Nanotechnology in Endodontics

Anil Kishen

Editor

Nanotechnology in Endodontics

Current and Potential Clinical Applications

 Springer

Editor
Anil Kishen
Department of Clinical Sciences
Discipline of Endodontics
University of Toronto
Faculty of Dentistry
Toronto, ON
Canada

ISBN 978-3-319-37924-1   ISBN 978-3-319-13575-5   (eBook)
DOI 10.1007/978-3-319-13575-5

Springer Cham Heidelberg New York Dordrecht London

Printed on acid-free paper

Springer International Publishing AG Switzerland is part of Springer Science+Business Media
(www.springer.com)

This book is dedicated to my teachers, students, colleagues and family for making a difference.

Anil Kishen

Preface

Current understanding emphasizes that endodontic disease is a biofilm-mediated infection. Therefore, elimination of biofilm bacteria from the root canal system remains to be the primary focus in the management of endodontic disease. Unfortunately, the endodontic environment is a challenging locale to eliminate surface-adherent biofilm bacteria. This advanced antimicrobial strategies are required for the optimal disinfection of previously infected root canal system. This step is crucial not only to predictably perform conventional root can treatments but also to develop tissue engineering–based strategies that can achieve organized repair or regeneration in previously infected teeth.

Nanotechnology is a rapidly advancing interdisciplinary field, and the use of nanomaterials is becoming more and more common for different healthcare purposes. Generally, nanomaterial-based approaches present distinct advantages over conventional approaches. At the nanometre scale, the characteristics of materials appear to be markedly different to those of the same material at the macro scale. Nanotechnology/nanoparticles in endodontics can offer different benefits, ranging from efficient miniaturization techniques to approaches that control the molecular assembly, all of which can create exciting opportunities for the prevention, diagnosis and treatment of endodontic disease.

This book is aimed to provide a comprehensive understanding of the current and potential application of nanoparticles in endodontics. Chapters 1, 2, and 3 cover the fundamental aspects of nanomaterials, with potential applications in endodontics. The basis of nanotechnology and nanomaterials in tissue engineering, methods for characterizing nanomaterials/nanoparticles, techniques to assess the cytotoxicity in nanomaterials/nanoparticles and the basis of nanomaterials/nanoparticles in drug and gene delivery are covered in these chapters. Chapters 4, 5, 6, and 7 deal with the applications of nanoparticles that are more specific to the field of endodontics. Antibacterial nanoparticles, nanoparticles for dentin tissue stabilization, nanoparticles in restorative/endodontic materials and the application of nanomaterials for minimally invasive treatment of dental caries are covered in these chapters.

This book is intended for graduate students and practising clinicians with interest in the new and exciting field of nanomaterials/nanoparticles.

Toronto, ON, Canada

Anil Kishen, BDS, MDS, PhD

Contents

Contributors

Ibtisam Al-Hashimi, BDS, MS, PhD Periodontics, Texas A&M University Baylor College of Dentistry, Dallas, TX, USA

Grace M. De Souza, DDS, MSc, PhD Clinical Sciences Department, Faculty of Dentistry, University of Toronto, Toronto, ON, Canada

Xuliang Deng, PhD Geriatric Dentisty, Peking University School and Hospital of Stomatology, Beijing, China

Prachi Desai, BTech Biomedical Engineering, Texas A&M University, College Station, TX, USA

Shimon Friedman, DMD Department of Endodontics, University of Toronto, Faculty of Dentistry, Toronto, ON, Canada

Akhilesh K. Gaharwar, PhD Department of Biomedical Engineering, Department of Materials Science and Engineering, Texas A&M University, College Station, TX, USA

Saji George, MSc, PhD Centre for Sustainable Nanotechnology, School of Chemical & Life Sciences, Nanyang Polytechnic, Singapore, Singapore

En-Tang Kang, PhD Department of Chemical and Biomolecular Engineering, National University of Singapore, Singapore, Singapore

Anil Kishen, PhD, MDS, BDS Department of Endodontics, Faculty of Dentistry, University of Toronto, Toronto, ON, Canada

Min Li, PhD Department of Chemical and Biomolecular Engineering, National University of Singapore, Singapore, Singapore

Raghavendra C. Mundargi, Msc, PhD School of Materials Science & Engineering, Nanyang Technological University, Singapore, Singapore

Koon Gee Neoh, ScD Department of Chemical and Biomolecular Engineering, National University of Singapore, Singapore, Singapore

Xu Wen Ng, PhD School of Materials Science & Engineering, Nanyang Technological University, Singapore, Singapore

Annie Shrestha, PhD, MSc, BDS Department of Endodontics, Faculty of Dentistry, University of Toronto, Toronto, ON, Canada

Venu Gopal Varanasi, PhD Department of Biomedical Sciences, Texas A&M University Baylor College of Dentistry, Dallas, TX, USA

Subbu S. Venkatraman, PhD Polymeric Biomaterials Group, School of Materials Science & Engineering, Nanyang Technological University, Singapore, Singapore

Yi Wu, MSc Department of Endodontics, School and Hospital of Stomatology, Tianjin Medical University, Tianjin, People's Republic of China

Janet P. R. Xavier, BTech Department of Biomedical Engineering, Texas A&M University, College Station, TX, USA

Xu Zhang, PhD Department of Endodontics, School and Hospital of Stomatology, Tianjin Medical University, Tianjin, People's Republic of China

Introduction

1

Shimon Friedman

Abstract

This new textbook on current and potential applications of nanotechnology in endodontics is offered to the endodontic community at a juncture when there is emerging understanding that traditional endodontic therapy may be limited in its ability to cure apical periodontitis and retain treated teeth. Improvements in treatment outcomes have been elusive, suggesting that "out of the box" approaches are needed beyond conventional endodontic therapy and restorative concepts.

Dr. Kishen's textbook offers the first focused glimpse at nanomaterials harnessed for root canal disinfection and stabilization of root dentin to overcome the microbial resilience, the ultimate challenge in endodontic therapy, and to enhance the resistance of root dentin to cracking. It provides an insight into how emerging nanotechnologies may benefit teeth and patients. Because of its focus on emerging innovative technologies, the content of this textbook and its detailed analysis are not yet found in any of the other endodontic textbooks. In this regard, it is a most timely addition to the endodontic texts which will become a valuable resource for endodontic clinicians and researchers.

This new textbook on current and potential applications of nanotechnology in endodontics by Dr. Anil Kishen is presented to the endodontic community at a juncture when there is emerging understanding that traditional endodontic therapy may be limited in its ability to cure apical peri-odontitis and retain treated teeth. Suffice it to observe that the clinical mid-term (4–10 year) treatment outcomes reported in literature from 50 to 60 years ago [1–5] and in current literature [6–8] are comparable [9, 10], to have one wonder whether they can indeed improve. This may be disappointing to many who consider that the elapsed decades, and especially the past 20 years, have brought along remarkable development of endodontic technologies, representing one of the most rapid and extensive technological evolutions in all of dentistry.

S. Friedman, DMD
Department of Endodontics,
University of Toronto Faculty of Dentistry,
124 Edward St., Toronto, ON M5G 1G6, Canada
e-mail: shimon.friedman@dentistry.utoronto.ca

A. Kishen (ed.), *Nanotechnology in Endodontics: Current and Potential Clinical Applications*,
DOI 10.1007/978-3-319-13575-5_1, © Springer International Publishing Switzerland 2015

The passage of time has allowed improved understanding of endodontic disease and its primary cause, the modalities applied to address that cause as well as the limitations of those modalities. Driven by an ever-increasing volume of research, this process has seen several shifts in focus that, in turn, have generated technological advances that improved almost every aspect of how endodontic therapy is delivered. The rapid pace of those advances challenged the endodontic community to keep up both with the changing focus and the technologies developed to improve endodontic therapy.

And yet, major breakthroughs in improving treatment outcomes have been elusive, suggesting that novel, 'out-of-the-box' approaches may have to be explored beyond the conventional concepts of canal disinfection and filling as well as tooth restoration. Dr. Kishen's textbook offers the endodontic community the first focused glimpse at nanotechnology that can be considered 'out of the box' when compared with existing endodontic therapeutic approaches.

Nanomaterials are being adapted to be harnessed in root canal disinfection and stabilization of root dentin in response to the emerged understanding that canal disinfection cannot be predictably achieved in all treated teeth even when using the most recent protocols and that root dentin in treated teeth may be vulnerable to cracks. The goals are to overcome the microbial resilience that represents the ultimate challenge in endodontic therapy and to enhance the resistance of root dentin to cracking and in so doing to circumvent the traditional reliance on just endodontic instruments and irrigation with topical antiseptics to reach anatomical intricacies of the root canal systems and dentinal tubules.

Development of nanomaterials towards potential applications in endodontics to address the above challenges is far from elementary. It requires specific knowledge to drive unique adaptations of nanotechnologies but also updating the clinicians on how nanotechnologies may be applied in practice. Compiling a textbook that captures emerging innovations and concepts is a challenge in itself. The biggest risk is that the emerging technologies will evolve rapidly, rendering the textbook obsolete shortly after it is published. Faced with this

risk, for the textbook to still provide value to readers, it needs to provide the comprehensive background on the novel technologies, what they are intended to achieve and the concepts of how they are expected to achieve those goals.

Dr. Kishen's textbook is designed specifically to provide the knowledge base, the detailed information on current and emerging nanotechnologies and insight into how those nanotechnologies may benefit teeth and patients. In this regard, this textbook is a most timely addition to the endodontic texts available to clinicians and researchers.

Dr. Kishen teamed up with a carefully selected panel of experts, allowing the reader to benefit from their cumulative expertise. With access to this collective expertise, the reader gains an in-depth insight into the current state of nanotechnologies as they may apply to disinfection of root canal systems and stabilization of root dentin. Because of its focus on emerging innovative technologies, the content of this textbook and its detailed analysis are not yet found in any of the other endodontic textbooks.

This textbook is organized in a logical sequence. Rather than assigning precedence to clinical applications in endodontics, the book first reviews the foundations of nanotechnology and nanomaterials as they evolved in the domain of tissue engineering. It is this domain after all that provided much of the inspiration to exploring potential benefits of nanotechnology in endodontics. This is then followed by characterization of nanomaterials with regard to their size, physical and chemical properties, their interactions with biomolecules such as proteins and membrane receptors and their cytotoxicity and its implications on safe application. The possibilities of drug and gene delivery are reviewed next, covering a range of applications in health care but also in periodontal and endodontic therapy. Only after this knowledge base is provided is the focus on applications in endodontic therapy introduced, primarily to enhance disinfection of the root canal system. Next reviewed is the potential application of nanomaterials in stabilization of root dentin with the aim of reducing its susceptibility to crack formation. The scope of the textbook is then further expanded to include applications of nanomaterials in restorative materials and in enhancing remineralization of carious dentin to enable less invasive caries removal procedures.

At once, this textbook will become a valuable resource for endodontic specialists, those engaged in both clinical practice and research. The clinicians may find in it some hope to see improved treatment outcomes in the future, which may encourage them to look for further developments in this dynamic area of nanotechnology applicable to endodontics. The researchers may find in it inspiration and guidelines to pursue research directions they may not have considered previously. All readers will find the textbook's content helpful in fostering understanding of the challenges that compromise the effectiveness of root canal disinfection and the long-term stability of root dentin in root-filled teeth. This understanding will potentially upgrade the readers' sophistication in their selected endeavours.

References

1. Strindberg LZ. The dependence of the results of pulp therapy on certain factors. An analytic study based on radiographic and clinical follow-up examination. Acta Odontol Scand. 1956;14:21.

2. Seltzer S, Bender IB, Turkenkopf S. Factors affecting successful repair after root canal therapy. J Am Dent Assoc. 1963;52:651–62.

3. Grossman LI, Shepard LI, Pearson LA. Roentgenologic and clinical evaluation of endodontically treated teeth. Oral Surg Oral Med Oral Pathol. 1964;17:368–74.

4. Engström B, Lundberg M. The correlation between positive culture and the prognosis of root canal therapy after pulpectomy. Odontol Revy. 1965;16:193–203.

5. Harty FJ, Parkins BJ, Wengraf AM. Success rate in root canal therapy. A retrospective study of conventional cases. Br Dent J. 1970;128:65–70.

6. De Chevigny C, Dao TT, Basrani BR, Marquis V, Farzaneh M, Abitbol S, Friedman S. Treatment outcome in endodontics: the Toronto study – phase 4: initial treatment. J Endod. 2008;34:258–63.

7. Ng Y-L, Mann V, Gulabivala K. A prospective study of the factors affecting outcomes of nonsurgical root canal treatment: part 1: periapical health. Int Endod J. 2011;44:583–609.

8. Ricucci D, Russo J, Rutberg M, Burleson JA, Spangberg LS. A prospective cohort study of endodontic treatments of 1,369 root canals: results after 5 years. Oral Surg Oral Med Oral Pathol Oral Radiol Endod. 2011;112:825–42.

9. Ng Y-L, Mann V, Rahbaran S, Lewsey J, Gulabivala K. Outcome of primary root canal treatment: systematic review of the literature – part 1. Effects of study characteristics on probability of success. Int Endod J. 2007;40:921–39.

10. Ng Y-L, Mann V, Rahbaran S, Lewsey J, Gulabivala K. Outcome of primary root canal treatment: systematic review of the literature – part 2. Influence of clinical factors. Int Endod J. 2008;41:6–31.

Advanced Nanomaterials: Promises for Improved Dental Tissue Regeneration

2

Janet R. Xavier, Prachi Desai, Venu Gopal Varanasi, Ibtisam Al-Hashimi, and Akhilesh K. Gaharwar

Abstract

Nanotechnology is emerging as an interdisciplinary field that is undergoing rapid development and has become a powerful tool for various biomedical applications such as tissue regeneration, drug delivery, biosensors, gene transfection, and imaging. Nanomaterial-based design is able to mimic some of the mechanical and structural properties of native tissue and can promote biointegration. Ceramic-, metal-, and carbon-based nanoparticles possess unique physical, chemical, and biological characteristics due to the high surface-to-volume ratio. A range of synthetic nanoparticles such as hydroxyapatite, bioglass, titanium, zirconia, and silver nanoparticles are proposed for dental restoration due to their unique bioactive characteristic. This review focuses on the most recent development in the field of nanomaterials with emphasis on dental tissue engineering that provides an inspiration for the development of such advanced biomaterials. In particular, we discuss synthesis and fabrication of bioactive nanomaterials, examine their current limitations, and conclude with future directions in designing more advanced nanomaterials.

J.R. Xavier, BTech • P. Desai, BTech
Biomedical Engineering, Texas A&M University,
College Station, TX, USA

V.G. Varanasi, PhD
Biomedical Sciences, Texas A&M University Baylor
College of Dentistry, Dallas, TX, USA

I. Al-Hashimi, BDS, MS, PhD
Periodontics, Texas A&M University Baylor
College of Dentistry, Dallas, TX, USA

A.K. Gaharwar, PhD (✉)
Department of Biomedical Engineering,
Department of Materials Science and Engineering,
Texas A&M University,
3120 TAMU, ETB 5024, College Station,
TX 77843, USA
e-mail: gaharwar@tamu.edu

2.1 Introduction

Nanotechnology is emerging as an interdisciplinary field that is undergoing rapid development and is a powerful tool for various biomedical applications such as tissue regeneration, drug delivery, biosensors, gene transfection, and imaging [1–4]. Nanotechnology can be used to synthesize and fabricate advanced biomaterials with unique physical, chemical, and biological properties [5–9]. This ability is mainly attributed to enhance surface-to-volume ratio of nanomaterials compared to the micro or macro counter-

A. Kishen (ed.), *Nanotechnology in Endodontics: Current and Potential Clinical Applications,*
DOI 10.1007/978-3-319-13575-5_2, © Springer International Publishing Switzerland 2015

parts. At the nanometer-length scale, materials behave differently due to the increased number of atoms present near the surface compared to the bulk structure.

A range of nanomaterials such as electrospun nanofiber, nanotextured surfaces, self-assembled nanoparticles, and nanocomposites are used to mimic mechanical, chemical, and biological properties of native tissues [8, 10]. Nanomaterials with predefined geometries, surface characteristics, and mechanical strength are used to control various biological processes [11]. For example, by controlling the mechanical stiffness of a matrix, cell–matrix interactions such as cellular morphology, cell adhesion, cell spreading, migration, and differentiation can be controlled. The addition of silica nanospheres to a poly(ethylene glycol) (PEG) network resulted in a significant increase in mechanical stiffness and bioactivity compared to PEG hydrogels [12]. These mechanically stiff and bioactive nanocomposite hydrogels can be used as an injectable matrix for orthopedic and dental applications. Apart from these, a range of nanoparticles is used to provide bioactive properties to enhance biological properties [12–15].

Dental tissue is comprised of mineralized tissues, namely, enamel and dentin, with a soft dental pulp as its core. Enamel is one of the hardest materials found in the body. It is composed of inorganic hydroxyapatite and a small amount of unique noncollagenous proteins, resulting in a composite structure [16]. Due to the limited potential of self-repair, once these dental tissues are damaged due to trauma or bacterial infection, the only treatment option that is available to repair the damage is the use of biocompatible synthetic materials [17]. Most of the synthetic implants are subjected to the hostile microenvironment of the oral cavity and thus have a limited lifespan and functionality. Thus, there is a need to develop biofunctional materials that not only aid in dental restoration but also mimic some of the native tissue functionally. One of the most important considerations in regeneration of dental tissue is engineering a bioactive dental implant that is highly resorbable with controlled surface structures, enhanced mechanical properties, improved

cellular environment, and effective elimination of bacterial infection [18].

While dental fillings and periodontal therapy are effective for the treatment of dental diseases, they do not restore the native tooth structure or the periodontium. Current technologies are focused on using stem cells from teeth and periodontium as a potential source for partial or whole tissue regeneration [16]; however, these approaches do not provide protection against future dental diseases. Recent advances in nanomaterials provide a wider range of dental restorations with enhanced properties, such as greater abrasion resistance, high mechanical properties, improved esthetics, and better controlled cellular environment [19, 20]. Currently, there is increasing interest in nanomaterials/nanotechnology/nanoparticles in dentistry as evident by the high number of publications; in fact, one-third of the tissue engineering publications are in the field of dentistry (Fig. 2.1).

This chapter reviews recent developments in the area of nanomaterials and nanotechnologies for dental restoration. We will focus on two major types of nanomaterials, namely, bioinert and bioactive nanomaterials. Bioactive nanomaterials include hydroxyapatite, tricalcium phosphate, and bioglass nanomaterials, whereas bioinert nanomaterials include alumina, zirconia, titanium, and vitreous carbon. Emerging trends in the area of dental nanomaterials and future prospects will also be discussed.

2.2 Anatomy and Development of the Tooth

To design advanced biomaterials for dental repair, it is important to understand the chemical and physical properties of native tissues. Teeth are three-dimensional complex structures consisting of the crown, neck, and root. Oral ectodermal cells and neural crest–derived mesenchyme are the primary precursors for mammalian teeth. Tooth development can broadly be divided into five major stages: dental lamina, initiation, bud stage, cap stage, bell stage and eruption (Fig. 2.2). Initially, ameloblasts and odontoblasts differentiate at the

Fig. 2.1 Number of (**a**) publications and (**b**) citations related to "nanomaterials" or "nanotechnology" or "nanoparticles" and tissue engineering/dental according to ISI Web of Science (Data obtained March 2014). A steady increase in the number of publication in the area of nanomaterials for dental research indicates growing interest in the field

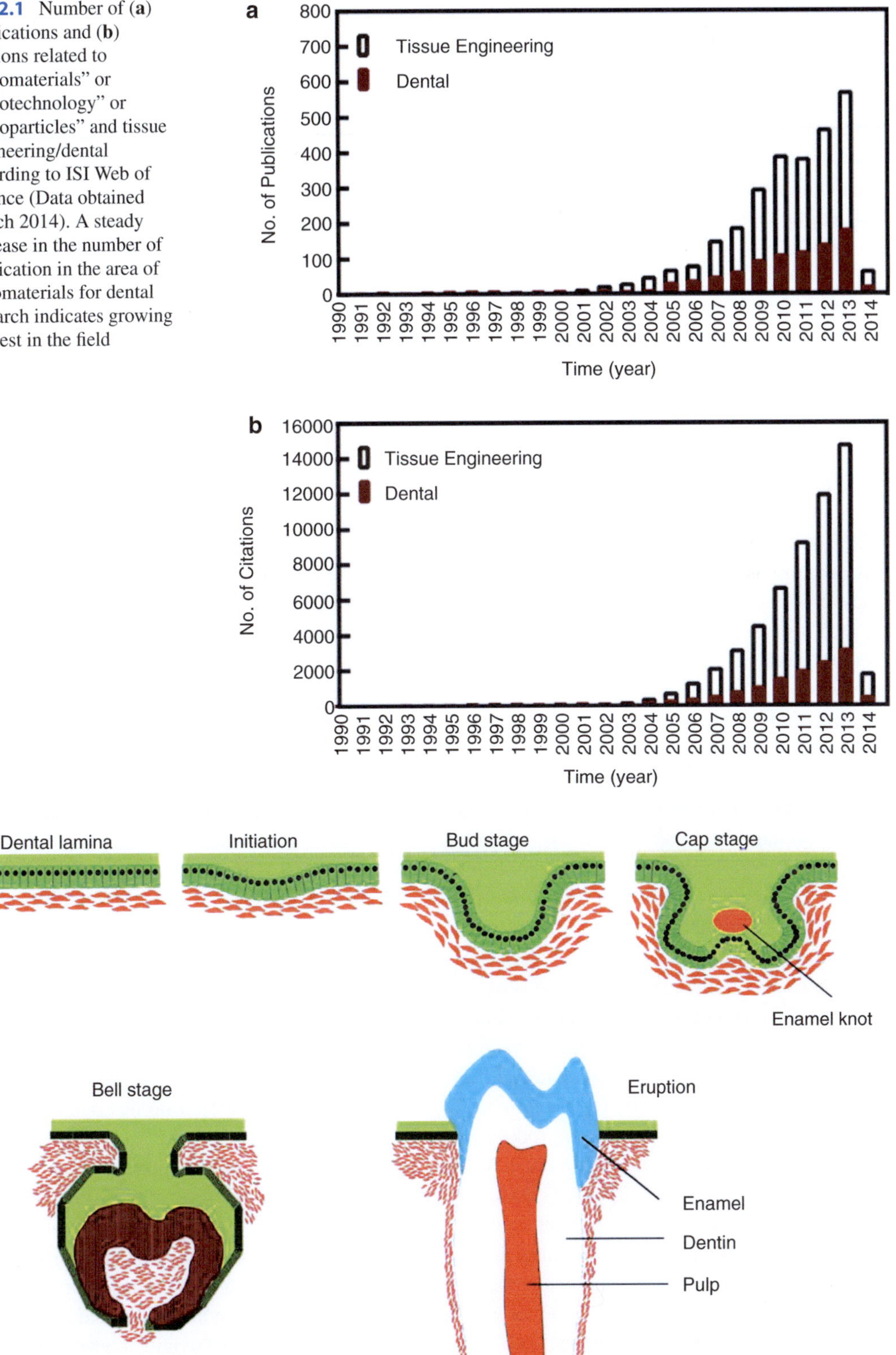

Fig. 2.2 Stages of tooth morphogenesis (Adapted from Nakashima and Reddi [97]. With permission from Nature Publishing Group)

junction between epithelium and mesenchyme to form enamel and dentin, which are the tooth-specific hard tissues. Following this, root formation is initiated by differentiation of cementoblasts from dental follicle mesenchyme to form cementum, which is the third hard tissue of the tooth. As teeth erupt into the oral cavity and the roots reach their final length, substantial amounts of epithelial cells are lost [21].

A range of soft and hard dental tissues is observed, depending on the anatomical location. For example, the tooth crown with a mechanical stiffness of 100 GPa is composed of the mineralized outer enamel layer with 0.6 % of organic matter and 0.36 % of proteins. Underlying the enamel layer, the mineralized dentin layer has moderate mechanical stiffness (80 GPa), and the inner pulp dental tissue is very soft (~65 GPa) [22]. The enamel layer is mechanically stiff as it is composed of mineralized tissue that is produced by specific epithelial cells called ameloblasts. This layer consists of specialized enamel proteins such as enamelin, amelogenin, and ameloblastin. These proteins participate in helping structural organization and biomineralization of the enamel surface [16]. The underlying dentin layer is 75 % mineralized tissue containing dental-derived mesenchymal cells called odontoblasts on the dental papilla [23, 24], while the inner dental pulp tissue present in the root canal consists of dentin, cementum, and periodontal ligament layers, which secures the tooth to the alveolar bone [4].

Most of the mineralized structure observed in dental tissue is composed of nonstoichiometric carbonated apatite known as nanocrystalline hydroxyapatite (HA) in a highly organized fashion from micrometer to nanometer-length scale. The nanocrystalline HA particles are rod-shaped particles 10–60 nm in length and 2–6 nm in diameter [25]. The second most important component of dental tissue is cells. Cells present in dental tissue include osteoclasts and osteoblasts present on the face of the alveolar bone, cementoblasts on the surface of the root canal, and mesenchymal tissue in the ligament tissue that is essential for the long-term survival of these dental tissues. Molecular signals initiated by these cells trigger a set of events that regulate tissue morphogenesis, regeneration, and differentiation.

Human teeth do not have the capacity to regenerate after eruption. Therefore, biomedical engineering of human teeth using nondental cells might be a potential alternative for functional dental restorations. Here, we will discuss various types of nanomaterials that are proposed to facilitate regeneration of dental tissues.

2.3 Nanomaterials for Dental Repair

2.3.1 Hydroxyapatite as a Biomaterial for Dental Restoration

Hydroxyapatite particle (HAp) is a naturally occurring mineral form of calcium apatite, which is predominately obtained in mineralized tissue [26–28]. Hydroxyapatite is also one of the major components of dentin. Due to its bone-bonding ability, hydroxyapatite has been widely used as a coating material for various dental implants and grafts. Additionally, HAp is highly biocompatible and can rapidly osteointegrate with bone tissue. Due to these advantages, HAp is used in various forms, such as powders [29], coatings [30], and composites [26, 31] for dental restoration. Despite various advantages, hydroxyapatite has poor mechanical properties (highly brittle) and hence cannot be used for load-bearing applications [28].

A range of techniques have been developed to improve the mechanical toughness of this HAp [32]. Such hybrid nanocomposites are used to design bioactive coatings on dental implants. Brostow et al. designed a porous hydroxyapatite (150 μm)–based material by selecting a polyurethane to fabricate porous material and improve the mechanical properties of the implants [32]. Polyurethane was used because of its tunable mechanical stiffness. These hybrid nanocomposites are shown to have enhanced mechanical strength along with an interconnected porous network. It was shown that a rigid polymer with 40 % alumina contained fewer smaller closed pores, which causes an increase in Young's modulus.

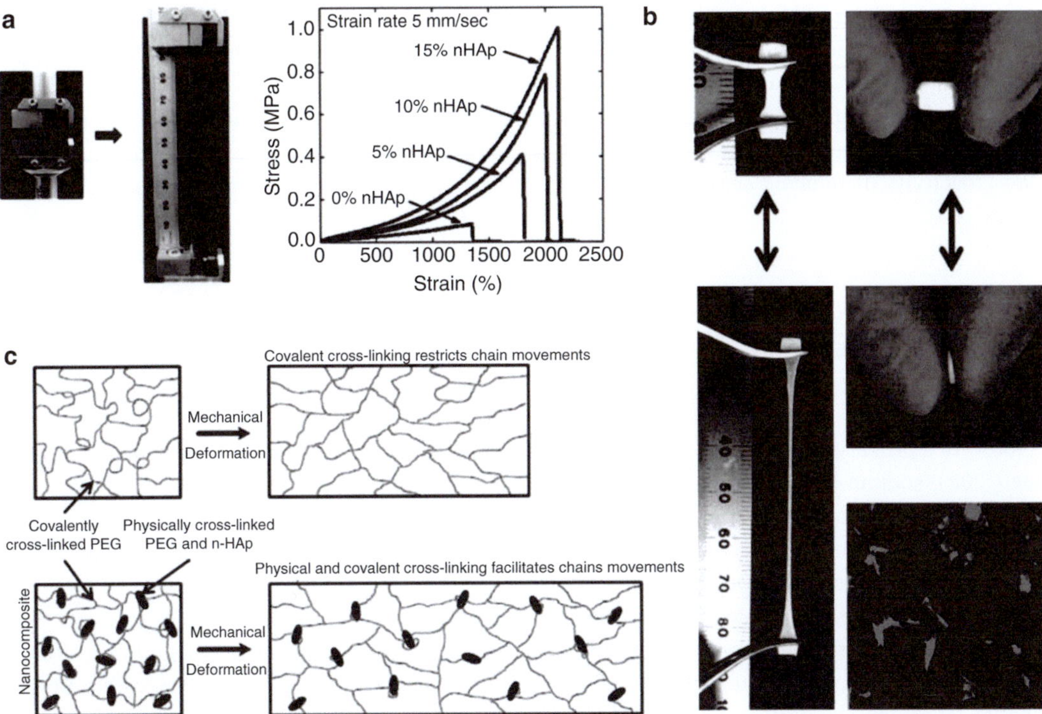

Fig. 2.3 Nanocomposite reinforced with nHAP. (**a**) Addition of nHAP to polymer matrix have shown to increase the mechanical strength by 5 folds. (**b**) The nHAP reinforced nanocomposite showed elastomeric properties indicating strong nHAP-polymer interactions at nano-length scale (**c**). The addition of nHAP also shown to promote cell adhesion properties of nanocomposite network (Reprinted from Gaharwar et al. [38]. With permission from American Chemical Society)

Earlier studies have highlighted the use of porous materials for dental fillings [33–35]. The increase in mechanical properties is mainly used to enhance surface interactions between nanoparticles and polymers. In a similar study, Uezono et al. showed that nanocomposites have significantly higher bioactivity when compared to microcomposites, as determined by an enhanced bone-bonding ability [36]. They fabricated titanium (Ti) rod specimens with a machined surface with nHAp coating and a nanohydroxyapatite/collagen (nHAp/Col) coating and placed it under the periosteum of a rat calvariu. After 4 weeks of implantation, they observed that nHAp/Col-coated titanium rods were completely surrounded by new bone tissue.

As compared to conventional microsized hydroxyapatite, nanophase hydroxyapatite displays unique properties such as an increased surface area, a lower contact angle, an altered electronic structure, and an increased number of atoms on the surface. As a consequence, the addition of hydroxyapatite nanoparticles to a polymer matrix result in enhanced mechanical strength. For example, Liu et al. showed that the addition of nHAp to chitosan scaffolds enhances the proliferation of bone marrow stem cells and an upregulation of mRNA for Smad1, BMP-2/4, Runx2, ALP, collagen I, integrin subunits, together with myosins compared to the addition of micron HAp [37]. The addition of nHAp significantly enhances pSmad1/5/8 in BMP pathways and showed nuclear localization along with enhanced osteocalcin production. In a similar study, nHAps are used to reinforce polymeric networks and enhance bioactive characteristics [38]. The addition of nHAps to a PEG matrix results in a significant increase in mechanical strength due to physical interactions between polymers and nanoparticles (Fig. 2.3). Additionally, this cell adhesion and spreading was also enhanced due to the nHAp addition. These bioactive nanomaterials can be used as an injectable matrix for periodontal regeneration and bone regrowth. Overall, nHAp-reinforced nanocomposites or surface coating improves mechanical stiffness and bioactivity of implants and can be used for dental restoration.

2.3.2 Dental Regeneration Using Bioactive Glass

Bioactive glass was developed by Hench et al. in 1960 with a primary composition of silicon dioxide (SiO_2), sodium oxide (Na_2O), calcium oxide (CaO), and phosphorous pentoxide (P_2O_5) in specific proportions. Bioglass is extensively used in the field of dental repair due to its bone-bonding ability [39–42]. When the bioglass is subjected to an aqueous environment, it results in the formation of hydroxycarbonate apatite/hydroxyapatite layers on the surface [43, 44]. Despite these advantages, bioglass is brittle and has a low wear resistance and thus cannot be used for load-bearing applications.

A range of techniques was developed to improve the mechanical properties of bioglasses. For example, Ananth et al. reinforced bioglass with yttria-stabilized zirconia. This yttria-stabilized zirconia bioglass is deposited on the titanium implant (Ti6Al4V) using electrophoretic deposition [45]. The yttria-stabilized zirconia bioglass (1YSZ-2BG) coating showed significantly higher bonding strength (72 ± 2 MPa) compared to yttrium-stabilized zirconia alone (35 ± 2 MPa). A biocompatibility test was performed to check for the ability to form an apatite layer on 1YSZ-2BG. A thin calcium phosphate film was observed on part of the 1 YSZ-2BG-coated surface after 7 days of immersion in simulated body fluid (SBF) without any precipitation. The apatite layer formation and size of calcium phosphate globules increased with an increase in the immersion time. The favored apatite formation on 1YSZ-2BG could be due to the Si-OH group arrangement on the bioglass (Fig. 2.4). Moreover, osteoblasts seeded on yttria-stabilized zirconia bioglass surfaces exhibited flattened morphology with numerous filopodial extensions. After 21 days of culture, the cells spread readily the on yttria-stabilized zirconia bioglass surface and resulted in the production of mineralized nodules. The enhanced mineralization of these bioactive surfaces is mainly attributed to the release of ions from the bioglass that facilitated the mineralization process.

Bioactive glass releases several ions that include sodium, phosphate, calcium, and silicon. Silicon in particular plays a central role as a bioactive agent. When released, it forms silanols in the near-liquid region above the glass surface. These silanols spontaneously polymerize to form a silica gel layer for the eventual nucleation and growth of a "bone-like" apatite. These same silanols also appear to influence collagen matrix synthesis by various cells types. In previous work, it was found that the ionic products from bioactive glass dissolution enhances the formation of a dense, elongated collagen fiber matrix, which was attributed to the presence of ionic Si in vitro [46, 47]. Moreover, it was found that these ions combinatorially effected the expression of genes associated with osteogenesis. The combinatorial concept of gene expression involves the use of gene regulatory proteins, which can individually control the expression of several genes, while combining these gene regulatory proteins can control the expression of single genes. Analogously, it was shown that Si and Ca ions can combinatorially regulate the expression of osteocalcin and this combinatorial effect can also enhance the mineralized tissue formation [48, 49]. Therefore, these ions can be used as an inductive agent to enhance the formation of mineralized tissue through the combinatorial of osteogenic gene expression.

2.3.3 Nanotopography Improves Biointegration of Titanium Implants

Titanium is one of the inert metals that is extensively used in the field of dental and musculoskeletal tissue engineering for load-bearing applications due to its excellent biocompatibility,

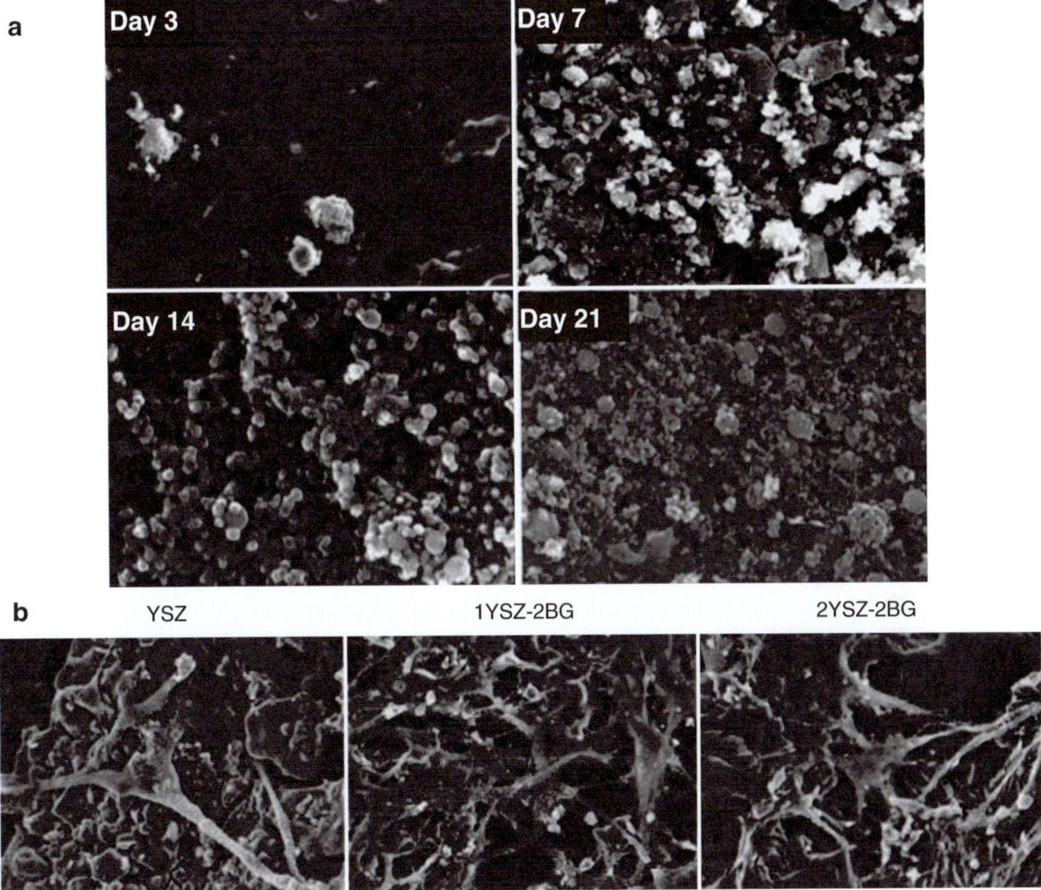

Fig. 2.4 Bioglass reinforced yttria-stabilized composite layer deposited on Ti6Al4V substrates. (**a**) SEM surface images of Ti6Al4V coated with 1YSZ–2BG immersed in simulated body fluid (SFB) for day 3, 7, 14, and 21. The formation CaP onto the surface of the 1YSZ–2BG indicates the catalytic effect of Si–OH, Zr–OH and Ti–OH groups on the apatite nucleation. (**b**) Morphological of osteoblast cells cultured on the Ti6Al4V surface coated with YSZ, YSZ–BG and 2YSZ-2BG (Adapted from Ananth et al. [45]. With permission from Elsevier)

high toughness, and excellent corrosion resistance and osteointegration [50–53]. When exposed to an *in vivo* microenvironment, a titanium surface spontaneously forms an evasive TiO_2 layer that is highly bioinert and provides excellent corrosion resistance [54]. Despite these favorable characteristics, the chemically inert surface of titanium implants is not able to form a firm and permanent fixation with the biological tissue to last the lifetime of the patient. Additionally, fibrous tissue encapsulation around the titanium implants leads to implant failure [51].

One of the possible techniques to reduce fibrous capsule formation and enhance tissue integration is to modulate surface structure and topography that directly aid in cell adhesion and mineralization [20]. Buser et al. observed that modifying the titanium implant surface with microtopography resulted in enhanced implant integration of underlying bone tissue *in vivo* [55, 56]. This is a widely accepted technique to enhance tissue integration of dental implants; however, these surface modification techniques at a micro length scale are far from satisfactory in preventing bone resorption and result in limited interaction with the natural tissues [57, 58].

Recent studies have shown that nanotopography of dental implants is far more effective in overcoming these problems. It is observed that nanoengineered surfaces are able to induce

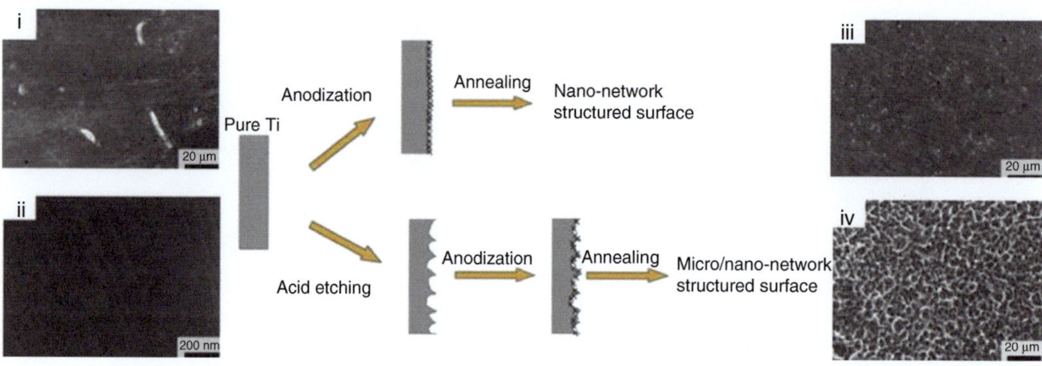

Fig. 2.5 Schematics showing surface modification of titanium implant using anodization and chemical surface etching. SEM images showing (*i, ii*) blank titanium, (*iii*) nano structured modified titanium surface and (*iv*) micro/nano modified surface (Adapted from Jiang et al. [50]. With permission from Elsevier)

osteoinductive signals to cells and facilitate their adhesion to the implant surface [20]. Nanotopography significantly modifies the biochemical and physiochemical characteristics of the implant surface and directly interacts with cellular components, favoring extracellular matrix deposition or formation of mineralized tissue at the dental implant surface [20]. Moreover, nanostructured titanium implants promote cell adhesion, spreading and proliferating as these nano surfaces directly interact with membrane receptors and proteins [20].

Apart from physical modification of titanium implant surfaces, chemical techniques to modify the surface characteristics have also been shown to influence dental tissue integration. For example, Scotchford et al. modified surface characteristics by functional groups (RGD) using molecular self-assembly to improve the osteointegration of a titanium implant surface [59, 60]. The RGD domains are immobilized on the titanium surface via a silanization technique using 3- aminopropyltriethoxysilane. Another method to impart nanotopography on the titanium implants includes chemical etching [61]. In this technique, the implant surface is treated with NaOH, which results in the formation of nanoetched structures and the formation of a sodium titanate layer. When this treated surface is subjected to simulated body fluid (SBF), nHAp crystals are deposited that aid in osteointegration of dental implants. In another study, Wang et al. showed that surface etching techniques to obtain nanostructure on the surface of implants can also improve mineralization on the implant surface [62]. They observed that the amorphous TiO_2 resulting due to the etching of the implant surface by H_2O_2 results in formation of mineralized tissue [63, 64].

Deposition of bioactive ceramic nanoparticles on the implant surface can also enhance the bonebonding ability [65, 66]. Nanosized calcium phosphate is deposited on the implant surface using the sol–gel transformation method to promote the formation of mineralized tissue on the implant surface [65, 66]. The implant surface modified with calcium phosphate promotes osteoblast attachment and spreading as shown by elongated filopodia. This resulted in interlocking of the implant with the adjoining bone in a rat model.

In a similar approach, Jiang et al. showed that combining nano- and microtopography, significantly enhanced osteointegration can be obtained (Fig. 2.5) [50]. They subjected the titanium implant surface to a dual chemical treatment consisting of acid etching followed by an NaOH treatment. The dual treatment resulted in nanostructured pores 15–100 nm in size and a microporous surface with a 2- to 7-μm size. The treated implant surface showed improved hydrophilicity, enhanced bioactivity, and increased corrosion resistance. Due to an increase in surface area, a significant increase in protein adsorption was observed on the nano/micro titanium implant. Thus, with the surface-functionalized and surface topography–modified titanium substrate, this

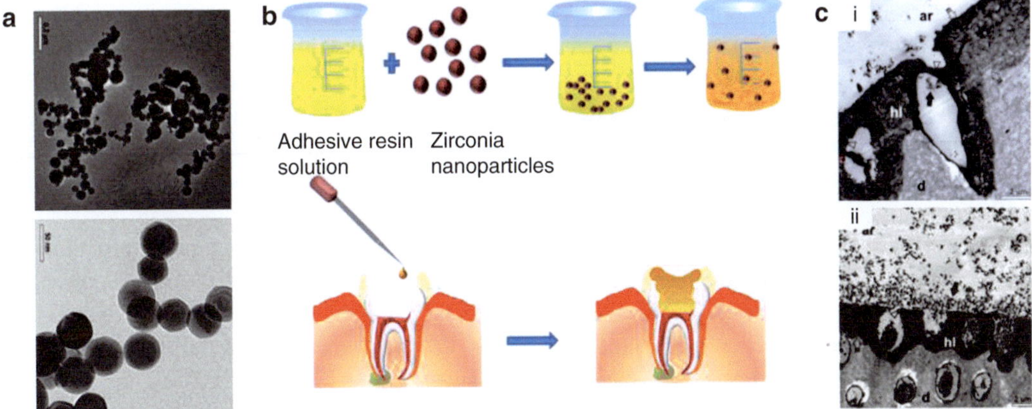

Fig. 2.6 (**a**) TEM images of zirconia nanoparticles prepared by laser vaporization. (**b**) Resin solution acting as adhesive is combined with zirconia nanoparticles. The composite resin loaded with zirconia nanoparticle showed uniform dispersion after ultrasonication. The adhesive dental resin loaded with nanoparticles are incorporated within to etched dentin with the implant. (**c**) TEM photographs showing dental resin composite with (**i**) 5 % wt and (**ii**) 20 % wt zirconia nanoparticles. The nanocomposite loaded with zirconia nanoparticles showed formation of submicron crystal enhancing remineralization and bioactivity at the implant-dentin interface. Adhesive resin (*ar*), hybrid layer (*hl*), demineralized dentin (*d*) are represented in image. Black arrow represents nanoparticles and open arrow represent formation of hybrid layer. (Adapted from Lohbauer et al. [77]. With permission from Elsevier)

ceramic--based material is clinically successful as a dental implant in regeneration of endosseous tissues.

2.3.4 Bioinert Zirconia Nanoparticles in Dentistry

Zirconia (or zirconium dioxide) is a polycrystalline biocompatible ceramic with low reactivity, high wear resistance, and good optical properties and thus extensively used in dental implantology and restorations [67, 68]. The mechanical properties of zirconia can be improved by phase transformation toughening using stabilizers such as yttria, magnesia, calcium, and ceria to stabilize the tetragonal phase of zirconia [69–72]. However, tetragonal zirconia is sensitive to low temperature degradation and results in a decrease in mechanical strength and surface deformation [68, 73]. By reducing the grain size of zirconia to nanoscale, phase modification can be arrested. Garmendia et al. showed that nanosized yttria-stabilized tetragonal zirconia (Y-TZP) can be obtained by spark plasma sintering [74, 75]. Nanosized Y-TZP showed significantly higher temperature stability and crack resistance compared to macrosized Y-TZP and zirconia [76].

Another approach to increase the mechanical properties of zirconia is to incorporate various nanoparticles such as carbon nanotubes and silica nanoparticles. For example, Padure et al. improved the toughness of nanosized Y-TZP by incorporating single-wall carbon nanotubes (SWCNTs) [76]. The addition of SWCNTs as a reinforcing agent to Y-TZP results in enhanced mechanical strength and making it attractive for dental restoration. In a similar approach, Guo et al. fabricated Y-TZP nanocomposite by reinforcing with silica nanofiber. The reinforced nanocomposite showed significant increase in the flexural modulus (FM), fracture toughness, flexural strength, and energy at break (EAB) compared to Y-TZP.

Nanosized zirconia can be used as a reinforcing agent in various dental fillers. Hambire et al. incorporated zirconia nanoclusters within a polymer matrix to obtain dental filler. The addition of zirconia nanoparticles resulted in significant increases in mechanical stiffness and enhanced tissue adhesion. Lohbauer et al. showed that a zirconia-based nanoparticle system can be used as dental adhesive (Fig. 2.6) [77]. Zirconia nanoparticles (20–50 nm) were prepared via a laser vaporization method. These nanoparticles are incorporated within the adhesive layer and have shown to increase tensile strength and promote mineralization after implantation.

Overall, nanostructured Y-TZP and nanosized TZP are extensively used in the field of dentistry due to the bioinert characteristic and aesthetic quality. The addition of zirconium nanoparticles to dental filler and incorporation into the dental tissue layers significantly enhance the mechanical stiffness and can promote bone bonding, restoring the dental defect and promoting accrual bone growth.

2.3.5 Antimicrobial Silver Nanoparticles for Dental Restoration

Over the centuries, silver has been used extensively in the field of medicine owing to its antimicrobial property, unique optical characteristic, thermal property, and anti-inflammatory nature [78]. Nano and micro particles of silver are used in conductive coatings, fillers, wound dressings, and various biomedical devices. Recently, there has been a growing interest in using silver nanoparticles in dental medicine, specifically for the treatment of oral cavities due to the antimicrobial property of silver nanoparticles [79]. The antimicrobial property of silver is defined by the release rate of silver ions. Although metallic silver is considered to be relatively inert, it gets ionized by the moisture, which results in a highly reactive state. This reactive silver interacts with the bacterial cell wall and results in structural changes by binding to the tissue protein and ultimately causing cell death [80].

Due to silver's antimicrobial properties, it is used in dental fillers, dental cements, denture linings, and coatings [81]. For example, Magalhaeces et al. evaluated silver nanoparticles for dental restorations [82]. They incorporated silver nanoparticles in glass ionomer cement, endodontic cement, and resin cement. The antimicrobial properties of these dental cements were evaluated against the most common bacterial species of *Streptococcus mutans* that are responsible for lesions and tooth decay. All these cements doped with silver nanoparticles showed significantly improved antimicrobial properties when compared to cement without any nanoparticles. In a similar study, Espinosa-Cristóbal et al. investigated the inhibition ability of silver nanoparticles toward *Streptococcus mutans* [83]. They evaluated the antimicrobial properties of silver on the enamel surface, which is commonly affected by primary and secondary dental caries. They evaluated three different sizes of silver nanoparticles (9.3, 21.3, and 98 nm) to determine minimum inhibitory concentrations (MICs) in *Streptococcus mutans* (Fig. 2.7). Due to the increase in surface area–to-volume ratio, silver nanoparticles with smaller size showed significantly enhanced antimicrobial properties.

Another application of these silver nanoparticles is evaluated in hard and soft tissue lining. Dentures (false teeth) are the prosthetic devices created to replace damaged teeth and are supported by hard and soft tissue linings present in the oral cavity [80]. These soft and hard tissue linings are subjected to higher mechanical stress during chewing and colonization of species, and this soft lining is one of the major issues due to invasion of fungi or plaque leading to mucosa infection. *Candida albicans* is the commonly found fungal colony near the soft linings. Chladek at al. modified these soft silicone linings of dentures with silver nanoparticles [84]. They showed that the addition of silver nanoparticles significantly enhances the antifungal properties and can be used for dental restorations. In a similar study, Torres et al. also evaluated antifungal efficiency of silver nanoparticles by incorporating it within denture resins [85]. They prepared denture resins by reinforcing silver nanoparticles within polymethylmethacrylate (PMMA). The addition of silver nanoparticles facilitates the release of silver ions and enhances antimicrobial activity. The denture resin loaded with silver nanoparticles showed improved inhibition of *C. albicans* on the surface along with an increase in flexural strength.

Apart from incorporating antimicrobial properties, silver nanoparticles also improve the mechanical strength of dental materials.

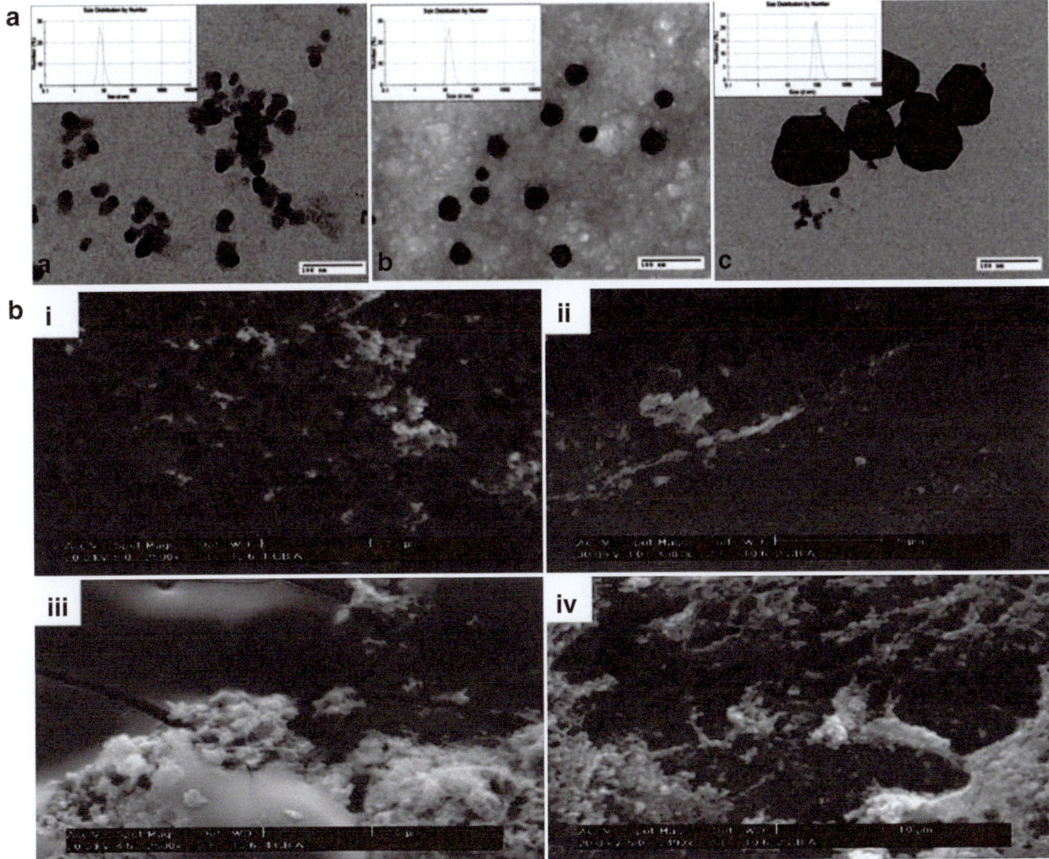

Fig. 2.7 (**a**) Different size of silver nanoparticles characterized using TEM and DLS. (**b**) SEM images showing the inhibition efficiency of silver nanoparticles towards *Streptococcus mutans* species (*i*) 9.3 nm (*ii*) 21.3 nm (*iii*) 98 nm (*iv*) negative control (Adapted from Espinosa-Cristóbal et al. [86]. With permission from Elsevier)

Mitsunori et al. showed that the addition of silver nanoparticles to porcelain significantly enhances mechanical toughness [86]. Additionally, an increase in hardness and fracture toughness of ceramic porcelain were observed with uniform distribution of silver nanoparticles. The addition of silver nanoparticles also arrests crack propagation on the implants that are developed due to the occlusal force. This is mainly due to the formation of smaller crystallites in porcelain ceramic with silver nanoparticles.

Overall, silver nanoparticles can be used to improve the mechanical properties such as modulus, fracture toughness, and hardness of dental biomaterials. Additionally, the release of silver ions from silver nanoparticles can significantly improve the antimicrobial properties of dental implants. Due to these unique property combinations, silver nanoparticles are extensively investigated in the field of dental research.

2.3.6 Bioactive Synthetic Silicates

Synthetic silicates (also known as layered clay) are disk-shaped nanoparticles 20–30 nm in diameter and 1 nm in thickness. These are composed of layers of $[SiO_4]$ tetrahedral sheets of Mg^{2+}, which complement their octahedral coordination by bridging with OH^- groups. The partial substitution of Mg^{2+} in the octahedral sheets by Li^+ charges the faces of the silicate nanoplatelets negatively, so the Na^+ ions are accommodated between the faces of the platelets for charge

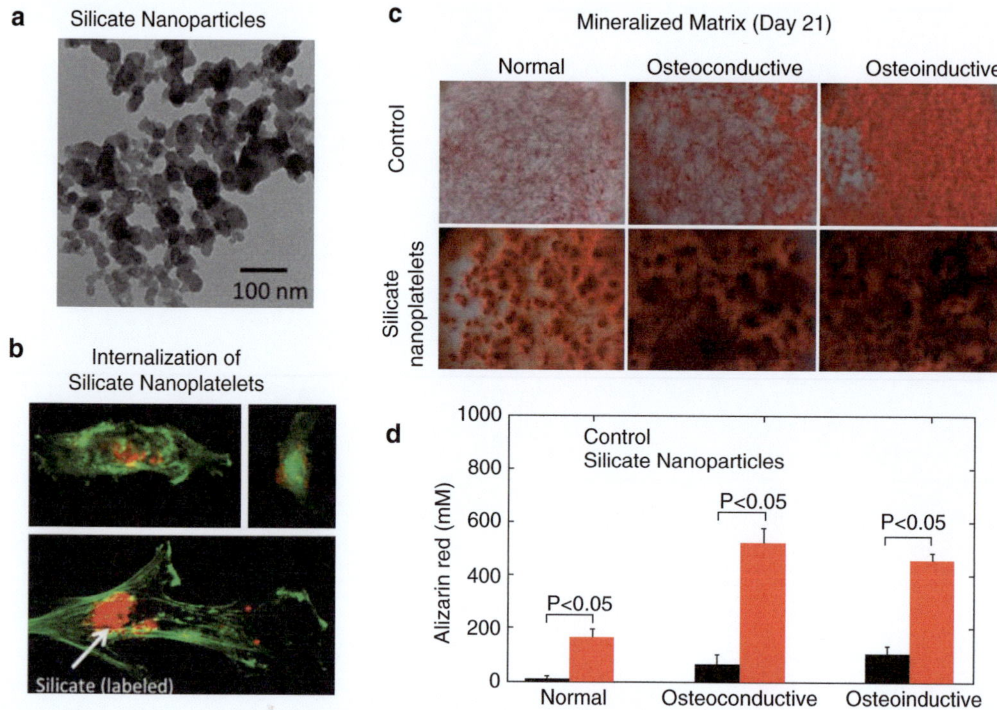

Fig. 2.8 (**a**) Synthetic silicates are 2D nanosheets with 20–30 nm in diameter and 1 nm in thickness. (**b**) Stem cells readily update these nansheets and differentiate into osteogenic lineages as determined by the production of mineralized matrix after 21 days of culture (**c**, **d**)

compensation, leading to a defined spatial distribution of charge on these nanoparticles [87]. Laponite, a type of synthetic silicate with empirical formula $Na^+_{0.7}[(Si_8Mg_{5.4}Li_{0.3})O_{20}(OH)_4]^-_{0.7}$, has been shown to be cytocompatible. These silicate nanoplatelets are shown to induce osteogenic differentiation of human mesenchymal stem cells (hMSCs) in the absence of osteoinductive factors, such as BMP-2 or dexamethasone (Fig. 2.8) [88]. A single dose of these silicate nanoparticles enhances the osteogenic differentiation of hMSCs when compared to hMSCs cultured in standard osteogenic differentiation conditions (in the presence of dexamethasone). Moreover, these synthetic silicates have shown to interact physically with both synthetic and natural polymers and can be used as injectable matrices for cellular therapies [13, 89, 90, 91]. These findings foster the development of new bioactive nanomaterials for repair and regeneration of mineralized tissue including bone and dental tissue.

Both natural and synthetic polymers are shown to physically interact with synthetic silicates. For example, long-chain poly(ethylene oxide) (PEO) has shown to form a physically cross-linked network with a shear thinning characteristic. The addition of silicates has shown to enhance the mechanical stiffness, structural stability, and physiological stability of a nanocomposite network. When these nanocomposites are dried in a sequential fashion, a hierarchical structure is formed [13]. These hierarchical structures consist of a highly organized layered structure composed of synthetic silicates and polymers. These structures are shown to have controlled cell adhesion and spreading characteristics [15]. The bioactive property of silicates helps in synthesizing such improved composites where silicates are acting as bioactive filters, triggering cues for specific regeneration approaches in bone-related tissues.

2.4 Future Outlook and Emerging Trends

In the future, the way forward in the field of dental regeneration is through the current knowledge in the field of tissue engineering, developmental biology, and cell and molecular biology. The regeneration capacity of these dental cells influences the extent for engineering the whole tooth. The regeneration potential for dental tissues is limited at the ameloblasts, which are mainly responsible for tissue regeneration, which is no longer present in adult dental tissue. Other dental tissues such as dentin consist of the neural crust cells, dentin mesenchymal cells that exihibit regenerative potential in response to injury [16], whereas cementum protecting the tooth has limited regeneration. Due to the limited regeneration capability of dental tissue nanotechnology, nanomaterials and stem cell–based approaches might facilitate regeneration of these tissues.

Several tooth structures have been regenerated in animal models using stem cell approaches. For enamel, epithelial cells of the Malassez (ERM) cells are used to regenerate enamel. These cells are located as a cluster near PDL at the tooth root. These cells remain in the G0 phase of their cell cycle and are a direct lineage of Herwig's epithelial root sheath, which are derived from enamel organs via cerival loop structures in developing enamel. These cells can differentiate into ameloblasts and can form the dentin–enamel matrix. The tissue structure system is then implanted onto the exposed dentin matrix to help rebuild the DEJ and enamel [92]. A similar procedure was used for tooth root regeneration, except that an entire regenerated tooth root structure is transplanted into the craniofacial pocket after in vitro development of a transplanted dentin matrix made from recovered dentin from extracted teeth and dentin follicle cells [93]. These stem cell approaches utilize existing tooth structures as scaffolds and have become very effective at repairing some of these structures in animal models.

For common dental deformities like periodontal disease or dental caries, focus has been drawn to two main approaches in tissue regeneration, the first being the introduction of filler materials into the point of defect with the goal being to induce bone regeneration, while the second approach focuses on instructing the cellular components present in the gums to take part in regeneration with the help of external cues [94]. Nanoparticles as injectable mixtures are suitable for filler-based approaches owing to their improved surface-to-volume ratio, reduced toxicity, and improved cellular response at the site of the dental defect (Fig. 2.9 and 2.10). The latter approach is by instructing the dental stem cells, which involves understanding of development of gums and cellular processes with the participation of key components of tissue engineering, which includes signals for development of morphological characters, stem cells to respond to morphogenesis, and scaffolds for mimicking an extracellular matrix (Fig. 2.11) [95]. Possible combination of these approaches that involves nanoparticles as scaffolds with stem cells and growth factors involving signaling molecules like BMP, FGFs, Shh, and Wnts will direct forward in a potential regenerative approach as a future vision for tooth biomedical engineering [19, 96].

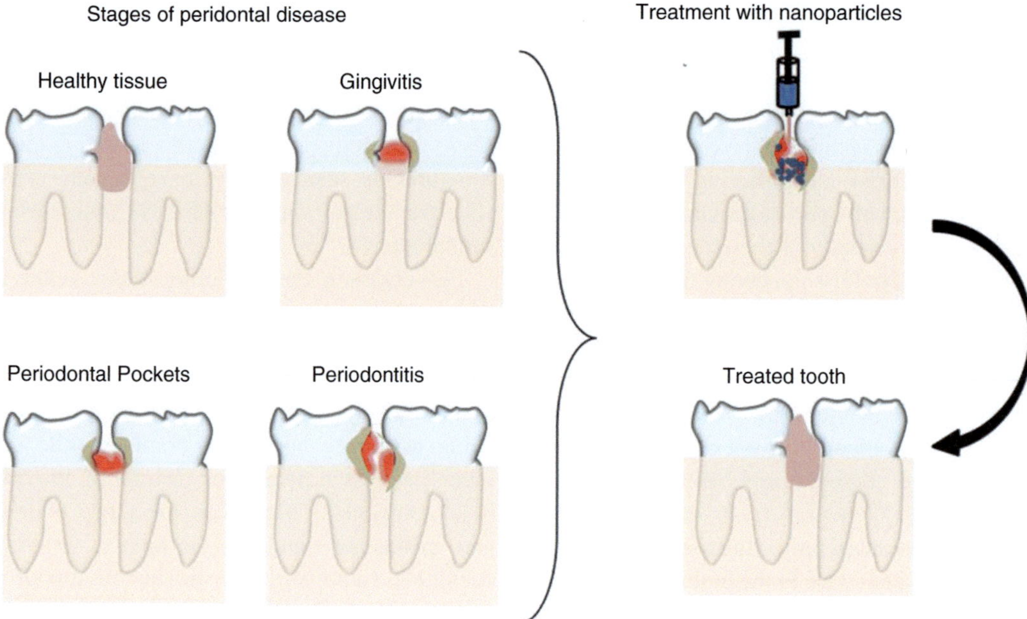

Fig. 2.9 Nanomaterials in the treatment of periodontal diseases. Various stages of periodontal diseases are shown in the figures starting with health tissue, gingivitis, formation of periodontal pockets, and periodontitis. These periodontitis can be treated by injecting nanoparticles for localized release of drug in the affected area

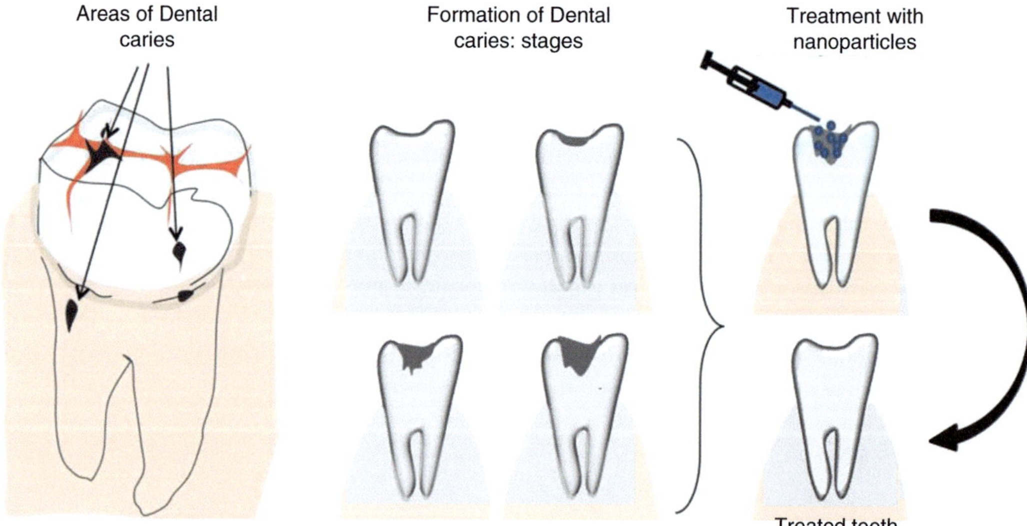

Fig. 2.10 Nanomaterials in the treatment of dental caries. Various stages of caries formation. These caries can be treated by injecting nanoparticles for localized release of drug in the affected area

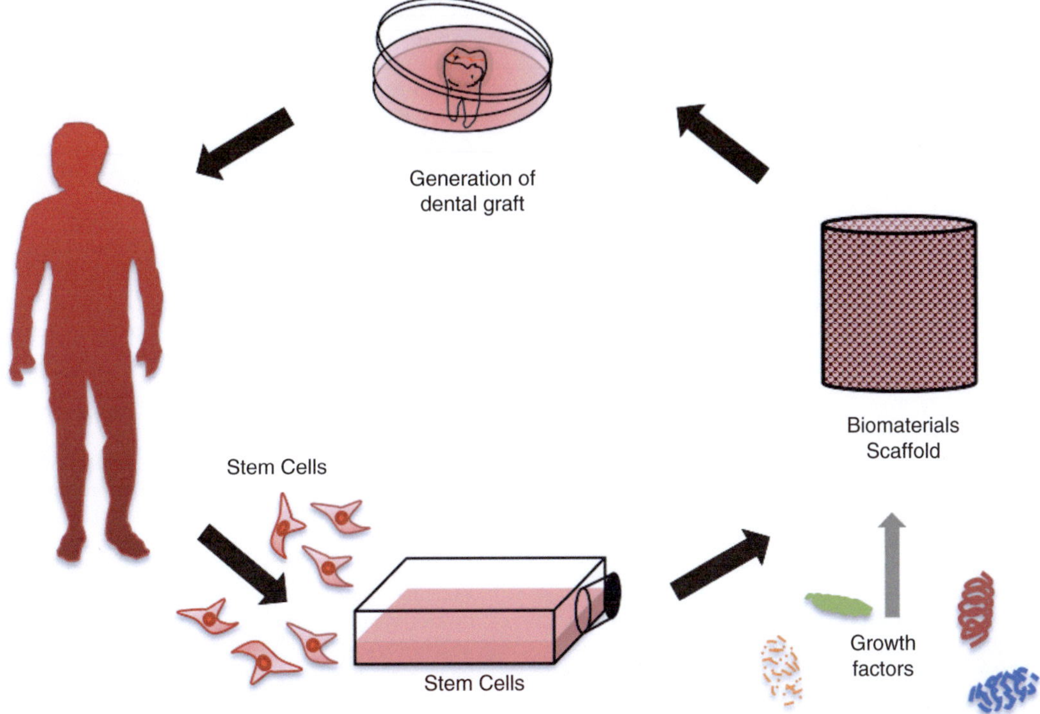

Fig. 2.11 The key components of dental tissue engineering are stem cells, growth factors and biomaterial scaffolds. Some of the key signaling molecules (growth factors) are BMPs, FGFs, Shh and Wnts. The stem cells used for dental regeneration include marrow derived stem cells (MSCs), dental pulp, and periodontal ligament (PDL) stem cells

2.5 Concluding Remarks

This paper presented the recent implementation of nanoscale materials from ceramic and metal particles to engineer dental tissue restoration. The synthesis and incorporation of these nanoparticles at the dental tissue repair site to bring persistent biological responses for tissue restoration are achieved by surface topography, modifications, and biological properties of these particles. Their effective stiffness, inert behavior, natural consistencies, biocompatibility, antimicrobial property, and improved protein–surface interaction for enhanced cell proliferation have rendered maximum responses for dental restoration. Also, their abilities in forming composites, injectable materials, and fibers waved a better efficiency in acting as a dental biomaterial. Possible limitations on controlling the nanoscale size of these materials in correspondence to their concentration at the implant surface and release rate cause toxicity, leading to the death of healthy cells, which needs more investigation in the dental field.

References

1. Peppas NA, Hilt JZ, Khademhosseini A, Langer R. Hydrogels in biology and medicine: from molecular principles to bionanotechnology. Adv Mater. 2006;18(11):1345–60.
2. Curtis A, Wilkinson C. Nantotechniques and approaches in biotechnology. TRENDS Biotechnol. 2001;19(3):97–101.
3. Gaharwar AK, Peppas NA, Khademhosseini A. Nanocomposite hydrogels for biomedical applications. Biotechnol Bioeng. 2014;111(3):441–53.
4. Venugopal J, Prabhakaran MP, Low S, Choon AT, Zhang Y, Deepika G, et al. Nanotechnology for nanomedicine and delivery of drugs. Curr Pharm Des. 2008;14(22):2184–200.
5. Carrow JK, Gaharwar AK. Bioinspired Polymeric Nanocomposites for Regenerative Medicine.

Macromolecular Chemistry and Physics. 2015; 261(3):248–64.

6. Langer R, Vacanti JP. Tissue engineering. Science. 1993;260(5110):920–6. PubMed PMID: 8493529.

7. Schexnailder P, Schmidt G. Nanocomposite polymer hydrogels. Colloid Polym Sci. 2009;287(1):1–11.

8. Thomas J, Peppas N, Sato M, Webster T. Nanotechnology and biomaterials. Boca Raton: CRC Taylor and Francis; 2006.

9. Goenka S, Sant V, Sant S. Graphene-based nanomaterials for drug delivery and tissue engineering. J Control Release. 2014;173:75–88.

10. Balazs AC, Emrick T, Russell TP. Nanoparticle polymer composites: where two small worlds meet. Science. 2006;314(5802):1107–10.

11. Xavier JR, Thakur T, Desai P, Jaiswal MK, Sears N, Cosgriff-Hernandez E, Kaunas R, Gaharwar AK. Bioactive Nanoengineered Hydrogels for Bone Tissue Engineering: A Growth-Factor-Free Approach. ACS Nano. 2015; DOI: 10.1021/nn507488s.

12. Gaharwar AK, Rivera C, Wu C-J, Chan BK, Schmidt G. Photocrosslinked nanocomposite hydrogels from PEG and silica nanospheres: structural, mechanical and cell adhesion characteristics. Mater Sci Eng C. 2013;33(3):1800–7.

13. Gaharwar AK, Kishore V, Rivera C, Bullock W, Wu CJ, Akkus O, et al. Physically crosslinked nanocomposites from silicate-crosslinked PEO: mechanical properties and osteogenic differentiation of human mesenchymal stem cells. Macromol Biosci. 2012;12(6):779.

14. Gaharwar AK, Rivera CP, Wu C-J, Schmidt G. Transparent, elastomeric and tough hydrogels from poly (ethylene glycol) and silicate nanoparticles. Acta Biomater. 2011;7(12):4139–48.

15. Gaharwar AK, Schexnailder PJ, Kline BP, Schmidt G. Assessment of using Laponite cross-linked poly (ethylene oxide) for controlled cell adhesion and mineralization. Acta Biomater. 2011;7(2):568–77.

16. Yen AH, Yelick PC. Dental tissue regeneration – a mini-review. Gerontology. 2011;57(1):85–94.

17. Ratner BD. Replacing and renewing: synthetic materials, biomimetics, and tissue engineering in implant dentistry. J Dent Educ. 2001;65(12):1340–7.

18. Lavenus S, Louarn G, Layrolle P. Nanotechnology and dental implants. Int J Biomater. 2010;915327:9.

19. Piva E, Silva AF, Nor JE. Functionalized scaffolds to control dental pulp stem cell fate. J Endod. 2014;40(4 Suppl):S33–40. PubMed PMID: 24698691.

20. Mendonça G, Mendonca D, Aragao FJL, Cooper LF. Advancing dental implant surface technology–from micron-to nanotopography. Biomaterials. 2008; 29(28):3822–35.

21. Thesleff I, Tummers M. Tooth organogenesis and regeneration. StemBook. Cambridge, UK: Harvard Stem Cell Institute; 2008–2009.

22. Baldassarri M, Margolis HC, Beniash E. Compositional determinants of mechanical properties of enamel. J Dent Res. 2008;87(7):645–9.

23. Lee SK, Krebsbach PH, Matsuki Y, Nanci A, Yamada KM, Yamada Y. Ameloblastin expression in rat incisors and human tooth germs. Int J Dev Biol. 1996;40(6):1141–50.

24. Brookes SJ, Robinson C, Kirkham J, Bonass WA. Biochemistry and molecular biology of amelogenin proteins of developing dental enamel. Arch Oral Biol. 1995;40(1):1–14.

25. Chen P-Y, McKittrick J, Meyers MA. Biological materials: functional adaptations and bioinspired designs. Prog Mater Sci. 2012;57(8):1492–704.

26. Sebdani MM, Fathi MH. Novel hydroxyapatite-forsterite-bioglass nanocomposite coatings with improved mechanical properties. J Alloys Compd. 2011;509(5):2273–6.

27. Shima T, Keller JT, Alvira MM, Mayfield FH, Dunsker SB. Anterior cervical discectomy and interbody fusion: an experimental study using a synthetic tricalcium phosphate. J Neurosurg. 1979;51(4):533–8.

28. Cao W, Hench LL. Bioactive materials. Ceram Int. 1996;22(6):493–507.

29. Fathi MH, Hanifi A. Evaluation and characterization of nanostructure hydroxyapatite powder prepared by simple sol–gel method. Mater Lett. 2007;61(18): 3978–83.

30. Sung Y-M, Lee J-C, Yang J-W. Crystallization and sintering characteristics of chemically precipitated hydroxyapatite nanopowder. J Crystal Growth. 2004; 262(1):467–72.

31. Sung Y-M, Kim D-H. Crystallization characteristics of yttria-stabilized zirconia/hydroxyapatite composite nanopowder. J Crystal Growth. 2003;254(3):411–7.

32. Brostow W, Estevez M, Lobland HEH, Hoang L, Rodriguez JR, & Vargar S. Porous hydroxyapatite-based obturation materials for dentistry. Journal of Materials Research 2008;23(06):1587–96.

33. Cottrell DA, Wolford LM. Long-term evaluation of the use of coralline hydroxyapatite in orthognathic surgery. J Oral Maxillofac Surg. 1998;56(8):935–41.

34. Ozgür Engin N, Tas AC. Manufacture of macroporous calcium hydroxyapatite bioceramics. J Eur Ceramic Soc. 1999;19(13):2569–72.

35. González-Rodríguez A, de Dios L-GJ, del Castillo JD, Villalba-Moreno J. Comparison of effects of diode laser and CO2 laser on human teeth and their usefulness in topical fluoridation. Lasers Med Sci. 2011;26(3):317–24.

36. Uezono M, Takakuda K, Kikuchi M, Suzuki S, Moriyama K. Hydroxyapatite/collagen nanocomposite-coated titanium rod for achieving rapid osseointegration onto bone surface. J Biomed Mater Res B Appl Biomater. 2013;101(6):1031–8.

37. Liu H, Peng H, Wu Y, Zhang C, Cai Y, Xu G, et al. The promotion of bone regeneration by nanofibrous hydroxyapatite/chitosan scaffolds by effects on integrin-BMP/Smad signaling pathway in BMSCs. Biomaterials. 2013;34(18):4404–17.

38. Gaharwar AK, Dammu SA, Canter JM, Wu C-J, Schmidt G. Highly extensible, tough, and elastomeric nanocomposite hydrogels from poly(ethylene glycol) and hydroxyapatite nanoparticles. Biomacromolecules. 2011;12(5):1641–50.

39. Kudo K, Miyasawa M, Fujioka Y, Kamegai T, Nakano H, Seino Y, et al. Clinical application of dental implant with root of coated bioglass: short-term results. Oral Surg Oral Med Oral Pathol. 1990;70(1):18–23.

40. Hench LL, Paschall HA. Direct chemical bond of bioactive glass-ceramic materials to bone and muscle. J Biomed Mater Res. 1973;7(3):25–42.

41. Stanley HR, Hench L, Going R, Bennett C, Chellemi SJ, King C, et al. The implantation of natural tooth form bioglasses in baboons: a preliminary report. Oral Surg Oral Med Oral Pathol. 1976;42(3):339–56.

42. Piotrowski G, Hench LL, Allen WC, Miller GJ. Mechanical studies of the bone bioglass interfacial bond. J Biomed Mater Res. 1975;9(4):47–61.

43. Oonishi H, Hench LL, Wilson J, Sugihara F, Tsuji E, Kushitani S, et al. Comparative bone growth behavior in granules of bioceramic materials of various sizes. J Biomed Mater Res. 1999;44(1):31–43.

44. Nganga S, Zhang D, Moritz N, Vallittu PK, Hupa L. Multi-layer porous fiber-reinforced composites for implants: *in vitro* calcium phosphate formation in the presence of bioactive glass. Dent Mater. 2012;28(11):1134–45.

45. Ananth KP, Suganya S, Mangalaraj D, Ferreira J, Balamurugan A. Electrophoretic bilayer deposition of zirconia and reinforced bioglass system on Ti6Al4V for implant applications: an in vitro investigation. Mater Sci Eng C. 2013;33(7):4160–6.

46. Varanasi VG, Owyoung JB, Saiz E, Marshall SJ, Marshall GW, Loomer PM. The ionic products of bioactive glass particle dissolution enhance periodontal ligament fibroblast osteocalcin expression and enhance early mineralized tissue development. J Biomed Mater Res A. 2011;98(2):177–84. PubMed PMID: 21548068.

47. Varanasi VG, Saiz E, Loomer PM, Ancheta B, Uritani N, Ho SP, et al. Enhanced osteocalcin expression by osteoblast-like cells (MC3T3-E1) exposed to bioactive coating glass (SiO2-CaO-P2O5-MgO-K2O-Na2O system) ions. Acta Biomater. 2009;5(9):3536–47. PubMed PMID: 19497391.

48. Varanasi VG, Leong KK, Dominia LM, Jue SM, Loomer PM, Marshall GW. Si and Ca individually and combinatorially target enhanced MC3T3-E1 subclone 4 early osteogenic marker expression. J Oral Implantol. 2012;38(4):325–36. PubMed PMID: 22913306.

49. Tousi NS, Velten MF, Bishop TJ, Leong KK, Barkhordar NS, Marshall GW, et al. Combinatorial effect of Si4+, Ca2+, and Mg2+ released from bioactive glasses on osteoblast osteocalcin expression and biomineralization. Mat Sci Eng C Mater. 2013;33(5):2757–65. PubMed PMID: WOS:000319630100037. English.

50. Jiang P, Liang J, Lin C. Construction of micro–nano network structure on titanium surface for improving bioactivity. Appl Surf Sci. 2013;280:373–80.

51. Wang H, Lin C, Hu R. Effects of structure and composition of the CaP composite coatings on apatite formation and bioactivity in simulated body fluid. Appl Surf Sci. 2009;255(7):4074–81.

52. Albrektsson TO, Johansson CB, Sennerby L. Biological aspects of implant dentistry: osseointegration. Periodontol 2000. 1994;4(1):58–73.

53. Suska F, Svensson S, Johansson A, Emanuelsson L, Karlholm H, Ohrlander M, et al. In vivo evaluation of noble metal coatings. J Biomed Mater Res B Appl Biomater. 2010;92(1):86–94.

54. Liu X, Chu PK, Ding C. Surface modification of titanium, titanium alloys, and related materials for biomedical applications. Mater Sci Eng R Rep. 2004;47(3):49–121.

55. Buser D, Schenk RK, Steinemann S, Fiorellini JP, Fox CH, Stich H. Influence of surface characteristics on bone integration of titanium implants. A histomorphometric study in miniature pigs. J Biomed Mater Res. 1991;25(7):889–902.

56. Petrie TA, Reyes CD, Burns KL, García AJ. Simple application of fibronectin-mimetic coating enhances osseointegration of titanium implants. J Cell Mol Med. 2009;13(8b):2602–12.

57. Cochran DL. A comparison of endosseous dental implant surfaces. J Periodontol. 1999;70(12):1523–39.

58. Shalabi MM, Gortemaker A, Van't Hof MA, Jansen JA, Creugers NHJ. Implant surface roughness and bone healing: a systematic review. J Dent Res. 2006;85(6):496–500.

59. Scotchford CA, Gilmore CP, Cooper E, Leggett GJ, Downes S. Protein adsorption and human osteoblast-like cell attachment and growth on alkylthiol on gold self-assembled monolayers. J Biomed Mater Res. 2002;59(1):84–99.

60. Germanier Y, Tosatti S, Broggini N, Textor M, Buser D. Enhanced bone apposition around biofunctionalized sandblasted and acid-etched titanium implant surfaces. Clin Oral Implants Res. 2006;17(3):251–7.

61. Sowa M, Piotrowska M, Widziołek M, Dercz G, Tylko G, Gorewoda T, Osyczka AM, Simka W. "Bioactivity of coatings formed on Ti–13Nb–13Zr alloy using plasma electrolytic oxidation." Materials Science and Engineering: C 49 (2015):159–173.

62. Wang XX, Hayakawa S, Tsuru K, Osaka A. A comparative study of in vitro apatite deposition on heat-, H(2)O(2)-, and NaOH-treated titanium surfaces. J Biomed Mater Res. 2001;54(2):172–8.

63. Uchida M, Kim H-M, Miyaji F, Kokubo T, Nakamura T. Apatite formation on zirconium metal treated with aqueous NaOH. Biomaterials. 2002;23(1):313–7.

64. Nanci A, Wuest JD, Peru L, Brunet P, Sharma V, Zalzal S, et al. Chemical modification of titanium surfaces for covalent attachment of biological molecules. J Biomed Mater Res. 1998;40(2):324–35.

65. Liu D-M, Troczynski T, Tseng WJ. Water-based sol–gel synthesis of hydroxyapatite: process development. Biomaterials. 2001;22(13):1721–30.

66. Xu W-P, Zhang W, Asrican R, Kim H-J, Kaplan DL, Yelick PC. Accurately shaped tooth bud cell-derived mineralized tissue formation on silk scaffolds. Tissue Eng Part A. 2008;14(4):549–57.

67. Deville S, Gremillard L, Chevalier J, Fantozzi G. A critical comparison of methods for the determination of the aging sensitivity in biomedical grade yttria-stabilized zirconia. J Biomed Mater Res B Appl Biomater. 2005;72(2):239–45.

68. Chevalier J. What future for zirconia as a biomaterial? Biomaterials. 2006;27(4):535–43.

69. Piconi C, Maccauro G. Zirconia as a ceramic biomaterial. Biomaterials. 1999;20(1):1–25.
70. Chevalier J, Deville S, Münch E, Jullian R, Lair F. Critical effect of cubic phase on aging in 3 mol% yttria-stabilized zirconia ceramics for hip replacement prosthesis. Biomaterials. 2004;25(24):5539–45.
71. Piconi C, Burger W, Richter HG, Cittadini A, Maccauro G, Covacci V, et al. Y-TZP ceramics for artificial joint replacements. Biomaterials. 1998;19(16):1489–94.
72. Bao L, Liu J, Shi F, Jiang Y, Liu G. Preparation and characterization of TiO2 and Si-doped octacalcium phosphate composite coatings on zirconia ceramics (Y-TZP) for dental implant applications. Appl Surf Sci. 2014;290:48–52.
73. Chevalier J, Gremillard L, Deville S. Low-temperature degradation of zirconia and implications for biomedical implants. Annu Rev Mater Res. 2007;37:1–32.
74. Marshall DB, Evans AG, Drory M. Transformation toughening in ceramics. Fract Mech Ceram. 1983;6:289–307.
75. Uzun G. An overview of dental CAD/CAM systems. Biotechnol Biotechnol Equip. 2008;22(1):530.
76. Oetzel C, Clasen R. Preparation of zirconia dental crowns via electrophoretic deposition. J Mater Sci. 2006;41(24):8130–7.
77. Lohbauer U, Wagner A, Belli R, Stoetzel C, Hilpert A, Kurland H-D, et al. Zirconia nanoparticles prepared by laser vaporization as fillers for dental adhesives. Acta Biomaterialia. 2010;6(12):4539–46.
78. Prabhu S, Poulose EK. Silver nanoparticles: mechanism of antimicrobial action, synthesis, medical applications, and toxicity effects. Int Nano Lett. 2012;2(1):1–10.
79. García-Contreras R, Argueta-Figueroa L, Mejía-Rubalcava C, Jiménez-Martínez R, Cuevas-Guajardo S, Sánchez-Reyna P, et al. Perspectives for the use of silver nanoparticles in dental practice. Int Dent J. 2011;61(6):297–301.
80. Chladek G, Barszczewska-Rybarek I, Lukaszczyk J. Developing the procedure of modifying the denture soft liner by silver nanoparticles. Acta Bioeng Biomech. 2012;14(1):23–9.
81. Hamouda IM. Current perspectives of nanoparticles in medical and dental biomaterials. J Biomed Res. 2012;26(3):143–51.
82. Magalhães APR, Santos LB, Lopes LG, Estrela CRA, Estrela C, Torres ÉM, et al. Nanosilver application in dental cements. ISRN Nanotechnol. 2012;2012 (365438):6.
83. Espinosa-Cristóbal LF, Martínez-Castañón GA, Téllez-Déctor EJ, Niño-Martínez N, Zavala-Alonso NV, Loyola-Rodríguez JP. Adherence inhibition of Streptococcus mutans on dental enamel surface using silver nanoparticles. Mater Sci Eng C. 2013;33(4): 2197–202.
84. Chladek G, Mertas A, Barszczewska-Rybarek I, Nalewajek T, Żmudzki J, Król W, et al. Antifungal activity of denture soft lining material modified by silver nanoparticles—a pilot study. Int J Mol Sci. 2011;12(7):4735–44.
85. Acosta-Torres LS, Mendieta I, Nuñez-Anita RE, Cajero-Juárez M, Castano VM. Cytocompatible antifungal acrylic resin containing silver nanoparticles for dentures. Int J Nanomedicine. 2012;7:4777.
86. Uno M, Kurachi M, Wakamatsu N, Doi Y. Effects of adding silver nanoparticles on the toughening of dental porcelain. J Prosthet Dent. 2013;109(4):241–7.
87. Lezhnina MM, Grewe T, Stoehr H, Kynast U. Laponite blue: dissolving the insoluble. Angewandte Chemie. 2012;51(42):10652–5. PubMed PMID: 22952053.
88. Gaharwar AK, Mihaila SM, Swami A, Patel A, Sant S, Reis RL, et al. Bioactive silicate nanoplatelets for osteogenic differentiation of human mesenchymal stem cells. Adv Mater. 2013;25(24): 3329–36.
89. Gaharwar AK, Schexnailder P, Kaul V, Akkus O, Zakharov D, Seifert S, et al. Highly extensible bionanocomposite films with direction-dependent properties. Adv Funct Mater. 2010;20(3):429–36.
90. Gaharwar AK, Schexnailder PJ, Dundigalla A, White JD, Matos-Pérez CR, Cloud JL, et al. Highly extensible bio-nanocomposite fibers. Macromol Rapid Commun. 2011;32(1):50–7.
91. Gaharwar AK, Mukundan S, Karaca E, Dolatshahi-Pirouz A, Patel A, Rangarajan K, et al. Nanoclay-enriched poly (ε-caprolactone) electrospun Scaffolds for osteogenic differentiation of human mesenchymal stem cells. Tissue Engineering Part A. 2014; 20(15–16):2088–2101.
92. Shinmura Y, Tsuchiya S, Hata K, Honda MJ. Quiescent epithelial cell rests of Malassez can differentiate into ameloblast-like cells. J Cell Physiol. 2008;217(3):728–38. PubMed PMID: 18663726.
93. Guo WH, He Y, Zhang XJ, Lu W, Wang CM, Yu H, et al. The use of dentin matrix scaffold and dental follicle cells for dentin regeneration. Biomaterials. 2009;30(35):6708–23. PubMed PMID: WOS:000271665300004. English.
94. Bartold P, McCulloch CAG, Narayanan AS, Pitaru S. Tissue engineering: a new paradigm for periodontal regeneration based on molecular and cell biology. Periodontol 2000. 2000;24(1):253–69.
95. Ferreira CF, Magini RS, Sharpe PT. Biological tooth replacement and repair. J Oral Rehabil. 2007;34(12):933–9. PubMed PMID: 18034675.
96. Dolatshahi-Pirouz A, Nikkhah M, Gaharwar AK, Hashmi B, Guermani E, Aliabadi H, et al. A combinatorial cell-laden gel microarray for inducing osteogenic differentiation of human mesenchymal stem cells. Scientific reports. 2014;4:3896.
97. Nakashima M, Reddi AH. The application of bone morphogenetic proteins to dental tissue engineering. Nat Biotechnol. 2003;21(9):1025–32.

Characterization of Nanomaterials/Nanoparticles

3

Koon Gee Neoh, Min Li, and En-Tang Kang

Abstract

In recent years, engineered nanoparticles have garnered increasing attention due to their potential for application in areas ranging from consumer and industrial products to medical diagnostics and therapeutics. This potential arises from the unique physical and chemical properties associated with the high surface-to-mass ratio and quantum phenomena of nanoparticles. Nanoparticles are in the same size range as many biomolecules such as proteins and membrane receptors, and their interactions with these biomolecules can be controlled by tuning the surface property/composition of the nanoparticles. Thus, nanoparticles can serve as useful imaging, diagnostic and therapeutic agents. On the other hand, these nanoparticles can also give rise to cytotoxic effects. Hence, it is imperative to carry out detailed characterization of engineered nanoparticles, especially those intended for medical applications, to predict their behavior in the *in vivo* environment. This chapter describes some methods that are useful for characterizing nanoparticles and their advantages, limitations, and challenges.

Abbreviations

AES	Auger electron spectroscopy
AFM	Atomic force microscopy
ATP	Adenosine 5′-triphosphate
ATR-FTIR	Attenuated total reflectance-Fourier transform infrared spectroscopy
CLSM	Confocal laser scanning microscope
DCS	Differential centrifugal sedimentation
DLS	Dynamic light scattering
DMSO	Dimethyl sulfoxide
ELS	Electrophoretic light scattering
ESCA	Electron spectroscopy for chemical analysis
FESEM	Field emission scanning electron microscopy

K.G. Neoh, ScD (✉) • M. Li, PhD • E.-T. Kang, PhD
Department of Chemical and Biomolecular
Engineering, National University of Singapore,
4 Engineering Drive 4, Singapore 117585, Singapore
e-mail: chenkg@nus.edu.sg

A. Kishen (ed.), *Nanotechnology in Endodontics: Current and Potential Clinical Applications*,
DOI 10.1007/978-3-319-13575-5_3, © Springer International Publishing Switzerland 2015

FTIR	Fourier-transform infrared spectroscopy
HPG	Hyperbranched polyglycerol
HRMAS NMR	High resolution magic angle spinning nuclear magnetic resonance spectroscopy
ICP	Inductively coupled plasma
ICP-MS	ICP-mass spectrometry
ICP-OES	ICP-optical emission spectrometry
INT	2-(4-iodophenyl)-3-(4-nitrophenyl)-5-phenyl tetrazolium chloride
ITC	Isothermal titration calorimetry
LEIS	Low-energy ion scattering
LDH	Lactate dehydrogenase
MTS	5-(3-carboxymethoxyphenyl)-2-(4,5-dimethylthiazolyl)-3-(4-sulfophenyl) tetrazolium inner salt
MTT	3-(4,5-dimethylthiazol-2-yl)-2,5-diphenyltetrazoliumbromide
MTX	Methotrexate
NAD^+	Nicotinamide adenine dinucleotide
NADH	Reduced form of nicotinamide adenine dinucleotide
NMR	Nuclear magnetic resonance
NTA	Nanoparticle tracking analysis
PAN	Polyacrylonitrile
PBS	Phosphate buffered saline
PEG	Polyethylene glycol
PDI	Polydispersity index
ppb	Parts per billion
ppm	Parts per million
ppt	Parts per trillion
RES	Reticuloendothelial system
RF	Radiofrequency
SEM	Scanning electron microscopy
STM	Scanning tunneling microscopy
SWCNTs	Single-walled carbon nanotubes
TEM	Transmission electron microscopy
TOF-SIMS	Time-of-flight secondary-ion mass spectrometry
WST-1	2-(4-iodophenyl)-3-(4-nitrophenyl)-5-(2,4-disulfophenyl)-2H-tetrazolium
WSTs	Water-soluble tetrazolium salts
XPS	x-ray photoelectron spectroscopy
XTT	2,3-bis(2-methoxy-4-nitro-5-sulfophenyl)-2H-tetrazolium-5-carboxanilide inner salt

3.1 Introduction

The rapid progress in nanotechnology has resulted in materials with nanoscale dimensions being used in a diverse range of applications in medicine and in the electronics, chemical, and aerospace industries and also in consumer products like cosmetics and sporting goods. The European Commission's Recommendation states that *'Nanomaterial' means a natural, incidental or manufactured material containing particles, in an unbound state or as an aggregate or as an agglomerate and where, for 50% or more of the particles in the number size distribution, one or more external dimensions is in the size range 1 nm–100 nm* [1]. Nanomaterials behave significantly differently from bulk materials due to two primary effects: (1) surface effects arising from the lower stability of surface atoms as compared to the bulk atoms since surface atoms have fewer neighbors, which result in lower coordination and unsatisfied bonds, and (2) quantum confinement effects in materials with delocalized electrons [2, 3].

Nanomaterials can be classified into three categories: zero-dimensional nanostructures such as nanoparticles, one-dimensional nanostructures such as nanorods, and two-dimensional nanostructures such as thin films. This article focuses on nanoparticles due to their importance in medicine and pharmacology. Nanoparticles are advantageous for application in biomedicine because they can be administered intravenously and be distributed to organs and tissues. With their nanoscale dimensions, the particles can interact closely with cells and cross biological membranes. Nanoparticulate systems with good potential for medical application include metal/inorganic/polymeric nanoparticles, liposomes, dendrimers, and polymer–drug conjugates [4, 5]. These nanoparticles may serve as drug delivery vehicles, diagnostic/imaging agents, or therapeutics, and they are often multicomponent and multifunctional.

The behavior of the nanoparticles in biological systems is highly dependent on their physicochemical properties. Nanoparticles, upon intravascular administration, will immediately encounter blood, which may induce agglomeration and sequestration [6]. Nanoparticles of 10–100 nm are considered optimal for *in vivo* delivery, as smaller ones (<10 nm) are rapidly removed by renal clearance while bigger ones (>200 nm) are quickly sequestered by the reticuloendothelial system (RES) [7]. In addition to size, the surface characteristics of nanoparticles also play an important factor in determining their life span and fate during circulation. Ideally, nanoparticles should have a hydrophilic surface to inhibit their interaction with plasma proteins and avoid uptake by the RES [8]. Nonspecific uptake by cells of the RES will result in a drastic reduction in the efficiency of nanoparticle-based diagnostics and therapeutics [9]. Nanoparticles intended for active targeting of specific receptors on cell surfaces would need to incorporate sufficient targeting ligands on their surfaces, and those serving as drug carriers would be required to release their payload at the intended site. Importantly, the potential side effects, in addition to the intended effects, of the nanoparticles on cells and tissues have to be understood. Thus, the characterization of the properties and effects of nanoparticles is a critical step in their formulation for medical use. The following sections will review the important current techniques for nanoparticle characterization and discuss the challenges that may be encountered.

3.2 Characterization of Physicochemical Properties of Nanoparticles

3.2.1 Size

The size of nanoparticles is a crucial factor in determining their interactions with cells and their distribution in the biological system. The ability of nanoparticles to extravasate from the vasculature and also their clearance from circulation depends on their size [6]. This is an important consideration for nanoparticulate drug carriers since long-circulating ones will have a higher chance of reaching their targets, resulting in improved treatment outcomes. The most commonly used tools for determining nanoparticle size are dynamic light scattering (DLS) and electron microscopy. DLS, also known as photon-correlation spectroscopy, measures the Brownian motion of nanoparticles in a dispersion and relates its velocity or translational diffusion coefficient to the size of the nanoparticles according to the Stokes–Einstein equation:

$$D_h = k_B T / 3\pi\eta D_t$$

where,

D_h is the hydrodynamic diameter
D_t is the translational diffusion coefficient
k_B is Boltzmann's constant
T is thermodynamic temperature
η is dynamic viscosity

In a DLS experiment, a laser beam is directed at the nanoparticle dispersion, and fluctuations in the intensity of the scattered light are monitored with a photon detector and related to the size of a hypothetical hard sphere that diffuses in the same fashion as the nanoparticles being measured. Thus, the equivalent-sphere hydrodynamic diameter obtained from DLS provides no information about the shape of the nanoparticles. The size distribution of the nanoparticle dispersion, as indicated by the polydispersity index (PDI), can also be obtained from the DLS measurement. The larger the PDI, the broader is the size distribution, and a PDI value from 0.1 to 0.25 indicates a narrow size distribution [10]. DLS is a very popular technique for measuring the size of nanoparticles because it requires minimal sample preparation and can be rapidly carried out (a few minutes per run) with small volumes of dilute dispersions. However, the results obtained from DLS can be skewed by the presence of aggregates or dust since the intensity of the light scattered by the nanoparticle varies as the sixth power of its diameter. Thus, for DLS measurement, it is critical to ensure that the nanoparticles are well dispersed. The usefulness of DLS is also limited if a multimodal particle size distribution is present. For example, when a mixture of 20 and 100 nm nanoparticles is measured, the signal of smaller particles is lost because the scattering intensity of small particles is masked by that of larger particles

[10]. The concentration of the dispersion may also affect the hydrodynamic size, as illustrated by Lim et al. using magnetic nanoparticles without surface coating dispersed in deionized water [11]. A sample that is too dilute may not result in sufficient scattering events for a proper measurement, while multiple scattering can occur in a highly concentrated sample. Furthermore, nanoparticles at high concentration have a higher tendency to aggregate. In view of the sensitivity of DLS to the presence of aggregates, this technique is useful for monitoring colloidal stability. This technique has been used by a number of investigators to assess the stability of nanoparticles in physiologically relevant media [12–14].

Electron microscopy can be carried out to complement the DLS technique for particle size measurements. Scanning electron microscopy (SEM) utilizes a high-energy electron beam to scan over the surface of the sample. The X-rays, backscattered electrons, and secondary electrons resulting from the interaction of the sample with the electron beam are collected and converted to provide information about the sample, including its surface features, size and shape of the features, and composition. The scanning electron microscope operates at high vacuum, and the specimen must be electrically conductive or sputter coated with a conductive layer (e.g., platinum). For SEM characterization of nanoparticles, the dry sample can be mounted directly on a sample holder. In conventional SEM, the electron beam is emitted from an electron gun with tungsten or LaB_6 filaments. By replacing these with a field emission gun, field emission scanning electron microscopy (FESEM) with higher resolution can be achieved. Transmission electron microscopy (TEM) also utilizes an electron beam to interact with the sample under high vacuum, but in this case, the transmitted electrons are detected. TEM provides higher resolution than SEM and provides more information at the atomic scale such as the crystal structure, but the sample must be thin enough to allow electron penetration [4]. For TEM characterization of nanoparticles, a dispersion of the nanoparticles is usually deposited directly onto support grids or films.

Sample preparation for electron microscopy may introduce artifacts, e.g., the dehydration of colloidal nanoparticles can affect the structure and morphology of the sample [15]. Some nanoparticles such as liposomes or polymers are invisible to TEM without heavy metal staining (e.g., with phosphotungstic acid) since they do not deflect the electron beam sufficiently [4]. As such, particle sizes determined by TEM and DLS can differ substantially. This is illustrated by a comparison of these two methods for measuring the size of magnetic nanoparticles either coated with oleic acid or grafted with a hydrophilic polymeric coating [16]. The TEM image of the oleic acid–stabilized magnetic nanoparticles dispersed in hexane shows that these nanoparticles are nearly monodispersed with an average diameter of about 12 nm (Fig. 3.1a). This value is close to the mean hydrodynamic diameter of ~15 nm as determined by DLS. The presence of electron-transparent oleic acid chains (chain length ~2 nm [17]) on the surface of the magnetic nanoparticles may have contributed to the slightly larger size as determined by DLS. In the case of magnetic nanoparticles grafted with hyperbranched polyglycerol (HPG) and methotrexate (MTX), the TEM image (Fig. 3.1b) indicates similarity in size with the oleic acid–stabilized magnetic nanoparticles even though its magnetic core constitutes less than 35 % of the total weight since in both cases TEM only shows the electron-dense magnetic nanoparticle core. On the other hand, the hydrodynamic diameter of these nanoparticles has increased to ~32 nm, and this significant increase in hydrodynamic diameter is attributed to the swelling of the hydrophilic HPG shell in water during the DLS measurement.

3.2.2 Surface Charge

Surface charge is another factor that plays an important role in determining how nanoparticles interact with proteins and cell membranes and their subsequent uptake into the cells [18, 19]. Furthermore, the surface charge of nanoparticles

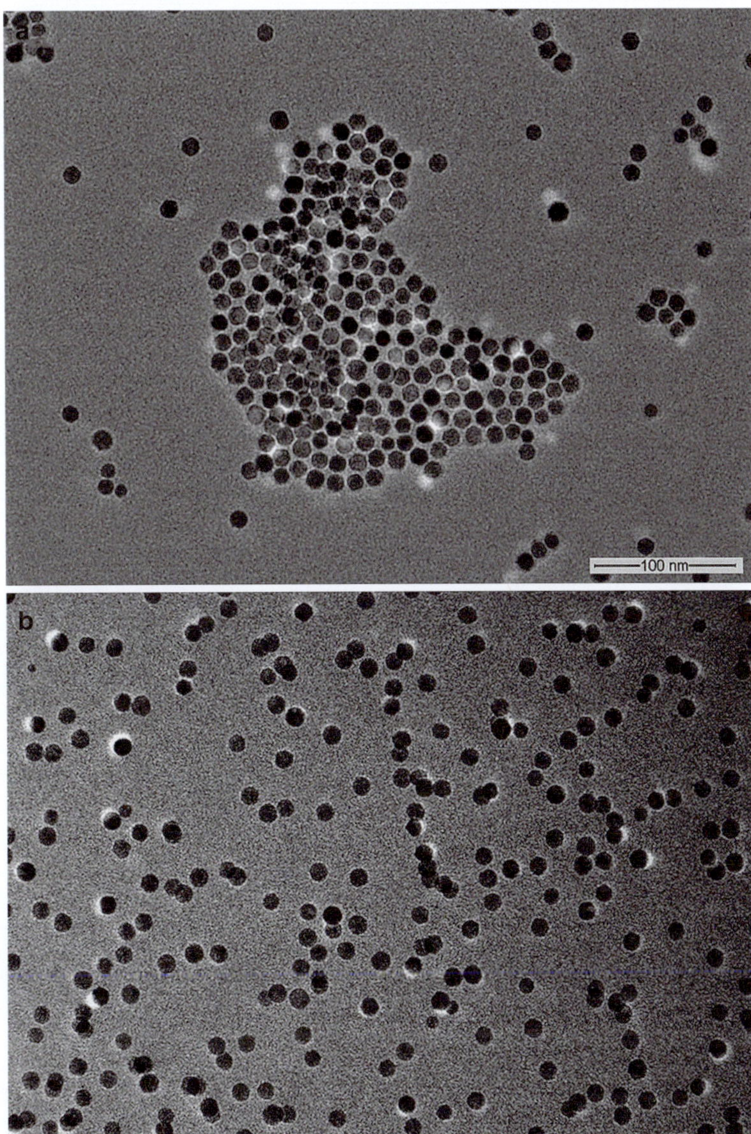

Fig. 3.1 TEM images of (**a**) oleic acid-stabilized magnetic nanoparticles dispersed in hexane, and (**b**) magnetic nanoparticles grafted with HPG and MTX dispersed in water (Reprinted from Li et al. [16]. With permission from Elsevier)

also affects their colloidal stability since it directly impacts the electrostatic repulsion of nanoparticles in dispersion [20]. The surface charge of a nanoparticle in dispersion affects the distribution of ions in the surrounding interfacial region, resulting in an increased concentration of counterions near the surface. The liquid layer surrounding the particle exists as two parts: the Stern layer, which is the region closest to the surface, where the ions are considered immobile, and an outer region that allows diffusion of ions. In this surrounding electric double layer, there is a notional boundary (slipping plane) within which the liquid moves together with the particle. The potential measured at this slipping plane is the zeta potential, which is not exactly the surface charge but is the potential of practical interest in dispersion stability because it determines the interparticle forces [21].

Zeta potential can be determined using electrophoretic light scattering (ELS), also known as

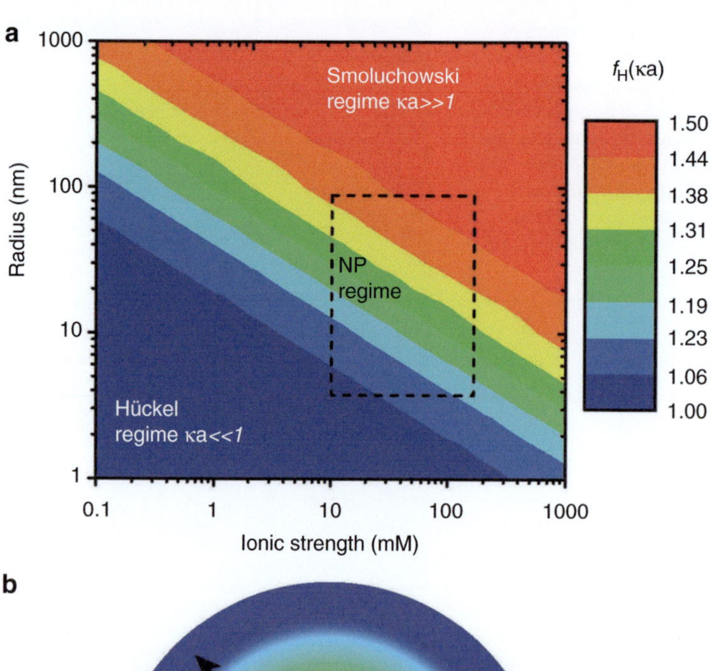

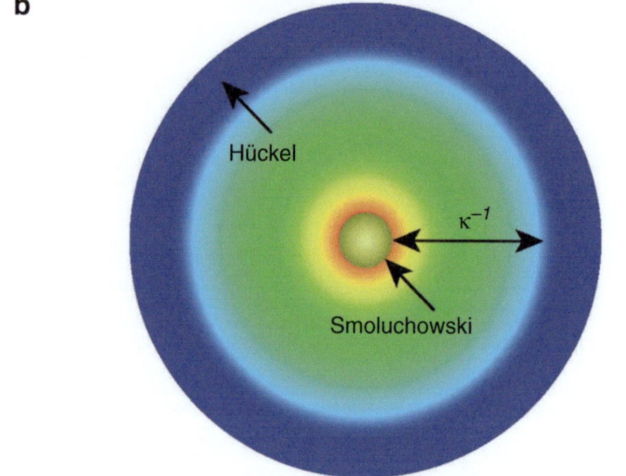

Fig. 3.2 (**a**) Henry's factor $f_H(\kappa a)$ colored according to the analyte radius and solution ionic strength (both with logarithmic scales): the Smoluchowski [$f_H(\kappa a) = 3/2$] and Hückel [$f_H(\kappa a) = 1$] limits are identified by *red* and *blue*, respectively, with the rectangle identifying particle radii and ionic strengths prevailing in common nanoparticle applications. (**b**) The Debye layer thickness κ^{-1} colored according to the accompanying Henry factor $f_H(\kappa a)$ in (**a**) (Reprinted from Doane et al. [22]. With permission from American Chemical Society)

laser Doppler electrophoresis or laser Doppler velocimetry. In this experiment, an electric field is applied across a colloidal dispersion. Charged particles in the dispersion will move toward the electrode of opposite polarity (electrophoresis) with a velocity or electrophoretic mobility that is proportional to the zeta potential. A laser beam is incident on the dispersion, and the electrophoretic mobility can be determined from the frequency shift of the light scattered by the moving particles. The measured electrophoretic mobility (U_E) is converted into zeta potential (ζ) through Henry's equation [22]:

$$U_E = 2\varepsilon\varepsilon_0 \zeta f_H(\kappa a) / 3\eta$$

where,

ε is the dielectric constant of the dispersant

ε_0 is the vacuum permittivity

$f_H(\kappa a)$ is the Henry's function

η is the viscosity

The dimensionless product κa describes the ratio of the particle radius (a) to the "thickness" of the double layer (κ^{-1}). At large κa (~100), the equation can be simplified to the Smoluchowski approximation with $f_H(\kappa a) = 1.5$ [23]. On the other hand, when $\kappa a \ll 1$, $f_H(\kappa a) = 1$, and this is known as the Hückel approximation. Doane et al. have provided a contour plot based on Ohshima's approximation to Henry's formula identifying the Smoluchowski and Hückel regimes for different nanoparticle sizes and solution ionic strengths, as shown in Fig. 3.2 [22].

A number of commercial instruments are available for the measurement of zeta potential. While the measurement is easily carried out with a dispersion of the nanoparticles, the effect of temperature, ionic strength, and pH of the dispersion on zeta potential has to be recognized [10, 24, 25]. A 5 °C variation in temperature can lead to significant changes in zeta potential [25]. High ionic strength decreases the electric double layer, resulting in a reduction of the absolute value of zeta potential, and divalent ions have a stronger effect than monovalent ions. At low pH, the nanoparticles acquire more positive charges, and the zeta potential decreases as the pH is increased. Thus, it is crucial to provide precise information on the dispersion when reporting zeta potential measurements.

3.2.3 Surface Composition

Nanoparticles possess a very high surface area–to-volume ratio, and the importance of the nanoparticle surface property and chemistry in determining their suitability for specific bioapplications cannot be underestimated. For example, there has been increasing interest in the use of superparamagnetic iron oxide nanoparticles for biomedical applications since these nanoparticles have low toxicity and are suitable for *in vivo* applications as they are biodegradable, with the iron product being recycled by cells. But mono-dispersed superparamagnetic iron oxide nanoparticles are often synthesized via a high-temperature decomposition process where oleic acid is used to stabilize the formed product [26, 27]. As a result, the nanoparticles disperse well in organic solvents but not in aqueous media. For biomedical applications, the surface of these nanoparticles has to be modified with a hydrophilic and biocompatible coating, and depending on the desired application, ligands may also be bound on the nanoparticle surface to promote interactions with cellular receptors. Thus, analysis of the surface composition of the nanoparticles is needed to confirm the success of the tailoring process.

X-ray photoelectron spectroscopy (XPS), also known as electron spectroscopy for chemical analysis (ESCA), is one of the most widely used techniques for analyzing the surface composition of nanoparticles. The popularity of XPS stems from its ability to (a) identify and quantify the elemental composition of the outer 10 nm or less of any solid surface for all elements from lithium to uranium on the assumption that the element of interest exists at > 0.05 atomic %. Hydrogen and helium are not detectable due to their extremely low photoelectron cross sections, (b) reveal chemical environment where the respective element exists, and (c) obtain the above information with relative ease and minimal sample preparation [28]. In XPS measurement, the sample is placed in a chamber under ultrahigh vacuum and subjected to x-rays, most commonly Al Kα x-rays (photon energy = 1486.7 eV). An electron is ejected from an atomic energy level by the X-ray photon, and its energy is analyzed by the spectrometer. By measuring the kinetic energy of the emitted photoelectrons, E_K, and knowing the photon energy $h\nu$, the binding energies of the photoelectrons, E_B, can be calculated using the Einstein equation:

$$E_B = h\nu - E_K - \varphi$$

where φ, the work function of the instrument, can be determined by calibration. It can be seen from the above equation that only binding energies lower than the exciting radiation can be probed.

Since the binding energies of the electron orbitals in atoms are known, the positions of the peaks in the XPS spectrum can be used to identify the atomic surface composition of the sample. Each element has a characteristic electronic structure and thus a characteristic XPS spectrum. Tables of binding energies for identifying and interpreting XPS spectra are available in the literature [29]. The usefulness of XPS for characterizing nanoparticles can be illustrated by using this technique for monitoring changes in the surface composition of magnetic nanoparticles subjected to surface modification, as shown in Fig. 3.3 [16]. Oleic acid–coated magnetic nanoparticles were first grafted with 3-(trimethoxysilyl)propylmethacrylate (MPS), and HPG was then grafted on the MPS-modified magnetic nanoparticles via thiol-ene click reaction.

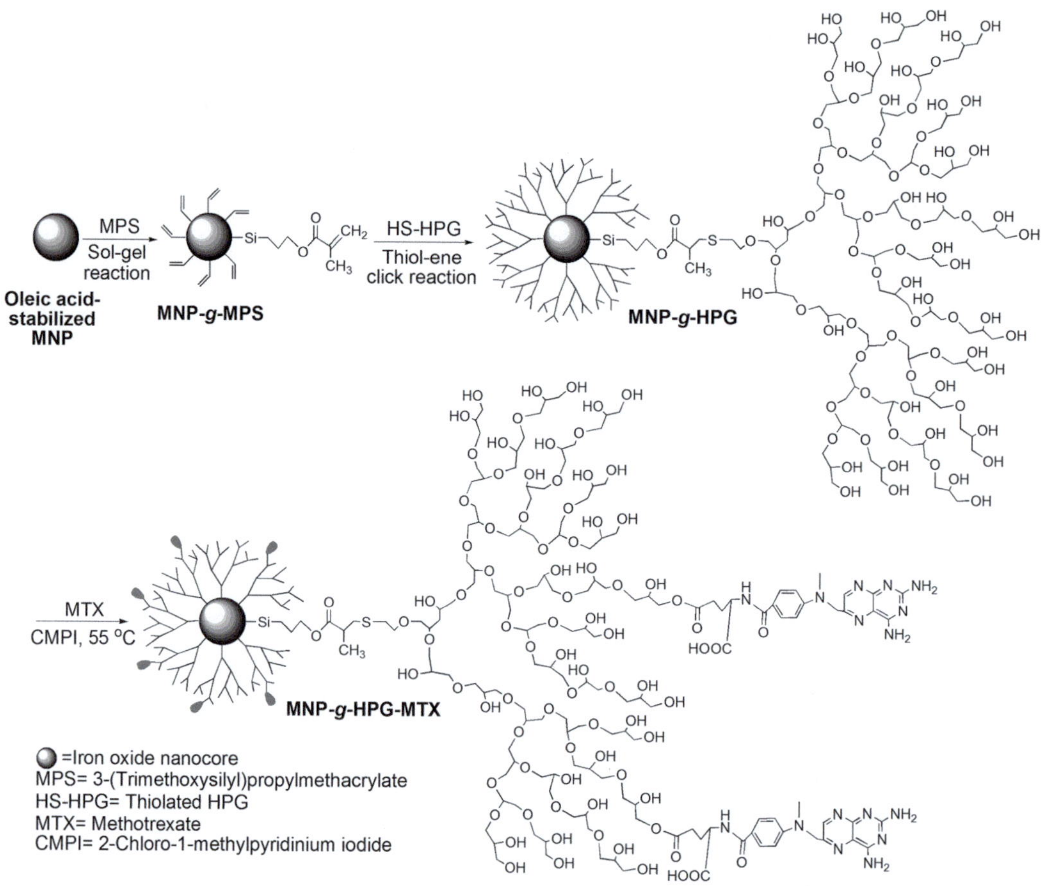

Fig. 3.3 Schematic illustration of the synthesis of HPG-grafted magnetic nanoparticles with conjugated MTX (Reprinted from Li et al. [16]. With permission from Elsevier)

Figure 3.4a shows the wide scan or survey scan obtained at low resolution of oleic acid–coated magnetic nanoparticles. The C 1 s peak associated with oleic acid, Fe 2p and Fe 3p peaks arising from the magnetic nanoparticles, and O 1 s peak from both oleic acid and magnetic nanoparticles can be identified. Figure 3.4b shows the corresponding spectrum of these nanoparticles after surface grafting with HPG. The increase in the O 1 s signal relative to the C 1 s signal is in line with the higher O/C ratio of HPG compared to oleic acid. The S 2p signal at a binding energy of about 168 eV (Fig. 3.4b) arises from the grafting of thiolated HPG via thiol-ene click reaction on the magnetic nanoparticles [16]. Thus, a comparison of Fig. 3.4b with Fig. 3.4a provides supporting evidence that HPG has been grafted on the surface of the magnetic nanoparticles. The

wide scan identifies the elements present, and quantification of the chemical composition as percentage atomic concentrations is possible from the peak area of each element corrected by the relative sensitivity factors (specific to the instrument) [30].

XPS not only allows the identification of the elements constituting the sample but also provides information on their chemical state based on the binding energy shift (or chemical shift). For example, the C 1 s high-resolution core-level spectrum of the oleic acid–coated magnetic nanoparticles in Fig. 3.4c shows a dominant peak at 284.6 eV attributable to C-C/C-H groups and a much smaller peak at 288.4 eV attributable to the COOH group of oleic acid. The shift of the COOH peak to a higher binding energy relative to the C-C/C-H peak arises from a decrease in

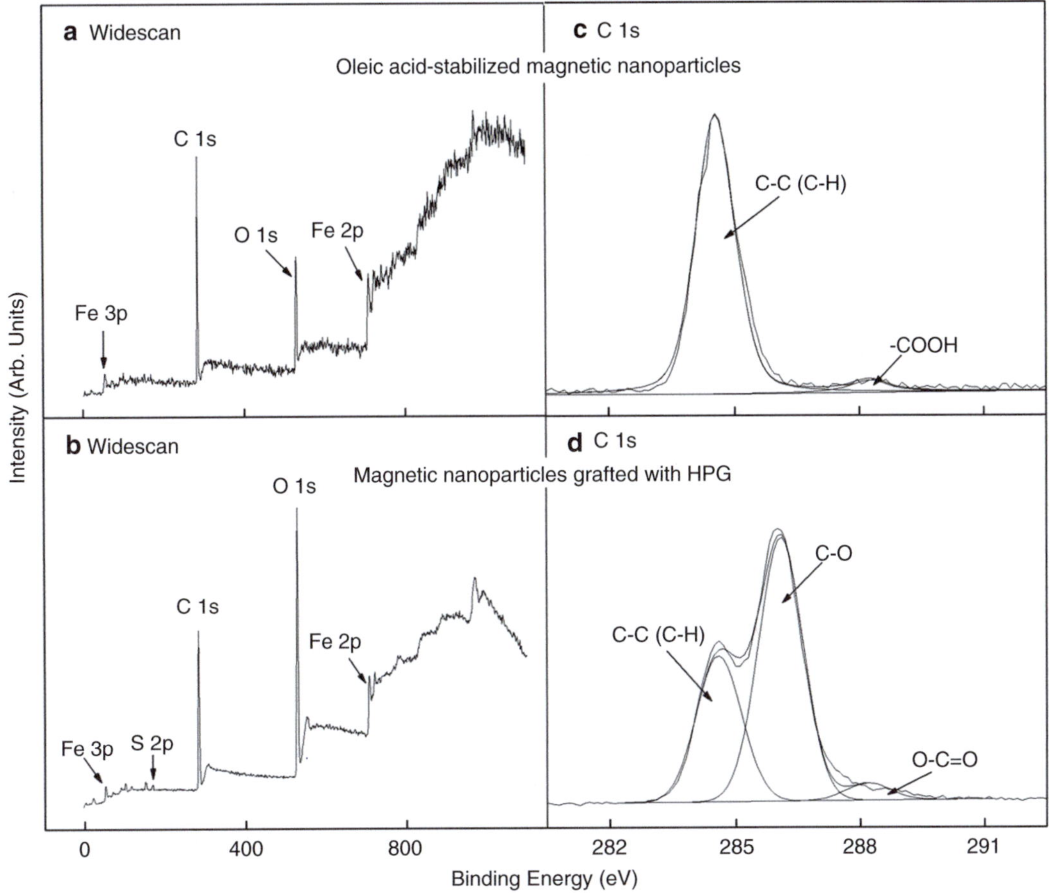

Fig. 3.4 XPS widescan spectra of (**a**) oleic acid-stabilized magnetic nanoparticles, and (**b**) magnetic nanoparticles grafted with HPG, and C 1 s core-level spectra of (**c**) oleic acid-stabilized magnetic nanoparticles, and (**d**) magnetic nanoparticles grafted with HPG

electron density in the C valence shell due to bonding to oxygen, which results in an increase in the binding energy of the C 1 s core orbital. The ratio of the peak areas gives the relative proportion of each component on the surface. In Fig. 3.4c, the area ratio of the C-C/C-H peak to the COOH peak is calculated to be 17.6:1, which is close to the theoretical ratio of 17:1 expected from the chemical structure of oleic acid. In Fig. 3.4d, the C 1 s peak envelope arising from C being present in multiple environments is more complex than in Fig. 3.4c, where two peak components can be clearly delineated. The peak envelope in Fig. 3.4d can be deconvoluted into three-component peaks at 284.6, 286.1, and 288.3 eV with the same full width at half maximum (FWHM). The peak at 284.6 eV is assigned to C-C/C-H groups as in Fig. 3.4c, while the dominant peak at 286.1 eV is attributed to C-O groups, consistent with the chemical structure of HPG. The minor O-C=O peak is likely to be from the MPS anchoring groups which did not react during the thiol-ene click reaction.

XPS is clearly a powerful technique capable of identifying a wide range of elements and their chemical states on surfaces. The analysis can be readily carried out, but the sample must be compatible with the ultrahigh vacuum environment. Since XPS is a surface-sensitive technique, cleanliness during sample preparation and during analysis is imperative to obtain the true surface composition. It must also be emphasized that although peak deconvolution and peak fitting tools are commercially available, curve fitting of peak envelopes must be undertaken with caution especially in the presence of overlapping peaks. Poorly fitted spectra with arbitrariness in peak width and shape can lead to incorrectly interpreted XPS data. More detailed discussions on curve fitting of core-level XPS data and the importance of combining curve fitting information with other supporting information and experiments are available in the literature [28, 31].

Auger electron spectroscopy (AES) is another common surface analysis tool for nanoparticles. Both AES and XPS detect electrons emitted from samples with kinetic energies typically below 2,000 eV, but in AES, the sample is irradiated with electrons instead of x-rays [32]. AES and XPS provide similar information, but AES gives a higher lateral resolution since the electron beam can be focused to a smaller size than x-rays. However, the electron beam can also impart more damage to the sample surface than x-rays. Baer et al. have provided a comparison of the types of information that can be extracted from nanoparticles using the surface analysis methods, AES, XPS, time-of-flight secondary-ion mass spectrometry (TOF-SIMS), low-energy ion scattering (LEIS), scanning tunneling microscopy (STM), and atomic force microscopy (AFM), as shown in Table 3.1. The ion-based techniques, TOF-SIMS and LEIS, can provide information on surface chemical composition to complement the electron spectroscopic techniques, but the scanning probe microscopy techniques, STM and AFM, are more suited for extracting information on physical properties like size, morphology, and surface heterogeneity.

Fourier transform infrared spectroscopy (FTIR) has been used for identifying functional groups on nanoparticles. In this technique, infrared radiation is passed through a sample, and some of the radiation is absorbed by the sample. The resulting spectrum shows a series of absorption bands characteristic of the frequency of vibrations of bonds between the atoms making up the sample. Since each sample comprises a unique combination of atoms, the spectrum provides a distinct identification of the sample akin to that of a molecular "fingerprint." The region between 2.5 and 25 μm (4,000–400 cm^{-1}) is most frequently used in modern infrared instruments [33]. The preparation of solid nanoparticle samples for FTIR analysis usually involves the milling of the sample with KBr and then compressing the fine powder into a thin pellet which is placed in the sample holder of the FTIR spectrometer. KBr is transparent to IR radiation and hence will not interfere with the absorption spectrum. However, since KBr is hygroscopic, precaution must be taken to ensure that the KBr is dry otherwise absorption bands from the adsorbed water may interfere with the sample's spectrum.

The use of FTIR for identifying the functional groups on nanoparticles is illustrated using the example of HPG-grafted magnetic nanoparticles discussed above. The FTIR spectrum of this sample shows dominant absorption bands at about 3,000–3,500 cm^{-1} attributed to the stretching vibrations of –OH groups, at 2,926 and 2,871 cm^{-1} due to CH_2 stretching vibrations, and at 1,086 cm^{-1} associated with the stretching vibrations of C–O–C ester groups (Fig. 3.5a). These groups are characteristic of the HPG graft layer. The HPG-grafted magnetic nanoparticles can be modified to be a drug delivery system for targeting cancer cells by conjugating MTX on the nanoparticle surface. MTX is one of the most widely used chemotherapy drugs for the treatment of many types of cancers [34, 35]. Furthermore, since MTX is an analogue of folic acid and the folate receptor is overexpressed on many types of human cancer cells, the MTX-functionalized nanoparticles can potentially act as a cancer-targeting agent in addition to being an anticancer drug [36]. The FTIR spectrum of the MTX-functionalized nanoparticles (Fig. 3.5b) can be compared with those of the HPG-grafted nanoparticles before conjugation with MTX (Fig. 3.5a) and free MTX (Fig. 3.5c). The doublet absorption bands at 1,646 and 1,605 cm^{-1}, which are the characteristic peaks of the amide groups of free MTX, appear in the spectrum of the MTX-functionalized nanoparticles but not the HPG-grafted nanoparticles, thus confirming the successful conjugation of MTX in the former.

Table 3.1 Types of information that can be extracted from nanoparticles using surface analysis tools, with several characteristics of the methods

	Information available	Probe	Detected	Lateral resolution	Information depth	Depth resolution
Electron spectroscopies						
Auger electron spectroscopy (AES)	Surface composition of individual large nanoparticles or distribution of smaller nanoparticles (depending on spatial resolution of specific instrument) Enrichment or depletion of elements at surface Presence and/or thickness of coatings and/or contaminants	Electrons (~3 to 20 keV)	Auger electrons	10 nm	~10 nm	~2 nm
X-ray photoelectron spectroscopy (XPS)	Analysis of a collection of particles deposited on a substrate or other support Surface composition and chemical state Presence and nature of functional groups on the surface Enrichment or depletion of elements at surface Presence and/or thickness of coatings or contaminants Nanoparticle size (when smaller than ~10 nm, can sometimes determine average particles size when too small to be detected by other methods or in complex matrix) Electrical properties of nanoparticles and coatings	X-rays	Photoelectrons	~2 μm	~10 nm	~2 nm

(continued)

Table 3.1 (continued)

	Information available	Probe	Detected	Lateral resolution	Information depth	Depth resolution
Incident ion methods						
Time-of-flight secondary-ion mass spectrometry (TOF-SIMS)	Usually analysis of a collection of particles or larger individual particles deposited on a supporting substrate Presence of surface coatings or contaminants on collections of nanoparticles Functional groups on surface	Ions (~3 to 20 keV)	Sputtered ions	~50 nm (inorganic); >200 nm (organic)	~1 nm (inorganic); ~1 nm (organic)	~1 nm (inorganic); ~10 nm (organic)
Low-energy ion scattering (LEIS)	Presence of ultra thin coating or contamination Effects of size	Ions (~2 to 10 keV)	Elastically scattered ions	~100 μm	~10 nm	~0.2 nm
Scanning probe microscopies						
Scanning tunneling microscopy (STM)	Electrical characteristics of individual nanoparticles Nanoparticle formation and/or size distribution of particles deposited or grown on a surface	Stylus	Tunneling current	~1 nm	~10 nm	
Atomic-force microscopy (AFM)	Shape, texture and roughness of individual particles and their distribution for an assembly of particles When particles structure is known, can provide information about crystallographic orientation	Stylus	Force or displacement	~1 nm	~10 nm	

Reprinted from Baer et al. [32]. With permission from Springer-Verlag

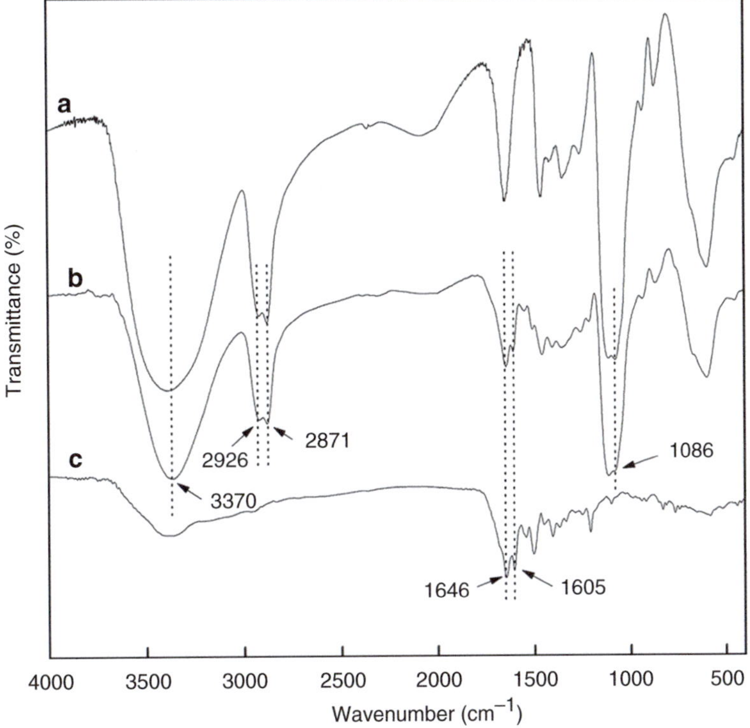

Fig. 3.5 FT-IR spectra of (*a*) magnetic nanoparticles grafted with HPG, (*b*) magnetic nanoparticles grafted with HPG and MTX and (*c*) free MTX

In addition to providing identification of functional groups on nanoparticles, FTIR can also be used for quantitative analysis. For quantitative analysis, calibration curves will first have to be established. Shan et al. used FTIR to quantify the amount of hemoglobin encapsulated in polymeric nanoparticles fabricated by the double emulsion method from poly(ethylene glycol)–poly(lactic acid)–poly(ethylene glycol) triblock copolymers [37]. Polyacrylonitrile (PAN) was chosen as an internal standard, and a constant amount was introduced to the calibration and test samples in order to avoid errors arising from the environment and laboratory manipulation. The calibration samples comprise hemoglobin, PAN, and blank polymeric nanoparticles. The IR absorbances at 1,540 cm^{-1} of hemoglobin (amide II) and at 2,241 cm^{-1} of PAN ($-C\equiv N$) were measured to establish the calibration equations relating the absorption band height ratio of hemoglobin and PAN as a function of the weight ratio of hemoglobin and PAN. From the absorbances of hemoglobin and PAN measured in the spectra of the test samples, these equations were then used to calculate the amount of encapsulated hemoglobin.

Although FTIR in the transmission mode is the most common, attenuated total reflectance–Fourier transform infrared spectroscopy (ATR-FTIR) has also been widely used. Unlike transmission FTIR, in which the IR beam travels in a straight path through a sample, in ATR, the infrared beam is directed into a high–refractive index crystal where it undergoes total internal reflection. The reflected IR beam creates an evanescent wave, which projects into the sample placed in contact with the crystal. Some of the energy of the evanescent wave is absorbed by the sample, and the attenuated radiation is passed to the detector. Tsai et al. used modified ATR-FTIR to carry out a quantitative study of competitive molecular adsorption on nanoparticles [38]. In these experiments, Au nanoparticles were deposited from a dispersion to form a film onto the germanium ATR crystal of the FTIR spectrometer. A flow cell was attached to the top of the ATR crystal, solutions with various concentrations of molecular adsorbates were then introduced into the flow cell, and the IR spectrum was collected. The surface density of the adsorbed molecules on the Au nanoparticles was computed by comparing the signal intensity of the IR band related to the adsorbate with the calibration curve established using the unbound molecule in deionized water.

Another method which is useful for characterizing organic molecules grafted on the surface of nanoparticles is high-resolution magic angle spinning nuclear magnetic resonance spectroscopy (HRMAS NMR). Nuclear magnetic resonance is based on the principle that all nuclei are positively charged and many nuclei, including ^{1}H and ^{13}C, have spin. If the number of neutrons and the number of protons in the nucleus are both even, it will not have a net spin. In nuclear magnetic resonance (NMR), the unpaired nuclear spins are of importance, and the behavior of the ^{1}H and ^{13}C nuclei has been exploited to deduce the structure of organic compounds. In NMR spectroscopy, the sample is placed in a magnetic field and excited via radiofrequency radiation to induce transitions between different nuclear spin states [39]. The precise resonance frequency of the energy transition is dependent on the effective magnetic field at the nucleus. Different atoms in a molecule experience slightly different magnetic fields due to their different local environments. Hence, information about the nucleus' chemical environment can be derived from its resonance frequency. In the NMR spectrum, the resonance frequency is expressed as chemical shift, which is the frequency of the signal relative to that of a standard molecule and divided by the spectrometer frequency. The unit of chemical shift is parts per million (ppm). In general, the more electronegative the nucleus is, the larger the chemical shift.

NMR has been the gold standard for characterization of small molecules, but conventional NMR signals of molecules attached to nanoparticles often lead to significant line broadening [40]. However, it has been known for over 40 years that if a solid sample is spun at an angle of 54.74° (the magic angle), the line width of signals in the NMR spectrum is significantly reduced [41]. Polito et al. employed HRMAS NMR to analyze carbohydrate-derivatized magnetic nanoparticles prepared via a surface diazo transfer/azide–alkyne click reaction [42]. They reported that the signals were sharp and well resolved, and the presence of a peak at ca. 8 ppm in the ^{1}H NMR spectra attributed to the triazole proton confirmed the success of the 1,3 dipolar cycloaddition. Zhou et al. used ^{1}H HRMAS NMR to fully characterize ligand structures on surfaces of gold nanoparticles [40]. They found that there are significant differences in detection sensitivity depending on the distance between the surface of the gold nanoparticles and protons in the ligand molecule, with a loss of sensitivity for protons closer to the nanoparticle surface.

3.2.4 Protein Adsorption

When nanoparticles are intravenously administered into the body, circulating plasma proteins can readily be adsorbed on the nanoparticle surface. The adsorbed proteins form a corona, and many of these proteins act as opsonins, which are capable of interacting with the specialized plasma membrane receptors on monocytes and macrophages [43, 44]. This leads to recognition and engulfment of the nanoparticles by these cells and the rapid removal of the nanoparticles (within minutes) from the blood circulation system. It is generally accepted that nanoparticles with neutral and hydrophilic surfaces will have a longer half-life [45, 46]. Thus, polymers like polyethylene glycol (PEG) or its derivatives [47, 48] and HPG [16, 49] have been grafted on nanoparticle surfaces to inhibit protein adosrption. Understanding the nanoparticle–protein interactions will provide insights into how the surface of nanoparticles should be modified for *in vivo* applications.

Isothermal titration calorimetry (ITC) has become one of the foremost tools for the study of the interaction of nanoparticles with proteins [50]. It can measure the quantity of protein adsorbed, the affinity of the protein for the surface, and the thermodynamics of the adsorption process in a single experiment [51]. In the ITC experiment, aliquots of protein are injected into a dispersion of nanoparticles, and the evolution of heat as a function of time is compared to that obtained with a control in the absence of the nanoparticles to account for the heat of dilution. At the completion of the experiment, the integrated heat is plotted as kJ/mol protein versus protein/nanoparticle ratio [51]. Using an equilibrium model, the binding constant for protein adsorption, the enthalpy of adsorption, and the number of binding sites on the nanoparticle can be estimated from the ITC isotherm as described by Becker et al. [51]. The recent studies on the interaction of bovine serum albumin with polyethyleneimine-functionalized ZnO nanoparticles [52] and with negatively and positively charged polystyrene nanoparticles [53] illustrate the usefulness of ITC.

Fluorescence spectroscopy is another method used to study nanoparticle–protein interaction. In this technique, a beam of light is incident on the sample, and the fluorescent light that is emitted is collected by a detector. In many of these studies, fluorescent-labeled proteins are used to allow visualization of the uptake of proteins by particles and to study the structural properties of the adsorbed proteins [50]. Alternatively, the technique may be used with nanoparticles, which are either fluorescent like Au nanoclusters [54] and quantum dots or modified with an organic fluorophore [55]. However, labeling of the protein or nanoparticle surface with a fluorescent tag may interfere with the subsequent protein–surface interactions. Bovine serum albumin, a model protein used in numerous studies on protein–nanoparticle interaction, possesses specific fluorescence properties due to the existence of two tryptophan residues in different positions (on the surface and in the hydrophobic pocket) [56]. Tryptophan is also highly sensitive to modifications of the local environment, and as such, its fluorescence emission spectrum is well suited for the study of protein conformational changes and binding to nanoparticles [56, 57]. Iosin et al. monitored the fluorescence spectra of bovine serum albumin solutions with different concentrations of added Au nanoparticles in the spectral region between 290 and 450 nm [56]. The fluorescence measurements were recorded using an excitation wavelength of 280 nm, which is far away from the plasmon resonance band of the Au nanoparticles. The intrinsic fluorescence band of bovine serum albumin centered at 336 nm (from emission of tryptophan residues) was quenched with increasing Au nanoparticle concentration. From the fluorescence spectra, the binding constant and the numbers of binding sites between Au nanoparticles and bovine serum albumin were determined.

There are a number of other methods besides ITC and fluorescence spectroscopy for studying the interaction between proteins and nanoparticles. Bell et al. quantified the amount of immunoglobulin G protein adsorbed on citrate-stabilized Au nanoparticles using three particle size measurement techniques, DLS, nanoparticle tracking analysis (NTA), and differential centrifugal sedimentation (DCS), as well as UV-visible spectroscopy [58]. For the DLS experiments, the thickness of the protein shell was calculated as half the difference between the diameter measured by DLS for the protein-coated particle and the corresponding diameter for the citrate-stabilized gold particle. The NTA method relates the rate of Brownian motion to particle size, and the equivalent-sphere hydrodynamic radius is calculated through the Stokes–Einstein equation. The thickness of the protein shell was calculated in the same manner as in DLS. DCS measures particle size based on the particle sedimentation rate, which depends upon its size and density. Interpretation of the results requires knowledge of the core particle size and the densities of the core particle, protein shell, and the surrounding fluid. In the UV-visible spectroscopy method, the extinction spectrum due to the gold-localized surface plasmon resonance was recorded. The position of the plasmon peak is sensitive to the size of the gold core and the refractive index of the media surrounding the core. For a protein shell, the peak position changes with both the density and thickness of the shell, and simulations of plasmon shift were carried out for particles with a range of shell thickness and refractive index. The authors concluded what with complete protein coverage, the thicknesses of the protein shell calculated from all these methods were consistent to within ~20 %.

3.3 *In Vitro* Assays of Cellular Uptake and Cytotoxicity of Nanoparticles

3.3.1 Cellular Uptake

As mentioned in Sect. 2.4, it is critical that nanoparticles introduced into the circulatory system are able to resist the adsorption of plasma proteins in order to avoid premature clearance by the body's immune system. On the other hand, for nanoparticles intended for target-specific imaging or drug delivery applications, it would be desirable for them to be taken in by the targeted cells with high efficacy. This is usually achieved by conjugating targeting ligands, which interact with specific cell receptors, on the nanoparticles, e.g., using folic acid [59–61] or MTX [62, 63] for targeting the folate receptor which is expressed at elevated levels on many human tumors but is generally absent in most normal tissues [64, 65]. Ideally, these nanoparticles can avoid capture by the RES and instead accumulate in the targeted cells [16, 66]. *In vitro* assays of cellular uptake are often carried out to predict the *in vivo* fate of the nanoparticles.

In *in vitro* assays of cellular uptake of nanoparticles, the cells of interest are first seeded on culture plates and then exposed to culture media containing different concentrations of the nanoparticles followed by a period of incubation. If the nanoparticles are fluorescent, the nanoparticles taken in by the cells can be visualized under a confocal microscope [67, 68]. In a confocal microscope, confocal pinholes allow only light from the plane of focus to reach the detector. This produces very clear images and enables the collection of "optical slices" to reconstruct three-dimensional representations of the sample. The most common type of confocal microscope is the confocal laser scanning microscope (CLSM). By labeling the various organelles of the cells such as cell membrane, nucleus, and lysosomes, the localization of the nanoparticles within the cells can be observed. An example is given in Fig. 3.6, which shows the CLSM images of lysosome, three types of chitosan-based nanoparticles with different surface charges, and their overlay signals after the incubation of these nanoparticles with HKC human proximal tubular epithelial cells for 6 h [69]. The results indicate that both the negatively charged nanoparticles (N-NPs) and the neutral nanoparticles (M-NPs) are highly co-localized with the lysosome, while some of the positively charged nanoparticles (P-NPs) escape from the lysosome into the cytoplasm.

For nanoparticles that are not inherently fluorescent, a fluorescent marker may be entrapped within the nanoparticle or covalently conjugated to the nanoparticle. Since the presence of a fluorescent marker may affect protein adsorption, it may also affect cellular uptake. On the other hand, a potential problem with physical entrapment is that the marker may leach out in the culture medium and subsequently contribute to the intracellular fluorescence. To evaluate the extent of the leaching problem, labeled nanoparticles can be subjected to dialysis, and the fluorescence curves of the labeled nanoparticles before and after dialysis can be measured. If the fluorescence curves are identical, it would rule out that the marker was released [70]. The contribution of any leached marker to intracellular fluorescence can be assessed by first carrying out a control experiment where fluorescent-labeled nanoparticles are suspended in the culture medium (without cells) for the period of the assay and then using the culture medium with the released marker for the incubation of the cells [71].

Another method for studying cell response to nanoparticles is flow cytometry [67, 68, 71, 72]. In this technique, the cells are first cultured with the nanopartcles as described above. After the incubation period, the cells are thoroughly washed, usually with phosphate-buffered saline (PBS), and harvested by trypsinization. The cells are injected into a stream that forces cells to travel individually to be interrogated by a laser beam in the flow cytometer. Information on the scattered light and fluorescence emission is gathered to generate information about the subpopulations within the sample. Imaging flow cytometry is a more

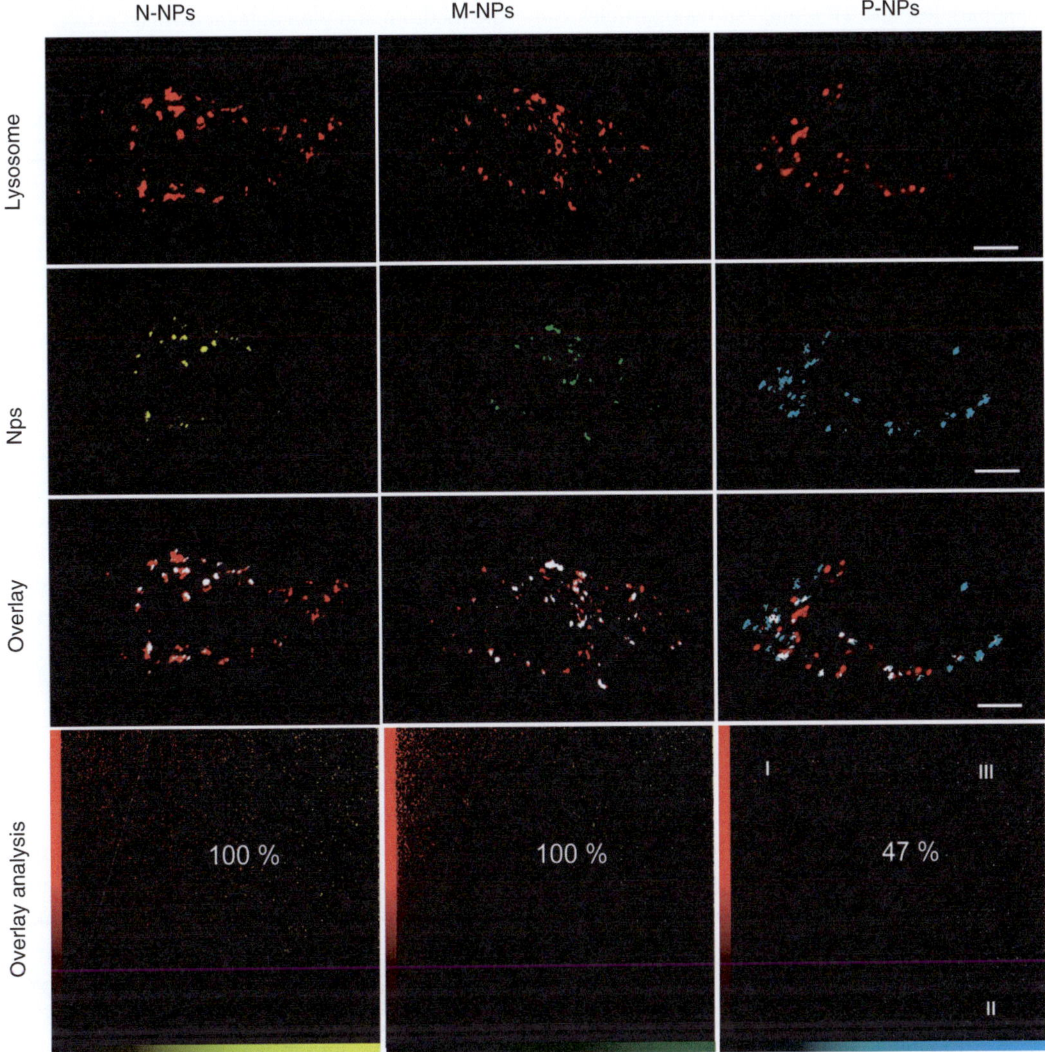

Fig. 3.6 CLSM images of lysosomes, nanoparticles (*NPs*), and their overlay signals after incubating different nanoparticles with HKC cells for 6 h. Lysosomes were labeled by LysoTracker Red. Nanoparticles were tested by their autofluorescence and shown in pseudocolor. In the overlay images, the signal of nanoparticles that were co-localized with the lysosomes was changed to white point by Leica LAS Co-localization software. In the overlay analysis panel, part I, II, and III indicated signals of the lysosomes, nanoparticles, and their co-localization points, respectively. The percentage number in each picture was the co-localization rate of nanoparticles calculated with Leica LAS Co-localization software. Scale bar: 10 μm (Reprinted from Yue et al. [69]. With permission from American Chemical Society)

advanced technique marrying the strengths of microscopy with those of flow cytometry. It combines a precise method of electronically tracking moving cells with a high-resolution multispectral imaging system to acquire multiple images of each cell in different imaging modes [73]. By combining confocal microscopy and imaging flow cytometry, Vranic et al. illustrated the possibility to distinguish between fluorescently labeled SiO_2 that were adsorbed on the cell surface and those internalized as well as nonfluorescent but light-diffracting TiO_2 nanoparticles [67].

Quantification of cellular uptake of many types of nanoparticles such as quantum dots, magnetic nanoparicles, and TiO$_2$ and Au nanoparticles can be carried out using inductively coupled plasma (ICP). The technique is based upon the use of a radiofrequency (RF)-induced argon plasma as the excitation source. While liquid and gas samples may be injected directly into the instrument, solid samples require extraction or acid digestion so that the analytes will be present in a solution [74]. In ICP optical emission spectrometry (ICP-OES), the light emitted from the plasma is separated into discrete component wavelengths and recorded. Each element emits at distinct wavelengths, and the amount of light emitted at a given wavelength is proportional to the concentration of the corresponding element in the solution injected into the instrument. Almost all naturally occurring elements, with the exception of hydrogen, oxygen, fluorine, and inert gases, can be determined. The elements that are not usually determined by ICP-OES fall into three categories: (1) those elements that occur either as trace contaminants in the argon gas used in ICP-OES (C from CO$_2$), constituents of the sample solvent (C, O, H), or contaminants from the environment or atmosphere (e.g., N), (2) those elements that require high excitation energy, such as the halogens, and (3) short-lived radioactive elements [74]. For most elements, the detection limits are 0.1–100 parts per billion (ppb). In ICP mass spectrometry (ICP-MS), the ions generated in the plasma are directed to a mass spectrometer, where they are separated on the basis of their mass-to-charge ratio before reaching the detector. The signal intensity for a given analyte ion is proportional to its concentration in the solution presented to the instrument. The ICP-MS instrument measures most of the elements in the periodic table, with exceptions such as hydrogen, helium, carbon, oxygen, nitrogen, fluorine, and neon, and the detection limits for ICP-MS are at or below the single parts per trillion (ppt) level for much of the periodic table [75].

To determine the cellular uptake of the nanoparticles, the cells are first cultured with the nanoparticles as described above. At the end of the incubation period, the cells are washed thoroughly with PBS to remove the nanoparticles in the medium and then detached via trypsinization. After counting with a hemocytometer, the cells are collected by centrifugation, and the cell pellets are digested in acid for the ICP analysis. A calibration curve using known amounts of the target element has to be generated to relate the ICP signals to quantitative amounts. The amount of nanoparticles taken in by the cells can then be expressed as pg/cell.

3.3.2 Cytotoxicity Assay

In vivo applications of nanoparticles require a thorough understanding of their potential cytotoxic effects on different mammalian cells. Nanoparticles may be inherently toxic, or they may be the carriers of toxic drugs. The level of cytotoxicity of nanoparticles is likely to be dose and time dependent. Thus, nanoparticulate drug carriers that can evade the RES and healthy cells and specifically target and accumulate in cancer cells will minimize harmful side effects associated with chemotherapy. *In vitro* assays are often used as an initial gauge of the potential cytotoxic effects of nanoparticles in the body. The two most commonly used *in vitro* assays rely on the measurement of the leakage of lactate dehydrogenase (LDH) from the cells and the metabolic activity in the cells, respectively. LDH is an enzyme localized in the cytosol, and its release into the cell culture medium is used as an indicator of loss of membrane integrity in compromised cells. The LDH assay can be carried out using commercial kits. After cells have been cultured with the nanoparticles for a desired period, the cell supernatant is collected and mixed with the assay solution. The released LDH is quantified using a coupled enzymatic reaction where LDH catalyzes the oxidation of lactate to pyruvate via reduction of NAD$^+$ (nicotinamide adenine dinucleotide) to NADH (reduced form of nicotinamide adenine dinucleotide), which in turn reduces 2-(4-iodophenyl)-3-(4-nitrophenyl)-5-phenyl tetrazolium chloride (INT) to a red, water-soluble formazan product [76, 77]. The absorbance of the formazan is measured using a spectrophotometer and is directly proportional to the amount of released LDH in the medium.

Unlike the LDH assay which gives a measure of the degree of cell death, assessment of metabolic activity provides information related to viable cell number. One of the most commonly used commercial kits for assessment of metabolic activity is based on the reduction of a yellow tetrazolium salt, 3-(4,5-dimethylthiazol-2-yl)-2,5-diphenyltetrazoliumbromide (MTT), by mitochondrial dehydrogenases to a purple, water-insoluble formazan product. After cells have been incubated with the nanoparticles for a desired period, the culture medium is removed, and the MTT solution is added and incubated with the cells. In order to solubilize the formed formazan product, organic solvents such as dimethyl sulfoxide (DMSO) or a combination of detergent and organic solvent may be used [78]. The absorbance of the solution is measured with a spectrophotometer, and the relative cell viability compared to control samples where the cells are incubated without nanoparticles is calculated from $[A]_{test}/[A]_{control}$, where $[A]_{test}$ and $[A]_{control}$ are the average absorbances of the test and control samples, respectively. More recently, tetrazolium reagents have been developed which can be reduced by viable cells to generate formazan products that are directly soluble in cell culture media [78]. For example, 5-(3-carboxymethoxyphenyl)-2-(4,5-dimethylthiazolyl)-3-(4-sulfophenyl) tetrazolium inner salt (MTS) can be used with an electron-coupling reagent, phenazine ethosulfate, instead of the MTT reagent. The absorbance of the formazan product can be directly recorded from the cell culture plate without a solubilization step, and thus, this protocol is simplified compared with the MTT assay. Other tetrazolium reagents that can be reduced by viable cells to generate formazan products that are directly soluble in cell culture media are the water-soluble tetrazolium salts (WSTs) and 2,3-bis(2-methoxy-4-nitro-5-sulfophenyl)-2H-tetrazolium-5-carboxanilide inner salt (XTT). Commercial kits based on MST, WSTs, and XTT are available from several vendors.

The Alamar Blue (resazurin) assay is another colorimetric assay to assess metabolic activity based on similar protocols as the tetrazolium compounds. In this assay, a blue dye with little intrinsic fluorescence is reduced to a pink fluorescent dye by cell metabolic activity. This provides the option of quantifying the results using fluorescence or absorbance measurements, but the latter is not often used because it is far less sensitive than measuring fluorescence [78]. Another viability assay based on metabolic activity is the measurement of the amount of adenosine 5′-triphosphate (ATP) in the cells since ATP has been widely accepted as a valid marker of viable cells [78, 79]. The assay protocol is simple and involves the addition of the ATP detection reagent from the commercial kit directly to cells in culture. The reagent contains a detergent to lyse the cells, ATPase inhibitors to stabilize the ATP released from the lysed cells, luciferin as a substrate, and the stable form of luciferase to catalyze the reaction that generates photons of light. The luminescent signal can then be measured. Since in the ATP assay the cells are lysed immediately upon addition of the reagent, information on the status of ATP is obtained at the instant the reagent is added. This is an advantage compared with the tetrazolium reduction or resazurin reduction assays that can require hours of incubation with viable cells to convert the substrate to a colored or fluorescent product [79].

It is important to recognize that there is a possibility the nanoparticles under evaluation may interfere with the reagents to be used in the assay for assessing the viability/cytotoxicity of cells. This is illustrated by the discrepancy in the results of different cytotoxicity assays carried out on single-walled carbon nanotubes (SWCNTs) [80]. The MTT assay indicated that after incubation of A549 cells with SWCNTs for 24 h, the cell viability is <50 %, whereas other assays based on LDH and WST-1 (2-(4-iodophenyl)-3-(4-nitrophenyl)-5-(2,4-disulfophenyl)-2H-tetrazolium) indicated that SWCNTs have minimal cytotoxicity. The authors postulated that the SWCNTs interacted with the insoluble formazan formed in the MTT assay but not with the reagents of the other assays. Thus, for new classes of nanoparticles, it would be advisable to verify cytotoxicity results with at least two independent assays.

3.4 Concluding Remarks

Nanoparticles offer great potential and opportunities in biomedical applications, and with the rapid expansion of the field of nanomedicine, many new nanoparticulate systems will be created in the future. Nanoparticles often have unique properties which are different from the bulk phase, and increasingly sophisticated nanoparticles are being designed to carry out multiple functions in the complex biological environment. Thus, the physicochemical and biological properties of the nanoparticles have to be well characterized and understood. However, the characterization of nanoparticles is complicated by a number of issues. Nanoparticles, with their large specific surface areas, are prone to contamination either during synthesis or sample preparation for analysis, which would skew the findings especially from surface-sensitive techniques such as XPS. Another complication is the disparity between the conditions used in the *in vitro* assays and the complex physiological environment. Significant knowledge gaps exist, and *in vitro* simulation of the cascade of events encountered by nanoparticles in their interaction with the physiological environment may not be possible. As such, animal model studies on the biodistribution, toxicology, and efficacy of the nanoparticles would be necessary. However, given the high costs and need to sacrifice animal lives in such studies, *in vitro* assays are essential, especially in offering preliminary insights into the suitability of the nanoparticles for the intended application. Thus, advances in technology to design and synthesize increasingly complex nanoparticles to widen their scope of applications should be accompanied by development of *in vitro* assays that can provide accurate predictive assessment of their performance in the human body.

References

1. European Commission (EU). Commission recommendation of 18 October 2011 on the definition of nanomaterial (2011/696/EU). Off J. 2011;L 275:38–40
2. Roduner E. Size matters: why nanomaterials are different. Chem Soc Rev. 2006;35:583.
3. Buzea C, Pacheco II, Robbie K. Nanomaterials and nanoparticles: sources and toxicity. Biointerphases. 2007;2:MR17–71.
4. Hall JB, Dobrovolskaia MA, Patri AK, McNeil SE. Characterization of nanoparticles for therapeutics. Nanomedicine. 2007;2:789–803.
5. Zhang L, Gu FX, Chan JM, Wang AZ, Langer RS, Farokhzad OC. Nanoparticles in medicine: therapeutic applications and developments. Clin Pharmacol Ther. 2008;83:761–9.
6. Veiseh O, Gunn JW, Zhang M. Design and fabrication of magnetic nanoparticles for targeted drug delivery and imaging. Adv Drug Deliv Rev. 2010;62:284–304.
7. Shubayev VI, Pisanic TR, Jin SH. Magnetic nanoparticles for theragnostics. Adv Drug Deliv Rev. 2009;61:467–77.
8. Moghimi SM, Szebeni J. Stealth liposomes and long circulating nanoparticles: critical issues in pharmacokinetics, opsonization and protein-binding properties. Prog Lipid Res. 2003;42:463–78.
9. Xie J, Xu C, Kohler N, Hou Y, Sun S. Controlled PEGylation of monodisperse Fe_3O_4 nanoparticles for reduced non-specific uptake by macrophage cells. Adv Mater. 2007;19:3163–6.
10. Cho EJ, Holback H, Liu KC, Abouelmagd SA, Park J, Yeo Y. Nanoparticle characterization: state of the art, challenges, and emerging technologies. Mol Pharm. 2013;10:2093–110.
11. Lim JK, Yeap SP, Che HX, Low SC. Characterization of magnetic nanoparticle by dynamic light scattering. Nanoscale Res Lett. 2013;8:381.
12. Wiogo HTR, Lim M, Bulmus V, Yun J, Amal R. Stabilization of magnetic iron oxide nanoparticles in biological media by fetal bovine serum (FBS). Langmuir. 2011;27:843–50.
13. Lazzari S, Moscatelli D, Codari F, Salmona M, Morbidelli M, Diomede L. Colloidal stability of polymeric nanoparticles in biological fluids. J Nanopart Res. 2012;14:920.
14. Lim JK, Majetich SA, Tilton RD. Stabilization of superparamagnetic iron oxide-gold shell nanoparticles in high ionic strength media. Langmuir. 2009;25:13384–93.
15. Mittal V, Matsko NB. Analytical imaging techniques for soft matter characterization. Berlin: Springer; 2012. p. 13–29.
16. Li M, Neoh KG, Wang R, Zong BY, Tan JY, Kang ET. Methotrexate-conjugated and hyperbranched polyglycerol-grafted Fe_3O_4 magnetic nanoparticles for targeted anticancer effects. Eur J Pharm Sci. 2013;48:111–20.
17. Zhang L, He R, Gu HC. Oleic acid coating on the monodisperse magnetite nanoparticles. Appl Surf Sci. 2006;253:2611–17.
18. Tatur S, Maccarini M, Barker R, Nelson A, Fragneto G. Effect of functionalized gold nanoparticles on floating lipid bilayers. Langmuir. 2013;29:6606–14.
19. Lin J, Zhang H, Chen Z, Zheng Y. Penetration of lipid membranes by gold nanoparticles: Insights into cellular uptake, cytotoxicity, and their relationship. ACS Nano. 2010;4:5421–9.

20. Hirsch V, Kinnear C, Moniatte M, Rothen-Rutishauser B, Clift MJD, Fink A. Surface charge of polymer coated SPIONs influences the serum protein adsorption, colloidal stability and subsequent cell interaction in vitro. Nanoscale. 2013;5:3723–32.

21. Jiang J, Oberdörster G, Biswas P. Characterization of size, surface charge, and agglomeration state of nanoparticle dispersions for toxicological studies. J Nanopart Res. 2009;11:77–89.

22. Doane TL, Chuang CH, Hill RJ, Burda C. Nanoparticle zeta-potentials. Acc Chem Res. 2012;45:317–26.

23. Kaszuba M, Corbett J, Watson FM, Jones A. High-concentration zeta potential measurements using light-scattering techniques. Phil Trans R Soc A. 2010; 368:4439–51.

24. Salgin S, Salgin U, Bahadir S. Zeta potentials and iso-electric points of biomolecules: the effects of ion types and ionic strengths. Int J Electrochem Sci. 2012;7:12404–14.

25. Kirby BJ, Hasselbrink EF. Zeta potential of microfluidic substrates: 1. Theory, experimental techniques, and effects on separations. Electrophoresis. 2004;25: 187–202.

26. Sun S, Zeng H, Robinson DB, Raoux S, Rice PM, Wang SX, et al. Monodisperse MFe_2O_4 (M = Fe, Co, Mn) nanoparticles. J Am Chem Soc. 2004;126:273–9.

27. Park J, An KJ, Hwang YS, Park JG, Noh HJ, Kim JY, et al. Ultra-large-scale syntheses of monodisperse nanocrystals. Nat Mater. 2004;3:891–5.

28. van der Heide P. X-ray photoelectron spectroscopy: an introduction to principles and practices. Hoboken: Wiley; 2012.

29. Chastain J, editor. Handbook of X-ray photoelectron spectroscopy. Eden Prairie: Perkin-Elmer Corporation; 1992.

30. XPS Spectra [Internet]. Casa Software Ltd.; c2013 [cited 2013 Nov 10]. Available from http://www.casaxps.com/ help_manual/manual_updates/xps_spectra.pdf.

31. Sherwood PMA. Curve fitting in surface analysis and the effect of background inclusion in the fitting process. J Vac Sci Technol A. 1996;14:1424–32.

32. Baer DR, Gaspar DJ, Nachimuthu P, Techane SD, Castner DG. Application of surface chemical analysis tools for characterization of nanoparticles. Anal Bioanal Chem. 2010;396:983–1002.

33. Coates J. Interpretation of infrared spectra, a practical approach. In: Meyers RA, editor. Encyclopedia of analytical chemistry. Chichester: Wiley; 2000. p. 10815–37.

34. Widemann BC, Adamson PC. Understanding and managing methotrexate nephrotoxicity. Oncologist. 2006;11:694–703.

35. Sun LC, Coy DH. Somatostatin receptor-targeted anti-cancer therapy. Curr Drug Deliv. 2011;8:2–10.

36. Cheng ZL, Elias DR, Kamat NP, Johnston ED, Poloukhtine A, Popik V, et al. Improved tumor targeting of polymer-based nanovesicles using polymer-lipid blends. Bioconjugate Chem. 2011;22:2021–9.

37. Shan X, Chen L, Yuan Y, Liu C, Zhang X, Sheng Y, et al. Quantitative analysis of hemoglobin content in polymeric nanoparticles as blood substitutes using Fourier transform infrared spectroscopy. J Mater Sci Mater Med. 2010;21:241–9.

38. Tsai DH, Davila-Morris M, DelRio FW, Guha S, Zachariah MR, Hackley VA. Quantitative determination of competitive molecular adsorption on gold nanoparticles using attenuated total reflectance-Fourier transform infrared spectroscopy. Langmuir. 2011;27:9302–13.

39. Carey FA. Chapter 13. Spectroscopy. In: Organic chemistry, 4th ed. [Internet]. McGraw Hill; c2000 [cited 2013 Nov 10]. Available from http://www. mhhe.com/physsci/chemistry/carey/student/olc/ ch13nmr.html.

40. Zhou H, Du F, Li X, Zhang B, Li W, Yan B. Characterization of organic molecules attached to gold nanoparticle surface using high resolution magic angle spinning ^1H NMR. J Phys Chem C. 2008;112: 19360–6.

41. Espinosa JF. High resolution magic angle spinning NMR applied to the analysis of organic compounds bound to solid supports. Curr Top Med Chem. 2011; 11:74–92.

42. Polito L, Monti D, Caneva E, Delnevo E, Russo G, Prosperi D. One-step bioengineering of magnetic nanoparticles via a surface diazo transfer/azide–alkyne click reaction sequence. Chem Commun. 2008;621–3.

43. Duguet E, Vasseur S, Mornet S, Devoisselle JM. Magnetic nanoparticles and their applications in medicine. Nanomedicine. 2006;1:157–68.

44. Walkey CD, Olsen JB, Guo H, Emili A, Chan WCW. Nanoparticle size and surface chemistry determine serum protein adsorption and macrophage uptake. J Am Chem Soc. 2012;134:2139–47.

45. Zahr AS, Davis CA, Pishko MV. Macrophage uptake of core-shell nanoparticles surface modified with poly(ethylene glycol). Langmuir. 2006;22:8178–85.

46. Vonarbourg A, Passirani C, Saulnier P, Benoit JP. Parameters influencing the stealthiness of colloidal drug delivery systems. Biomaterials. 2006;27:4356–73.

47. Fan QL, Neoh KG, Kang ET, Shuter B, Wang SC. Solvent-free atom transfer radical polymerization for the preparation of poly(poly(ethyleneglycol) monomethacrylate)-grafted Fe_3O_4 nanoparticles: synthesis, characterization and cellular uptake. Biomaterials. 2007;28:5426–36.

48. Moghimi SM, Hunter AC, Murray JC. Long-circulating and target-specific nanoparticles: theory to practice. Pharmacol Rev. 2001;53:283–318.

49. Wang L, Neoh KG, Kang ET, Shuter B, Wang SC. Superparamagnetic hyperbranched polyglycerol-grafted Fe_3O_4 nanoparticles as a novel magnetic resonance imaging contrast agent: An in vitro assessment. Adv Funct Mater. 2009;19:2615–22.

50. Welsch N, Lu Y, Dzubiella J, Ballauff M. Adsorption of proteins to functional polymeric nanoparticles. Polymer. 2013;54:2835–49.

51. Becker AL, Welsch N, Schneider C, Ballauff M. Adsorption of RNase A on cationic polyelectrolyte

brushes: a study by isothermal titration calorimetry. Biomacromolecules. 2011;12:3936–44.

52. Chakraborti S, Joshi P, Chakravarty D, Shanker V, Ansari ZA, Singh SP, et al. Interaction of polyethyleneimine-functionalized ZnO nanoparticles with bovine serum albumin. Langmuir. 2012;28:11142–52.

53. Baier G, Costa C, Zeller A, Baumann D, Sayer C, Araujo PHH, et al. BSA adsorption on differently charged polystyrene nanoparticles using isothermal titration calorimetry and the influence on cellular uptake. Macromol Biosci. 2011;11:628–38.

54. Shang L, Brandholt S, Stockmar F, Trouillet V, Bruns M, Nienhaus GU. Effect of protein adsorption on the fluorescence of ultrasmall gold nanoclusters. Small. 2012;8:661–5.

55. Röcker C, Pötzl M, Zhang F, Parak WJ, Nienhaus GU. A quantitative fluorescence study of protein monolayer formation on colloidal nanoparticles. Nat Nanotech. 2009;4:577–80.

56. Iosin M, Toderas F, Baldeck PL, Astilean S. Study of protein–gold nanoparticle conjugates by fluorescence and surface-enhanced Raman scattering. J Mol Struct. 2009;924–926:196–200.

57. Shi XJ, Li D, Xie J, Wang S, Wu ZQ, Chen H. Spectroscopic investigation of the interactions between gold nanoparticles and bovine serum albumin. Chin Sci Bull. 2012;57:1109–15.

58. Bell NC, Minelli C, Shard AG. Quantitation of IgG protein adsorption to gold nanoparticles using particle size measurement. Anal Methods. 2013;5:4591–601.

59. Park EK, Lee SB, Lee YM. Preparation and characterization of methoxy poly(ethylene glycol)/poly(epsilon-caprolactone) amphiphilic block copolymeric nanospheres for tumor-specific folate-mediated targeting of anticancer drugs. Biomaterials. 2005;26:1053–61.

60. Sudimack J, Lee RJ. Targeted drug delivery via the folate receptor. Adv Drug Deliv Rev. 2000;41:147–62.

61. Wang L, Neoh KG, Kang ET, Shuter B. Multifunctional polyglycerol-grafted Fe_3O_4@SiO_2 nanoparticles for targeting ovarian cancer cells. Biomaterials. 2011;32:2166–73.

62. Kohler N, Sun C, Fichtenholtz A, Gunn J, Fang C, Zhang M. Methotrexate-immobilized poly(ethylene glycol) magnetic nanoparticles for MR imaging and drug delivery. Small. 2006;2:785–92.

63. Rosenholm JM, Peuhu E, Bate-Eya LT, Eriksson JE, Sahlgren C, Lindén M. Cancer-cell-specific induction of apoptosis using mesoporous silica nanoparticles as drug-delivery vectors. Small. 2010;6:1234–41.

64. Weitman SD, Lark RH, Coney LR, Fort DW, Frasca V, Zurawski VR, et al. Distribution of the folate receptor Gp38 in normal and malignant-cell lines and tissues. Cancer Res. 1992;52:3396–401.

65. Ross JF, Chaudhuri PK, Ratnam M. Differential regulation of folate receptor isoforms in normal and malignant-tissues in-vivo and in established cell-lines – physiological and clinical implications. Cancer Am Cancer Soc. 1994;73:2432–43.

66. Lu S, Neoh KG, Huang C, Shi Z, Kang ET. Polyacrylamide hybrid nanogels for targeted cancer chemotherapy via co-delivery of gold nanoparticles and MTX. J Colloid Interface Sci. 2013;412:46–55.

67. Vranic S, Boggetto N, Contremoulins V, Mornet S, Reinhardt N, Marano F, et al. Deciphering the mechanisms of cellular uptake of engineered nanoparticles by accurate evaluation of internalization using imaging flow cytometry. Part Fibre Toxicol. 2013;10:2.

68. Lerch S, Dass M, Musyanovych A, Landfester K, Mailänder V. Polymeric nanoparticles of different sizes overcome the cell membrane barrier. Eur J Pharm Biopharm. 2013;84:265–74.

69. Yue ZG, Wei W, Lv PP, Yue H, Wang LY, Su ZG, Ma GH. Surface charge affects cellular uptake and intracellular trafficking of chitosan-based nanoparticles. Biomacromolecules. 2011;12:2440–6.

70. Gottstein C, Wu G, Wong BJ, Zasadzinski JA. Precise quantification of nanoparticle internalization. ACS Nano. 2013;7:4933–45.

71. Davda J, Labhasetwar V. Characterization of nanoparticle uptake by endothelial cells. Int J Pharm. 2002;233:51–9.

72. Patra HK, Banerjee S, Chaudhuri U, Lahiri P, Dasgupta AK. Cell selective response to gold nanoparticles. Nanomedicine. 2007;3:111–9.

73. Basiji DA, Ortyn WE, Liang L, Venkatachalam V, Morrissey P. Cellular image analysis and imaging by flow cytometry. Clin Lab Med. 2007;27:653–70.

74. Hou X, Jones BT. Inductively coupled plasma/optical emission spectrometry. In: Meyers RA, editor. Encyclopedia of analytical chemistry. Chichester: Wiley; 2000. p. 9468–85.

75. Perkin Elmer Technical note. The 30-minute guide to ICP-MS [Internet]. Waltham: PerkinElmer, Inc.; c2004–2011 [cited 2013 Nov 10]. Available from http://www.perkinelmer.com/PDFs/Downloads/tch_icpmsthirtyminuteguide.pdf.

76. Korzeniewski C, Callewaert DM. An enzyme-release assay for natural cytotoxicity. J Immunol Methods. 1983;64:313–20.

77. Kroll A, Dierker C, Rommel C, Hahn D, Wohlleben W, Schulze-Isfort C, et al. Cytotoxicity screening of 23 engineered nanomaterials using a test matrix of ten cell lines and three different assays. Part Fibre Toxicol. 2011;8:9.

78. Riss TL, Moravec RA, Niles AL, Benink HA, Worzella TJ, Minor L. Cell viability assays. In: Sittampalam GS, Gal-Edd N, Arkin M, Auld D, Austin C, Bejcek B, et al., editors. Assay guidance manual [Internet]. Bethesda: Eli Lilly & Company and the National Center for Advancing Translational Sciences; 2004. [cited 2013 Nov 10]. Available from: http://www.ncbi.nlm.nih.gov/books/NBK144065/.

79. Niles AL, Moravec RA, Riss TL. Update on *in vitro* cytotoxicity assays for drug development. Expert Opin Drug Discov. 2008;3:655–69.

80. Wörle-Knirsch J, Pulskamp K, Krug H. Oops they did it again! Carbon nanotubes hoax scientists in viability assays. Nano Lett. 2006;6:1261–8.

Nanomaterial Properties: Implications for Safe Medical Applications of Nanotechnology

4

Saji George

Abstract

Nanomaterials with exciting functional properties are being increasingly used for therapeutic applications. Research studies over the past decade have shown that properties that make nanomaterials useful for their therapeutic application may also give rise to potential hazardous outcomes. Experimental evidences suggest that variations in nanomaterial size, shape, aspect ratio, surface chemistry, dispersal state and bio-persistence in addition to material 'specific' properties could all contribute individually or combinatorially to biological injury *via* toxicological pathways. While nanomaterial-induced injury pathways could be associated with human disease conditions, it should be noted that currently, there is no human disease that can be directly related to the exposure to engineered nanomaterials. Therefore, in this book chapter, the term 'toxicity' is referring to nanomaterial-induced injury mechanisms from experimental studies and details the major properties of nanomaterials in relation to their potential adverse outcomes. In addition, topics on strategies to elucidate property–activity relationships, merits and demerits of cellular models and animal models of toxicology and the use of high-throughput screening for rapid identification of 'danger signal' has been emphasized to provide the reader with a comprehensive understanding on the discipline of 'nanotoxicology'.

4.1 Introduction

4.1.1 Increasing Nanotechnology Applications in Medicine

Among the different sectors of nanotechnology applications, health and fitness is the most proficient sector with a fast growing list of nanotechnology applications (Fig. 4.1) [1]. The widespread application of nanotechnology for diagnostics, treatment and monitoring of the progression of

S. George, MSc, PhD
Centre for Sustainable Nanotechnology,
School of Chemical & Life Sciences, Nanyang
Polytechnic, 180 Ang Mo Kio Ave 8,
Room Q 607B, Singapore 569830, Singapore
e-mail: saji_george@nyp.edu.sg

A. Kishen (ed.), *Nanotechnology in Endodontics: Current and Potential Clinical Applications*,
DOI 10.1007/978-3-319-13575-5_4, © Springer International Publishing Switzerland 2015

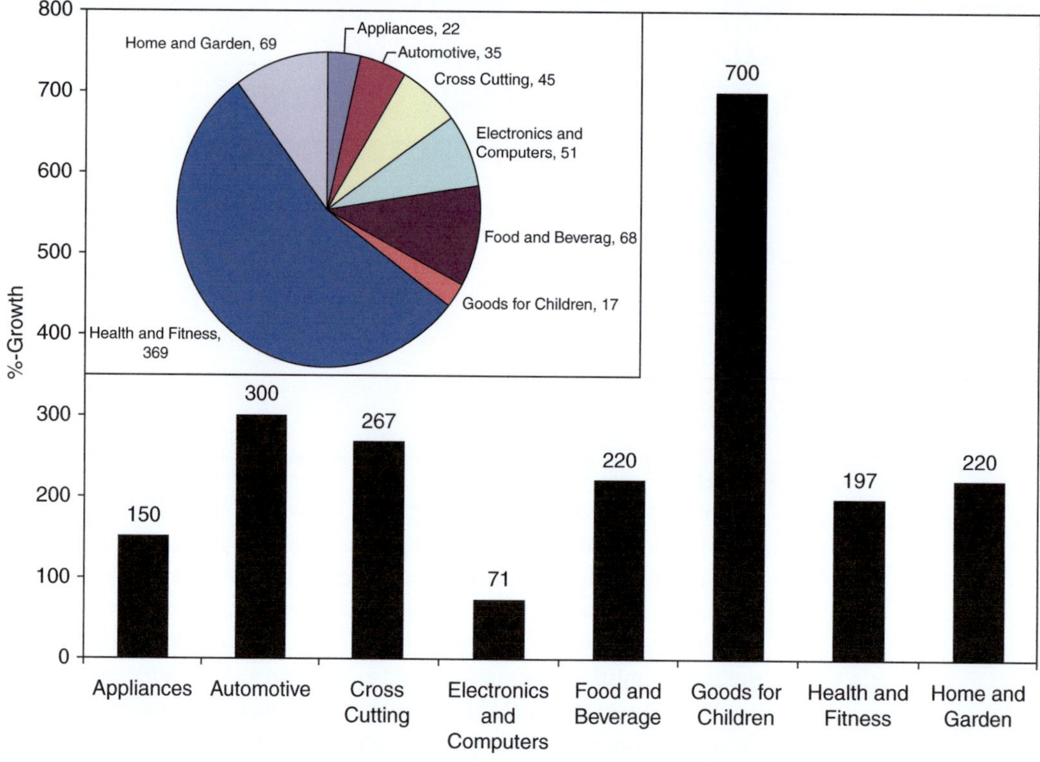

Fig. 4.1 Distribution (*inset*) and growth of consumer goods containing nanomaterials based on the identification of eight commercial sectors (Reprinted from Meyer et al. [1]. With permission from American Chemical Society)

disease has led to the emergence of a separate discipline called 'nanomedicine'. The advantages of using nano-sized systems for therapeutic applications can never be overestimated given their tunable functionalities and ability to access biological compartments and interact with biomolecules more effectively. While physical dimensions in the nano range present many advantages in terms of transport, passive targeting and increased surface availability of molecules, the chemistry of core particles and capping agents provides additional platforms for customizing target specificity, cellular trafficking and sustained release of therapeutics. More exciting is the possibility of embedding these multiple functionalities into a single entity which distinguishes nanotechnology from other recent developments in the field of medicine.

The major areas of nanotherapeutic applications in health care are cancer treatment, infection control, vaccine delivery, scaffold for tissue engineering and immunotherapy. To date, nanomaterials of metals, carbon, silica, liposomes, polymers and composite materials are developed with varying purposes. Although most of the advanced therapeutic interventions using nanotechnology are yet to reach the hands of medical practitioners, the medical benefits of nanotechnology has already been experienced with the advent of new-generation cancer medicine, diagnostic systems and antimicrobials. The disruptive growth of nanotechnology applications in diverse sectors and medicine in particular has raised questions on its safety to humans and the environment. In fact, ensuring the safety of nanotherapeutics has been recognised as one of the major roadblocks in the translation of exciting research findings.

Some of the fundamental properties of materials begin to shift as their size approaches 'nano', which has been the primary reason for the excitement about nanomaterials. Therefore, it is reasonable to speculate that the biological responses to nanomaterials could also be different from their bulk counterparts. In fact, the differential response of biological systems to nanomaterials has been well captured in nanomedicine. On the flip side, however, the same properties that make nanomaterials attractive for their therapeutic applications such as ability to penetrate biological compartments, biological persistence required for slow and sustained release of therapeutic agents, altered redox potential and ability to generate reactive oxygen species, dissolution and release of chemical components, *etc.* could generate negative outcomes at the nano–bio interface. Research studies over the years have shown that the toxicology of nanomaterials is different from the toxicology of the bulk material or the chemical entity that constitutes these materials. Since the field of studying toxicology of nanomaterials – nanotoxicology – has advanced over the past decade, there has been significant improvement in our understanding on the properties of nanomaterials that could make them potentially dangerous.

4.2 General Physico-Chemical Properties of Nanomaterials in Relevance to Their Toxic Potential

Over the past decade, there have been many studies addressing the hazard potential of nanomaterials. As our understanding on this subject is evolving, it is reasonable to state that there are '*general*' mechanisms and material '*specific*' mechanisms that determine the toxicity of nanomaterials. Physical parameters such as extremely small size and increased surface area, shape and aspect ratio, dispersion state, *etc.* constitute the general properties applicable to all nanomaterials regardless of their chemical composition. Chemical properties such as dissolution chemistry, band-gap energy, surface defects, *etc.* are some of the material-specific properties that govern toxic potentials of nanomaterials. The following section will detail some of the important physical and chemical factors that in general shape the toxicity of nanomaterials.

4.2.1 Size of Nanomaterials in Relevance to Toxicity

Among the physico-chemical properties, the primary size is the defining criterion for nanomaterials. The National Nanotechnology Initiative (NNI), USA, defines nanotechnology as "*the understanding and control of matter at dimensions of roughly 1 to 100 nm, where unique phenomena enable novel applications. Encompassing nanoscale science, engineering and technology, nanotechnology involves imaging, measuring, modeling, and manipulating matter at this length scale*" [2]. The US Food and Drug Administration adopts this definition of nanotechnology and nanomaterials. When all three dimensions (length, width and height) or radius are in the range of 1–100 nm, the materials is generally referred to as 'nanoparticle', and a material with one or two dimension(s) within 1–100 nm is generally referred to as 'nanomaterial'. Examples of nanoparticles include spheres or pseudospheres of metal (Ag, Au, Pt), metal oxides (Al_2O_3, TiO_2, CeO_2, ZnO, CuO, Fe_2O_3, etc.), silica particles (SiO_2), spherical carbon nanoparticles (C_{60}, C_{70} etc) and quantum dots (CdSe, CdSe/ZnS). Carbon nanotubes (SWCNTs, MWCNTs), carbon nanowhiskers, nanopods, nanorods, nanobelts and the fibrous structure of metal and metal oxides are some of the classic examples of nanomaterials. Although the terms 'nanoparticle' and 'nanomaterial' are used interchangeably, it should be noted that the material properties as well as the toxicity of the same chemical composition could vary between its 'nanoparticle' or 'nanomaterial' form, which will be discussed later.

Experimental and epidemiological evidences suggest that the biological activity of a given material could increase with the decrease in particle size. Earlier indications on the relevance of particle size on the biological activity of materials came from studies on inadvertent particle and fibre exposures [3, 4]. The adverse health effects from inhaled ultrafine particles of aerodynamic diameters of <100 nm were comparatively higher than larger coarse (aerodynamic diameter <10 μm, PM10) and fine (aerodynamic diameter

<2.5 µm, PM2.5) air pollution particles [3, 5]. In the case of inhalation exposure, 'ultrafine particles' of size 1–100 nm get deposited in the alveolar region impeding their removal via physiological function of lungs as well as through macrophage clearance mechanisms [6]. These particles get translocated through circulatory and lymphatic systems to distant tissues and organs. Nanoparticles in circulation can be taken up by the organs and tissues including the brain, liver, heart, kidneys, spleen, bone marrow and nervous system. At the cellular level, however, the entry of particles is governed by biological and thermodynamic principles. The binding of the nanoparticle–ligand conjugate to the receptor produces a localized decrease in the Gibbs free energy, which induces the membrane to wrap around the nanoparticle to form a closed-vesicle structure [7]. The vesicle eventually buds off the membrane and fuses with other vesicles to form endosomes, which fuse with lysosomes where degradation occurs. The size-dependent uptake of nanoparticles is likely related to the membrane-wrapping process, where it was observed that optimum size for the spherical particles for uptake through endocytosis is 25 nm [8]. One of the most documented and well-known endocytic pathways is clathrin-mediated endocytosis, wherein the physical interaction of nanomaterials with cell surfaces prompts assembly of clathrin molecules on the interior side of lipid bilayers to form a membrane vesicle that wraps around the nanomaterial. Particles interacting with cell surfaces could also be internalized *via* non-receptor-mediated pathways, among which pinocytosis is the most frequently reported in literature. These routes of uptake, which include macropinocytosis, can potentially allow uptake of materials up to 300 nm in diameter [9].

The ability of nanoparticles to translocate to distant organs is a double-edged sword. While this is advantageous for delivery of therapeutics to distal organs, it could impede the clearance of nanoparticles from different tissues and organs. Here, it is worth noting that more than the ability to access biological compartments, the toxicity of particles increases with decreasing size because the proportion of the surface-exposed atoms/mol-

ecules increases with decreasing size. The higher proportion of surface-exposed atoms/molecules increases their chance of interacting with biomolecules present in their vicinity. Thus, for a given mass of materials, the effective 'reactive surface' is higher for smaller particles contributing to the potential to cause higher toxicity in comparison to their bulkier counterpart. In the case of gold nanoparticles, it was observed that the toxicity increases tremendously when the particle size is in the range of 1–2 nm [10]. Nonetheless, it is reasonable to summarize that the ability to trespass biological compartments and increased proportion of surface-exposed atoms/molecules of the 'extremely small' nanomaterials are the two most relevant general properties that makes nanomaterials potentially more dangerous. Since surface area of a nanomaterial is related to its primary particle size, surface area as a determining factor for toxicity will not be discussed under separate topics. However, it should be noted that surface area is an important parameter to consider for the dosimetry of nanomaterials for toxicological assessment.

The properties related to primary particle size and increased surface areas of nanomaterials are greatly influenced by the agglomeration behaviour of particles. Particles suspended in media or biological fluids agglomerate depending on the dominant thermodynamic forces between particles, biomolecules and ionic entities in the suspending media. Since agglomeration reduces the effective surface area and increases the overall size, it will be relevant to understand the factors influencing agglomeration particles and methods to prevent/minimize particle agglomeration. Therefore, strategies aimed at stabilizing particle agglomeration are often required when investigating the toxicity of nanomaterials.

Agglomeration Behaviour of Nanomaterials in Body Fluids and Cell Culture Media

Body fluids and cell culture media (used for the growth of cells under laboratory conditions) often contain salts and mineral ions besides biomolecules. Through a series of studies, Allouni et al. showed that the stability of nanoparticles in cell culture media is determined by multiple factors such as particle concentration, ionic strength and the presence of proteins [11]. Mineral ions and salts could interact with nanomaterials according to their charge compatibility, often leading to agglomeration and precipitation of nanomaterials. As proposed in the Derjaguin, Landau, Verwey and Overbeek (DLVO) theory, the stability of a colloidal suspension is based on the net balance of two forces: the electrostatic repulsion which prevents aggregation and a universal attractive van der Waals force which acts to bind particles together [12]. In electrolytes containing solutions such as cell culture media, the charged surface of nanomaterials attracts counter-ion charge clouds. This electrostatic double layer extends far from the particle surface contributing to the long-range interactions. In low–ionic strength solutions, the electrostatic repulsion between similarly charged nanoparticles is relatively higher than van der Waals forces of attractions, keeping particles separated from each other, thereby stabilizing the nanoparticle suspension. However, in high–ionic strength media such as biological fluids and cell culture media, the electrical double layer is compromised, and the magnitude of the repulsive electrostatic forces decreases, whereby van der Waals attraction dominates. Therefore, the net interaction potential becomes purely attractive at long and short range, leading to nanoparticle agglomeration and precipitation of agglomerated particles.

Agglomeration of particles could lead to increase in size with a concomitant decrease in effective surface area. Increase in size of the particles could negatively affect the uptake of particles into tissue compartments and cells. Therefore, it is important to ensure adequate dispersion of particles for desired and more accurate outcome in biological studies. Therefore, many dispersion strategies to suit the specific application scenarios have been developed to ensure the stability of nanomaterials in biological media.

Dispersion of Nanomaterials in Biological Media

It is evident from the previous discussions that in order to prevent agglomeration of particles and to achieve stabilization in biological media/fluids, the attractive inter-particle forces should be overcome by equivalent repulsive forces. Most dispersion strategies are based on the electrostatic repulsion [13] between the particles and steric hindrances imparted by the polymeric molecules decorating the material surface or combinations of these two effects [11, 14]. Polymeric coating materials can be either synthetic or natural in origin. Synthetic polymers often employed for stabilizing nanoparticle suspension include poly(ethylene-co-vinyl acetate), poly(vinylpyrrolidone) (PVP), poly(lactic-co-glycolic acid) (PLGA), poly(ethyleneglycol) (PEG) and poly(vinyl alcohol) (PVA), while natural polymers with a desired property include serum albumin, dextran, chitosan, *etc.* [15]. For example, surface coating with citric acid or humic acid is used to stabilize the iron oxide nanomaterial because of the electrostatic repulsion and steric hindrances provided by these coating materials [16]. Although such attempts are reasonable from the standpoint of targeted drug delivery, the purposeful surface modifications especially with synthetic chemical compounds could yield misleading toxicological interpretation about the nanomaterial in question. Therefore, when toxicity of nanomaterials is attempted, a rational approach to disperse nanomaterials could be the use of proteins or phospholipids that are part of the biological fluids at the site of action. Such an approach has been employed by many workers where serum albumin and dipalmitoyl phosphatidylcholine (DPPC) has been used to disperse and stabilize nanomaterial suspension in cell culture media when studying the cytotoxicity of nanoparticles [17–20]. Taking TiO_2 nanoparticles as an example, Ji et al. showed that the dispersion of nanoparticles can be improved by supplementation of cell culture media with albumin [21]. However, they observed that the degree of dispersion could vary from medium to medium mostly because of the varying concentrations of phosphate. They also showed that fetal bovine serum (FBS) is an effective dispersing agent for TiO_2 nanoparticles, which was due to synergistic effects of its multiple protein components, such as albumins, γ-globulin and apo-transferrin. George et al. detailed a protocol involving the use and sequence of application of ultra-sonication and albumin to ensure stability of nanomaterial suspension (Fig. 4.2) [22]. Similarly, Portel et al. showed the dispersion capability and biocompatibility of media containing albumin protein and DPCC in PBS for nanomaterial suspension [20]. Although FBS is an excellent dispersion agent, the serum components could interfere with the toxic potential, leading to a drop in the toxicity of nanoparticles. This could lead to gross underscoring of the possible toxic potential of nanoparticles, as it has been observed with nanomaterials of heavy metals. Thus, while the use of FBS in cell culture media for cytotoxicity studies in mammalian systems is reasonable, the application of FBS as dispersing agent may not be appropriate when studying toxicity to environmental organisms. Alternate dispersion strategies, for example, use of Alginic acid as environmentally relevant dispersion agent, should be applied in such cases [23].

The relevance of dispersal states of nanomaterials in influencing their biological properties is exemplified by a series of studies conducted with Multi-Walled Carbon Nanotubes (MWCNTs) by Wang et al. They induced differential agglomeration states of MWCNTs in exposure media by suspending the particles in media with or without supplementation of dispersion agents. Suspending MWCNTs in exposure media resulted in agglomeration, while supplementation of the media with bovine serum albumin (BSA) and a lung surfactant DPPC dispersed these materials. The results showed that well-dispersed, relatively hydrophobic MWCNTs induce more robust pro-fibrogenic effects under *in vitro* and *in vivo* test conditions [24, 25]. Well-dispersed MWCNTs induced more prominent TGF-β1 and IL-1β production *in vitro* as well as TGF-β1, IL-1β and PDGF-AA production *in vivo* than non-dispersed tubes [25].

4.2.2 Shape of Nanomaterials in Relevance to Toxicity

Nanomaterials cover a broad range of shapes, most of which fall in sphere, pyramid, cube, rod, tube or wire. Nanomaterials with distinct physico-chemical properties can be made from the same chemical composition by altering the geometry of the material. Shape of the nanomaterial is relevant with regard to its toxic properties because it determines

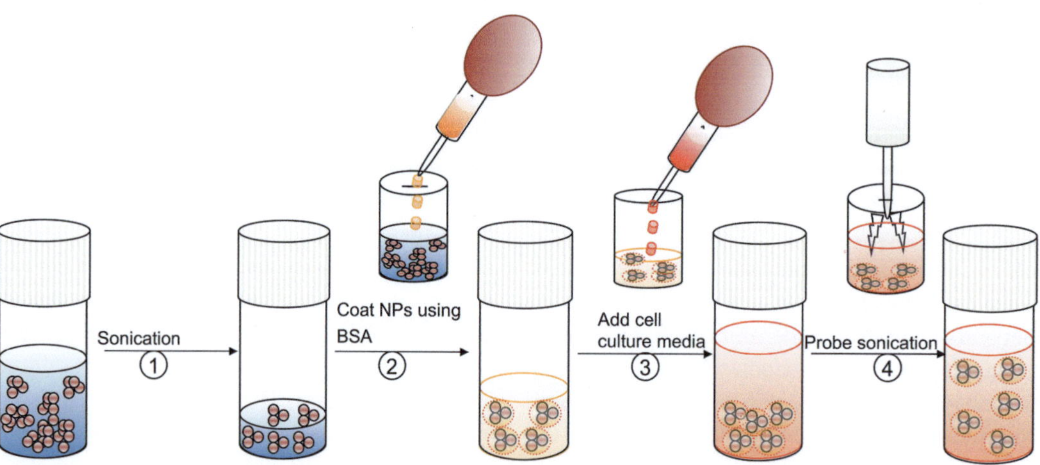

Fig. 4.2 Preparation of nanoparticle dispersion for cytotoxicity studies. (*1*) Sonicate (water bath) the stock solution of nanomaterial. (*2*) Aliquote the quantity required to make working concentration and add with dispersing agent (BSA) (e.g., 2 % BSA for serum free cell culture media) (*3*) Add cell culture media. Serum free cell culture media should be supplemented with BSA at 2 mg/mL concentration (*4*) Sonicate the nanomaterial suspension using probe sonicator and use the fresh nanomaterial suspension for cytotoxicity studies (Reprinted from Ji et al. [21]. With permission from American Chemical Society)

factors such as the area of contact and the internalization pathway of nanomaterials into tissue compartments and cells, leading to different biological outcomes. In addition to these direct effects, shapes of nanomaterials could also influence their interaction with each other in suspensions. Nanomaterials of symmetrical shapes, for example, can give rise to self-assembled structures exhibiting periodicity in assembly of the building blocks (single nanocrystal). These interactions in turn shape nanomaterials' macroscopic-like behaviour such as their locomotion in circulatory systems. The influence of size and shape of nanomaterials on their movement through blood vessels and accumulation in different body compartments has been highlighted by research works on silica-based drug delivery nanomaterials. Decuzzi et al. studied a combinatorial library of silica nanomaterials with differing shapes – sphere, disc and cylinder – in mice models of tumours in order to study their accumulation in the major organs and within the tumour [26]. They observed that relatively bigger particles get accumulated in the reticuloendothelial system (RES). Discoidal particles were observed to accumulate more than others in most of the organs and in tumours, whereas cylindrical particles were deposited in liver [26]. Similarly, Meng et al. compared the cellular uptake of mesoporous silica nanoparticles (MSNPs) of different aspect ratio [27]. Their studies showed that the aspect ratio (ratio of length to diameter) of rod-shaped particles determines the rate and abundance of MSNP uptake by a macropinocytosis process in HeLa and A549 cancer cell lines. MSNPs with an aspect ratio of 2.1–2.5 were taken up in larger quantities compared to shorter- or longer-length rods [27]. While these examples provide examples of the influence of material shape on their transport and compartmentalization in the body, there are studies showing direct evidence of how shape of material influences its toxic outcome.

4.2.3 Fibre Pathogenicity Paradigm

The discussion about shape in determining toxicity of materials will be incomplete without mentioning about '*fibre pathogenicity paradigm*'. Fibre pathogenicity paradigm deals with the toxicology of high–aspect ratio materials, primarily to respiratory system. Epidemiological studies over the past 40 years or so had revealed the occupational hazard associated with inhalation of high–aspect ratio materials such as asbestos, vitreous and organic fibres [28]. Similar pathological conditions are implied from emerging studies on nanomaterials of high aspect ratio. Although the fibre pathogenicity paradigm is established for inhalation toxicology, the essential factor that determines fibre toxicity such as aspect ratio and bio-persistence is also expected to determine the harmful outcomes from high–aspect ratio nanomaterials used for drug delivery and tissue engineering applications. Therefore, we will examine the material properties and molecular mechanisms involved in the fibre pathogenicity paradigm in this chapter.

The pathogenicity of any fibre-shaped material is determined by factors such as diameter of fibre, length of fibre, rigidity of fibre and its bio-persistence. Length is the main determinant of fibre toxicity. Length-dependent pathogenicity of materials has been demonstrated from epidemiological and experimental studies, and this paradigm can be generalized to high–aspect ratio materials, including those of emerging nanomaterials such as carbon nanotubes. The accessibility of fibre into distal compartments of lungs is determined by its diameter. Fibres with submicron diameter can be deposited beyond the ciliated airways [29]. The removal of materials deposited in alveolar regions is slow as it is mediated through macrophages. However, bio-persistence – the retention of the structural integrity of the material inside organs/tissues – is largely determined by the chemical composition of the fibre. A fibre that is made up of bio-soluble material may degrade and form shorter fibres or solubilize completely into its molecular entities, which are incapable of inducing a fibre pathogenicity response.

The primary mechanism by which high–aspect ratio materials induce toxicity is by activating inflammatory cells recruited to the site of their deposition. Alveolar macrophages recruited to the site of their deposition in lungs try to clear these materials through phagocytosis. However, materials with length exceeding the 'engulfing' capacity of macrophages (>15 μm) give rise to a persistent macrophage activation and chronic inflammation, generally referred to as *'frustrated phagocytosis'* (Fig. 4.3) [29]. The chronic inflammation accompanied by the release of oxidants and cytokines attracts inflammatory cells to the site of deposition, leading to an inflammatory response. In the case of inhaled particles, as experimental evidence suggests, the constant stress induced by these persistent particles could lead to 'mesothelioma', or cancer of the mesothelium [30–32]. Apart from the hindrance in clearing these particles, the mechano-transduction mechanisms, including the cyclic stretching of tissues and local shear forces exerted by nanomaterials of high aspect ratio, is also suspected to modulate toxicological responses [33, 34].

While the influence of aspect ratio in influencing the toxic potential of nanomaterials such as carbon nanotubes and TiO_2 nanowires has been established, there was considerable uncertainty on the critical lengths and aspect ratios for inducing biological perturbations. However, a study using a combinatorial library of CeO_2 demonstrated that at lengths ≥200 nm and aspect ratios ≥22, CeO_2 nanorods could induce progressive cytotoxicity and pro-inflammatory effects. The relatively low 'critical' length and aspect ratio were associated with small nanorod/nanowire diameters (6–10 nm), which facilitates the formation of stacking bundles due to strong van der Waals and dipole-dipole attractions [35]. Thus, this *in vitro* study suggests that both length and diameter components of aspect ratio should be considered when addressing the cytotoxic effects of high–aspect ratio materials [35].

4.2.4 Surface Functionality in Relation to Toxicity

The surface chemistry of a nanomaterial or chemical entity that coats nanomaterials can influence the biological outcomes of nano–bio interactions in many ways. The surface functionality determines the half-life of nanomaterials in circulation and its accumulation in specific body compartments. Capping nanomaterials with poly(ethylene) glycol (PEG) has been observed to increase their blood half-life [36]. The circulation time of PEGylated nanomaterials increases

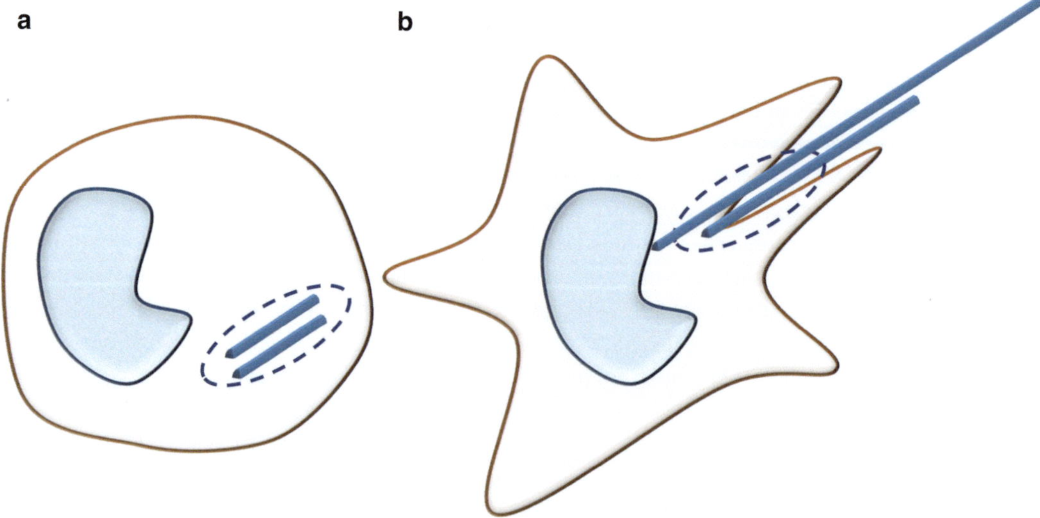

Fig. 4.3 Cartoon showing the phagocytic responses to short (**a**) and long fibres (**b**). Short fibres are completely internalized by phagocytosis whereas long fibres initiate the process of phagocytosis. The long aspect ratio impedes the complete internalization of fibres leading to frustrated phagocytosis

because PEG protects nanomaterials from phagocytosis by circulating macrophages. In fact, the second generation of nanomaterials for medical application uses the strategy of modifying the surface chemistry in order to achieve longer circulation time and active targeting. Surface functionalization using a co-polymer strategy using PEG and poly(ethyl) imine was also shown to increase the accumulation of nanomaterials in tumours [37]. These studies invariably show that functional groups play a major role in determining the ADME of nanomaterials. While the prolonged half-life in blood and site-specific accumulation of nanomaterials could have indirect effects on its potential toxicity, there are direct evidences on the influence of surface chemistry on the toxicity of nanomaterials.

In a study by Ryman-Rasmussen et al. using human epidermal keratinocytes cells (HEKs), the surface coating of QDs (commercially available CdSe core/ZnS shell-(QD 565) and QD 655) was found to cause differential effects on the cellular uptake, pro-inflammatory cytokine release and cytotoxicity [38]. Cytotoxicity was observed for quantum dots without any coating (QD 565) and QD 655 coated with carboxylic acids or PEG-amine, while no significant cytotoxicity was observed for PEG-coated QDs. Only carboxylic acid–coated QDs significantly increased release of IL-1b, IL-6 and IL-8, indicating that QD surface coating is a primary determinant of cytotoxicity and immunotoxicity in HEKs. This study also suggested the relevance of surface charge of the QDs to elicit biological responses where the cationic QDs showed the highest potential. The cationic quantum dots or dendrimers gain entry into the cytoplasm of cells by forming transient pores on the cell membranes which often result in cytotoxicity [39]. However, studies from Andre Nel's group from UCLA showed that cytotoxicity is determined not only by the surface charge but also by the cationic density of the polymer coating the nanomaterial. Mesoporous silica nanoparticles (MSNP) with potential use as carriers for hydrophobic cancer drugs and nucleic acids (for gene therapy) showed high efficiency of toxicity when polyethylenimine (PEI) of molecular size 25 kD was coated to

the surface by non-covalent bonds. However, by reducing the molecular weight (MW) of the attached polymers to 0.6–10 KD, it was possible to retain high cellular uptake and drug delivery while reducing or even eliminating cytotoxicity. Apart from the type of surface coating, the pattern on the particle surface could also influence the uptake of nanomaterials and their ability to cause cytotoxicity. Verma et al. showed that gold nanomaterials with sub-nanometre striations of alternating anionic and hydrophobic groups could penetrate the plasma membrane without causing disruption to lipid bilayers, whereas the particles with random distribution of anionic and hydrophobic groups on the particle surface were mostly trapped in endosomes [40].

Chemical modification of nanomaterial surfaces with the intention of reducing their toxicity has revealed that surface functionality of nanomaterials could modify the toxic potential of nanomaterials. There have been many efforts to reduce the toxicity of nanomaterials mainly used for drug delivery purpose by chemical modification of their surfaces. Stasko et al. observed elimination of cytotoxicity of dendrimers in HUVEC cells when the surface amines of parental dendrimers were chemically modified to neutral acetamide or Polyethylene glycol (PEG) functionalities [41]. Similarly, functionalizing the fullerene and single-walled carbon nanotubes (SWNTs) with caroboxyl or hydroxyl groups was reported to decrease cytotoxic potential [42, 43].

4.3 Material-Specific Properties of Nanomaterials in Relation to Its Toxicity

The above-mentioned factors such as size, shape and surface functionalization are the major properties in general that influence toxicity of nanomaterials. In addition, there are material-specific properties that are primarily associated with the chemical composition of nanomaterials that shape their interactions at nano–bio interfaces. Some of the material–specific properties of nanomaterials of relevance to human exposure because of their application in oral care products, food or therapeutics will be detailed in the following section. The strategy of elucidating property–activity relationships will be emphasized in an attempt to elaborate how nanotoxicology tools and studies are incrementally adding to our knowledge of factors that shape nano–bio interactions.

4.3.1 Dissolution as the Primary Mechanism of Toxicity of ZnO Nanoparticles

Nanomaterials of zinc oxide (ZnO) have been used extensively in cosmetics, personal care products, semiconductor devices, etc. Traditionally, ZnO in combination with eugenol has been used in restorative dentistry as a filling material in root canal therapy. The broad-spectrum antimicrobial activity of ZnO holds promising dental applications for infection control. Recent studies on endodontic pathogens warrant the potential use of ZnO nanomaterials to control dental infections [44]. Thus, while ZnO nanomaterials show promising applications in restorative dentistry as a cementing and antimicrobial material, the safety of this material for clinical application cannot be overlooked.

A comparative cytotoxicity screening of a compositional library of nanomaterials had shown that ZnO is relatively more toxic [45]. ZnO is a material with high dissolution kinetics in aquatic media, where dissociative adsorption of water layers results in the release of highly reactive Zn^{2+} ions. However, ZnO nanoparticles with systematic variation in dissolution kinetics were required to establish dissolution as a paradigm of ZnO toxicity. In a study reported by George et al., the challenge in creating ZnO nanomaterials with differential dissolution kinetics was overcome by doping Fe into the crystal matrix of ZnO [22]. Doping of ZnO with different levels of Fe was achieved by changing the concentrations of precursor solutions during the synthesis of these nanoparticles by flame spray pyrolysis to create a compositional library of ZnO nanoparticles with 0, 1, 2, 4, 6, 8 and 10 % Fe doped into the crystal matrix [22]. Dissolution kinetics studies showed that doping of a ZnO matrix with Fe could decrease the dissolution and release of Zn^{2+} depending on the level of atomic percentage of Fe doped into ZnO. Cytotoxicity studies using this library of ZnO nanomaterials demonstrated that decreasing the dissolution and rate of Zn^{2+} release could reduce the cytotoxicity of ZnO [22]. In other words, the decrease in cytotoxicity was proportionate to the dopant level of Fe in ZnO crystal matrix, demonstrating dissolution and release of Zn^{2+} ions as the primary mechanism of toxicity in ZnO nanoparticles [22]. The true predictive power of this cellular study was demonstrated by studies in different animal models [46]. The decreasing toxicity along with decreasing dissolution kinetics of ZnO was demonstrated by (a) the decreasing mortality rate in zebrafish embryos and (b) the decrease in acute inflammatory effects in the rodent lung [46]. Thus, these studies – on the identification of ZnO as a potentially toxic nanomaterial, identifying dissolution and shedding of Zn ions as the mechanism of toxicity, reengineering the material to reduce the toxic material property, ensuring the decreased toxic potential of reengineered materials – demonstrate the possibility of developing safe-by-design strategies through a multidisciplinary approach.

Many studies have recently been reported on the antimicrobial and cytotoxic properties of ZnO or composite materials containing ZnO nanomaterials. Sahu et al. reported that variation in ZnO content in a HA-ZnO composites can cause significant differences in its cytotoxic potential and suggested tailoring the ZnO content in HA composites to improve its biocompatibility [47]. An *in vivo* study in guinea pigs where ZnO nanoparticles were implanted in a mandible in the region of the symphysis showed only mild signs of inflammatory response [48]. While these cytotoxicity studies and short-term animal experiments support the possible use of ZnO nanomaterials for dental applications, a definite answer on its biocompatibility is yet to be sought.

4.3.2 Dangling Bonds' Surface Chemistry in Determining the Potential Toxicity of Silica Nanoparticles

Silicates or silicate-based composite materials are frequently used in dentistry as filling materials such as glass ionomer cements, compomers, composites and adhesive systems [49]. In addition, they also form components of dental ceramics, for instance, for veneers, inlays and onlays, for denture teeth and for full-ceramic crowns [49]. Along with the advancement in nanotechnology, there are studies showing the potential use of silica nanomaterials to improve the functional properties of silicate-based composite materials. The silica chemistry which is well utilized for improved mechanical properties of these composite materials has also been implied in its potential to cause adverse biological outcomes.

Silica nanoparticles can exist in crystalline and amorphous forms. It has been well known that the crystalline silica, such as quartz and cristobalite, are capable of inducing silicosis, lung cancer and autoimmune diseases [50–52]. Although the toxicity mechanism of quartz remains unclear [53], their surface properties, such as free radicals, surface charge and silanol density, have been envisaged to play crucial roles in the pathogenesis [54]. The relative toxicity of amorphous, fumed, mesoporous and other crystalline silica polymorphs (e.g., silicalite) still remains debatable.

Using a silica nanoparticle library, Zhang et al. identified key material properties that govern the potential toxicity of silica nanomaterials [55]. The library of silica particles used in their study consisted of amorphous colloidal or Stober silica, fumed silica, mesoporous silica, silicalite and nanosized Min-U-Sil (quartz). These silica nanoparticles have different surface displays of silanol groups and surface defects capable of generating adverse biological effects such as oxidative stress and cell membrane damage. Moreover, the differential distribution of strained and unstrained siloxane rings in these particles allowed the determination of the relationship between silanols and hydroxyl groups to the biological outcomes. Cytotoxicity assessment showed that fumed silica and nano-sized Min-U-Sil induced cell membrane damage, increased intracellular calcium flux and increased superoxide generation, while other types of silica showed low or no toxicological responses. Closer examination of material surfaces using Raman spectroscopy revealed that fumed silica had relatively higher density of '*highly strained*' three-member siloxane group, while other silica nanoparticles had considerably less of these three-member siloxane groups, exhibited on their surface. Highly strained three-member siloxane groups presented higher incidence of dangling bonds that was proposed to generate reactive oxygen species. In fact, electron paramagnetic resonance (EPR) analysis showed that hydroxyl radical generation was relatively higher in fumed silica, suggesting the role of dangling bond surface chemistry in contributing to the toxic potential of silica nanoparticles. Thus, oxidative stress and cell membrane damage mediated by the dangling bond surface chemistry was found to be a mechanism of silica nanoparticle toxicity, which otherwise was correlated to the crystalline phase alone [55].

4.3.3 Role of Surface Defects in Determining Toxicity of Silver Nanomaterials

Silver nanotechnology has been extensively applied in personal care products, mainly because of its antimicrobial properties. More than 30 % of all nano-enabled consumer products currently in market are using silver nanotechnology, making it one of the high-volume production-engineered nanomaterials. The potential use of silver nanomaterials for control of dental biofilm has been highlighted by a recent surge of reported studies [56, 57].

Thus far, dissolution and release of Ag ions has been attributed as the primary mechanism of silver nanoparticle toxicity [58–60]. However, studies conducted in bacteria have indicated that critical design parameters such as size and shape of Ag nanoparticles may also influence the bactericidal effect of Ag nanomaterials [61, 62]. Morone et al. reported that nano-prisms (also referred to as 'nano-plates') are more toxic compared to other shapes [62]. George et al. tested the differential toxicity of a Ag nanomaterial library comprising of Ag nanospheres (10, 20 and 40 nm), Ag nanoplates and Ag nanowires [23]. They noticed that while most of the Ag nanoparticles induced N-acetyl cysteine-sensitive oxidative stress effects in vertebrate cellular models, Ag nanoplates were considerably more toxic than other particle shapes [23]. Interestingly, Ag ion shedding and bioavailability failed to explain the high toxicity of the nanoplates. Since surface reactivity is one of the mechanisms of nanoparticle-mediated toxicity, they conducted an erythrocyte lysis assay – a well-established assay to measure the surface reactivity of particles. The involvement of surface reactivity in modulating the toxicity of Ag nanoplates was evident from the relatively higher erythrocyte lysis by Ag nanoplates. Examination of the particle surface using high-resolution transmission electron microscopy (HRTEM) revealed high levels of crystal defects (stacking faults and point defects) on the nanoplate surfaces. Crystal defects that are rich in dangling bonds are reported to enhance the surface reactivity of nanomaterials [63]. Consequently, passivation of the surface defects by coating with cysteine reduced the toxicity of Ag nanoplates, demonstrating the important role of crystal defects in contributing to Ag nanoparticle toxicity in addition to the established roles of Ag ion shedding. Figure 4.4 depicts the shape-dependent toxicity of silver nanomaterials.

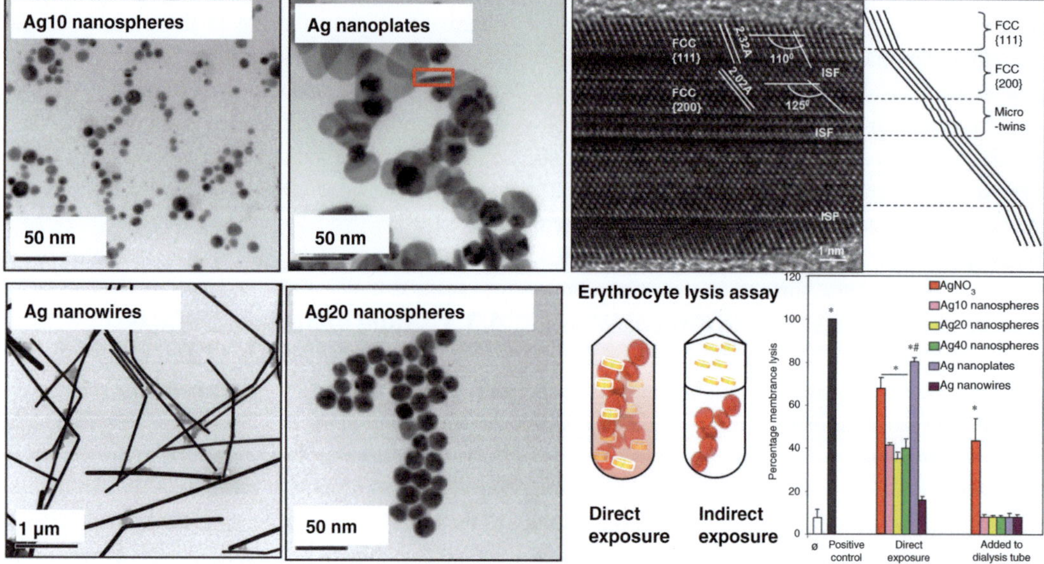

Fig. 4.4 Shape dependent toxicity of silver nanomaterials. *Left panel* shows the Transmission Electron Microscopy (TEM) of a panel of silver nanomaterial used in this study that involves spheres of different sizes and shapes. *Upper right panel.* High-resolution image of a vertically oriented {110} plane (*i.e.*, along the plane of the image) of a silver nanoplate to show its structural defects. *Lower right panel.* Measurement of erythrocyte lysis with Ag nanoparticles and $AgNO_3$ added directly to or physically segregated from the red blood cells (RBC). Heparinized mouse RBC were treated with 10 μg/mL silver nanoparticles or $AgNO_3$ (2.5 ppm) for 4 h. The samples were centrifuged and the hemoglobin absorbance of the supernatants measured at 540 nm in a microplate reader. Triton X-100 (5 %), leading to 100 % cell lysis, was used as the positive control to normalize all the values. While direct contact with silver nanoparticles or $AgNO_3$ induced RBC lysis, the effect was significantly higher for silver nanoplates. However, when the silver nanoparticles were introduced in dialysis tubing, the lytic effect was dramatically reduced. $AgNO_3$ leaching from the dialysis bag could also induce RBC lysis (Reprinted from George et al. [23]. With permission from American Chemical Society). * Statistically significant from control ($p < 0.05$). # Statistically significant differences between Ag nanoplates and other nanoparticle types ($p < 0.05$). Error bars represent the standard deviations from average values

4.3.4 Physico-Chemical Properties That Determine the Potential Toxicity of Hydroxyapatite Nanomaterials

Hydroxyapatite $(Ca_{10}(PO_4)_6(OH)_2)$ is the principal inorganic component of tooth structure. Microscopic examinations of tooth structure have revealed that dental enamel and dentine have hierarchical structures and surface features at the nanometre scale. The dimensions of enamel crystallites are 26.3 ± 2.2 nm thick, 68.3 ± 13.4 nm wide and between 100 and 1,000 nm long [64], while in the case of dentine, the hydroxyapatite crystallites are of ~20 nm in size [65]. The natural occurrence of hydroxyapatite nanomaterials in biological structures has been the rationale for the application of their synthetic version in a wide range of medical [66] and personal care products [67, 68]. In general, nanomaterials of hydroxyapatite are perceived as safe for medical and dental applications because of their relatively high biocompatibility.

A study by Chen et al. suggested the role of surface charges on the cytotoxicity of hydroxyapatite nanomaterials. They used carboxylic acids 12-aminododecanoic acid and dodecanedioic acid as surface functionalizing agents to generate positively and negatively charged hydroxyapatite nanomaterials, respectively [69]. They noticed that cellular uptake and cellular proliferation was relatively higher for positively charged hydroxyapatite nanomaterials. In a study conducted in HepG2 cells, the size-dependent cytotoxicity was noticed where the cytotoxicity followed an order of 45-nm > 26-nm > 78-nm > 175-nm [70]. The effect of size on the cell proliferation and cytotoxicity is far from resolved as another study by Shi et al. showed that hydroxyapatite nanoparticles of 20 nm induced higher cell proliferation and lower cellular apoptosis [71].

4.4 Safety Testing of Nanomaterials

So far, we have been discussing about the general and material specific properties that contribute towards potential toxicity of nanomaterials. In the following part of this chapter, we will go through the aspects of safety testing of nanomaterials. This section will detail *in vivo* and *in vitro* approaches for testing safety of nanomaterials with emphasis on the merits and demerits of these models. Different aspects of *in vitro* testing have been elaborated since animal-free test models are gaining popularity among researchers for toxicity testing of chemicals and nanomaterials. Finally, the utility of high-throughput screening will be discussed for the readers to gain some understanding on this relatively new tool for toxicity testing.

4.4.1 In Vivo Testing Using Animal Models

Animal models have served as the 'gold standard' for evaluating toxicity of chemical compounds as well as particles for over a century. *In vivo* studies help in evaluating not only the lethal effects of nanomaterials but also their long-term effects, tissue localization, biodistribution and retention/excretion of nanomaterials. Several animal studies have helped in elucidating the importance of size and chemical composition in manifesting the biodistribution of nanomaterials. Unlike *in vitro* studies using cell models, *in vivo* studies could yield valuable information on the organ-specific retention time and fate of nanomaterials from the body. For example, Cho et al. showed that SiO_2 nanomaterials of 100 and 200 nm were retained in kidney, liver and spleen of mice injected with these particles, while 50 nm NPs were cleared in urine [72]. Fitzpatrick et al. showed the differential retention time of quantum dots injected in mice [73]. Rapid clearings of QDs were observed in reticuloendothelial systems, while the retention time varied from 2 to 5 days in liver. However, the lymph nodes retained QDs for a period up to 2 years. Most importantly, their work indicated the possible breakdown of QDs under *in vivo* conditions where the mechanism of degradation is influenced by subcellular location (i.e. vesicle physiology, presence of peroxide, pH level) or simply a result of long-term infinite dilution. Earlier, Chen et al. had shown that the aggregation state of QDs is an important factor for distribution and excretion in the body. QDs that remained nano sized *in vivo* excreted via the kidney, and quantum dots that aggregated into larger particles remained in liver tissue until 5 days after a single intravenous injection [74]. Their result also points towards the importance of physicochemical information such as agglomeration status of nanomaterials in biologically relevant microenvironments in explaining their potential toxicity and retention in the body. In a study that examined the effect of size of gold nanomaterials on their *in vivo* toxicity in mice, it was shown that mice injected with 3, 5, 50 and 100 nm gold NPs behaved normally and survived throughout the experimental period, while those injected with 8, 17, 12 and 37 nm gold NPs exhibited symptoms of toxicity [75]. In short, it is evident that toxicity testing in animal models merits from the capability of understanding the relative distribution, retention and fate of nanomaterials in the body that would have been difficult to evaluate using *in vitro* cell-based models.

Although an animal-based model gives valuable information on the possible fate and metabolism in a living organism, the huge cost and time required to elaborate toxicity limits its usefulness in screening a large number of nanomaterials. It has been estimated that about \$2.8 billion are spent worldwide every year for toxicological studies using animals [76]. The estimated economic burden of safety evaluation of nanomaterials in the United States for testing existing nanomaterials ranges from \$249 million for primary screening using simpler screening assays to \$1.18 billion for a more comprehensive long-term *in vivo* testing [77]. In Europe, regulatory agencies are concerned about the huge estimated cost of \$ 13.6 billions for toxicological testing of chemicals (including nanomaterials) over the next decade to comply with the European Union's REACH (Registration, Evaluation, Authorisation and Restriction of Chemicals) legislation, which came into force since 2007. Moreover, toxicity screening at this level requires sacrificing of about 54 millions animals, contributing to multi-dimensional issues related to animal ethics [78]. The time taken to complete the toxicity profiling of all existing nanomaterials could be between 34

and 53 years [77]. Besides the huge cost and time involved, animal-based toxicity testing also suffers from the poor correlation to toxicity in humans because of the differences in the physiology (evolutionary adaptations) and likelihood of being exposed to multitudes of other environmental factors [76]. The need of an alternate toxicity screening because of the inadequacy of current models to accommodate all newer materials being synthesized in a time- and cost-effective way is discussed more elaborately by Hartung [76]. Accordingly, the commission communiqué issued in Rome in the month of August 2009 at the World Congress on Alternatives and Animal Use in the Life Sciences observed that a 'Faster, cheaper and more reliable alternative methods will contribute to increased safety' while reducing the use of animals [79].

4.4.2 In Vitro Nanotoxicity Testing

As the lowest functional unit of an organism, cells represent the ultimate target for an intruding material, and from a biochemical perspective, cells respond to foreign materials essentially similar to that of the whole organism. Cell-based assays in the context of nanotoxicity can be regarded as an analytical procedure that assesses the biological outcomes resulting from the interaction of nanomaterials with viable cells. Right choice of cellular responses as the *in vitro* end points could yield valuable information about the potential impact of nanomaterials in the whole organism. Unlike in the case of whole organisms, the manifestations of biological outcomes is faster (often in the order of minutes to hours) in cells, thereby providing rapid read-out of relevant biological outcomes. Cell-based assays make it possible to screen for nanomaterials with potential adverse effects on human health because they can provide rapid but relevant biological responses. There are numerous inter-related biological and molecular events pre-stage or accompanying the induction of cell death in response to engineered nanomaterials. In order to achieve the true toxicological significance of a cellular injury response, the choices of material characterization techniques, cell type and cytotoxicity events are critical. Therefore, we will detail these aspects in the following section.

Characterization of Nanomaterials for Cytotoxicity Testing

The first type of characterization involves identification of the intrinsic properties of materials as initially acquired or synthesized, including chemical composition, size, shape, purity, crystal structure and surface area. Some of the commonly used techniques for this purpose include (1) transmission electron microscopy (TEM) and scanning electron microscopy (SEM) primarily to obtain particle size, shape and aspect ratio; (2) X-ray diffraction (XRD) to identify crystal structure of ENMs and to determine the percentage of crystallinity of each phase; (3) inductively coupled plasma mass spectroscopy (ICP-MS) or elemental analysis to ensure chemical purity of engineered nanomaterials (ENMs) (4) Brunauer–Emmett–Teller (BET) gas absorption/desorption method to analyse surface area as well as porosity of particles [80]. These constitute some of the basal material properties that need to be reported in nanotoxicity studies. However, in some specific cases, additional material characterization using high-resolution transmission electron microscopy (HRTEM), X-ray photon spectroscopy (XPS) and thermogravimetric analysis (TGA) to understand crystallography, surface chemical composition and organic and water content may be required.

In addition to the basic 'as produced' physicochemical characterization, the change of nanomaterial properties when dispersed in biological media warrants for additional characterization in their 'as dispersed' state. More often, the additional characterization includes determination of zeta potential, hydrodynamic diameter, stability of the dispersion and dissolution of particles in the media. In general, cell culture and biological media contain high ionic strength, which may lead to charge shielding and electric double-layer compression on the material surface, thereby facilitating nanomaterial agglomeration. For example, when TiO_2 nanoparticles are dispersed in different biological exposure media depending on concentration, they can form agglomerates that vary from tens of nanometers to over a micron in size during assessment of their hydrodynamic diameter [21]. This exemplifies the need to consider dispersing agents to achieve stable particle suspensions to perform toxicity studies. For example, Ji et al. observed that, when TiO_2 nanoparticles are dispersed in cell culture media at 50 μg mL^{-1} concentration, only less than 10 μg mL^{-1} remain in the supernatant after 24 h [21]. However, improved stability was achieved following the introduction of bovine serum albumin (BSA), a combination of dipalmitoyl phosphatidylcholine (DPPC) and BSA or adding fetal bovine serum (FBS). The FBS was the best dispersant, leaving more than 40 μg mL^{-1} TiO_2 nanoparticles in the supernatant by 24 h [21]. Similar dispersion methods and stability evaluation techniques have been introduced for other classes of ENMs used in cellular and vertebrate model studies [45, 46, 81].

Choice of Cell Types for Cytotoxicity Testing

Accidental or therapeutic exposure of nanomaterials may not necessarily be restricted to a single tissue type in a human body, and often, it becomes necessary to test an array of cell types for cytotoxicity evaluation. Different cell lines representing almost entire cell types in a human body have been successfully cultivated under defined growth conditions. The versatility in the selection of cell types and culture conditions has been of great advantage in evaluating the potential health hazards due to the intentional and unintentional exposure to nanomaterials. Therefore, among various factors that may affect the reliability of conclusions drawn from an *in vitro* cytotoxicity evaluation, the right choice of the cell line is of outmost importance. Knowledge on the route of exposure of the nanomaterial and their accumulation in specific tissues in the body would enable one to choose the right cell line to be selected for the study. It has been demonstrated that TiO_2 that is either injected intraperitoneally or ingested orally could transcytosis across epithelial lining or across endothelial cells into the blood circulation to reach the final destination [82, 83]. The importance of the right choice of cell lines for a cytotoxicity study of nanomaterials has been well recognized. For example, Committees on Toxicity, Mutagenicity and Carcinogenicity of Chemicals in Food, Consumer Products and the Environment, UK, in their 'Joint statement on nanomaterial toxicology' state that the experimental model for the safety/toxicity assessment of nanomaterials should be based on the probable sites of contact, for example, primary cultures of human skin cells or other organs as (http://cot.food.gov.uk/cotstatements 2005/307429).

Choice of Cytotoxic Events

Measuring phenotypic and/or genotypic markers by using *in vitro* assays developed based on mechanisms of toxicity or injury pathways are important because the mechanisms of toxicity are the only common thread underlining both *in vitro* cellular toxicity and *in vivo* disease outcomes. Using mechanism-based *in vitro* assays will have more predictive power if human diseases are also caused by the same mechanisms, as it has been demonstrated in the toxicity of air pollution particles [5, 84]. Air pollution particles could induce lung inflammation, asthma and cardiovascular diseases with increased morbidity and mortality wherein oxidative stress has been shown to play a major role in these disease processes [85].

When nanomaterials interact with cells, they can evoke a multitude of responses depending upon the nature of the assault. These include several proximate injury responses that are directly related to the unique physicochemical properties of nanomaterials as well as more downstream injury responses, including cytotoxicity. Nanomaterials have been found to mediate cell injury through various mechanisms [86] including induction of oxidative stress [87], heavy metal ions released by dissolution of nanomaterials [88], photoactivated generation of free radicals, high cationic charge density [89], which may result in one or more of protein denaturation, foreign body granulomas, frustrated phagocytosis, triggering of the protein-unfolding response [90] and triggering of immune cell 'danger signal', activation of signaling cascades [91]. Injury response due to oxidative stress has been documented extensively in literature and is one of the best understood mechanisms of Nanomaterial-mediated toxicity. Nanomaterial-induced oxidative stress involves an incremental series of cellular responses that are encapsulated in the *'hierarchical oxidative stress paradigm'* [3, 84] (Fig. 4.5). The cellular responses to oxidative stress are discrete but sequential events involving antioxidant defence, pro-inflammatory effects and finally cytotoxic responses leading to cell death. Each of these response levels is initiated by specific biological sensors and activation mechanisms. When cells are subjected to mild oxidative stress, the transcription factor Nrf2 is activated, leading to the up-regulation of the expression of phase II enzymes. These cellular events that attempt to restore the redox equilibrium of cells are grouped as tier-1 response. If the level of oxidant injury increases, cells express pro-inflammatory cytokines by activating signalling pathways such as the mitogen-activated protein kinase (MAPK) and nuclear factor-kappa B (NF-κB) cascades (Tier 2). These inflammatory effects contribute to disease processes such as asthma and atherosclerosis. At the highest level of oxidative stress, interference in mitochondrial inner-membrane electron transfer or changing open/closed status of permeability transition pore could lead to a drop in ATP synthesis and release of pro-apoptotic factors, ultimately leading to cell death (Tier 3). Although the above-mentioned cellular responses are explained primarily based on the oxidative stress induced by nanomaterials, other nanomaterial properties may also propel cells through the same sequential events. Accordingly, pro-inflammatory responses and cytotoxicity can also be manifested as a downstream injury response to particle effects involving surface shedding of toxic metal ions (e.g. interference in mitochondrial electron transduction by cationic nanoparticles or dissolved Zn^{2+}), electronically active surfaces (e.g. semiconductors) and presence of redox cycling chemical compounds (e.g. transition metal impurities in carbon nanotubes) [3, 84]. Hence, for a comprehensive mechanistic cytotoxicity evaluation, often it is necessary to incorporate multiple biological end points that capture cellular events in Tier 1–3 of the hierarchi-

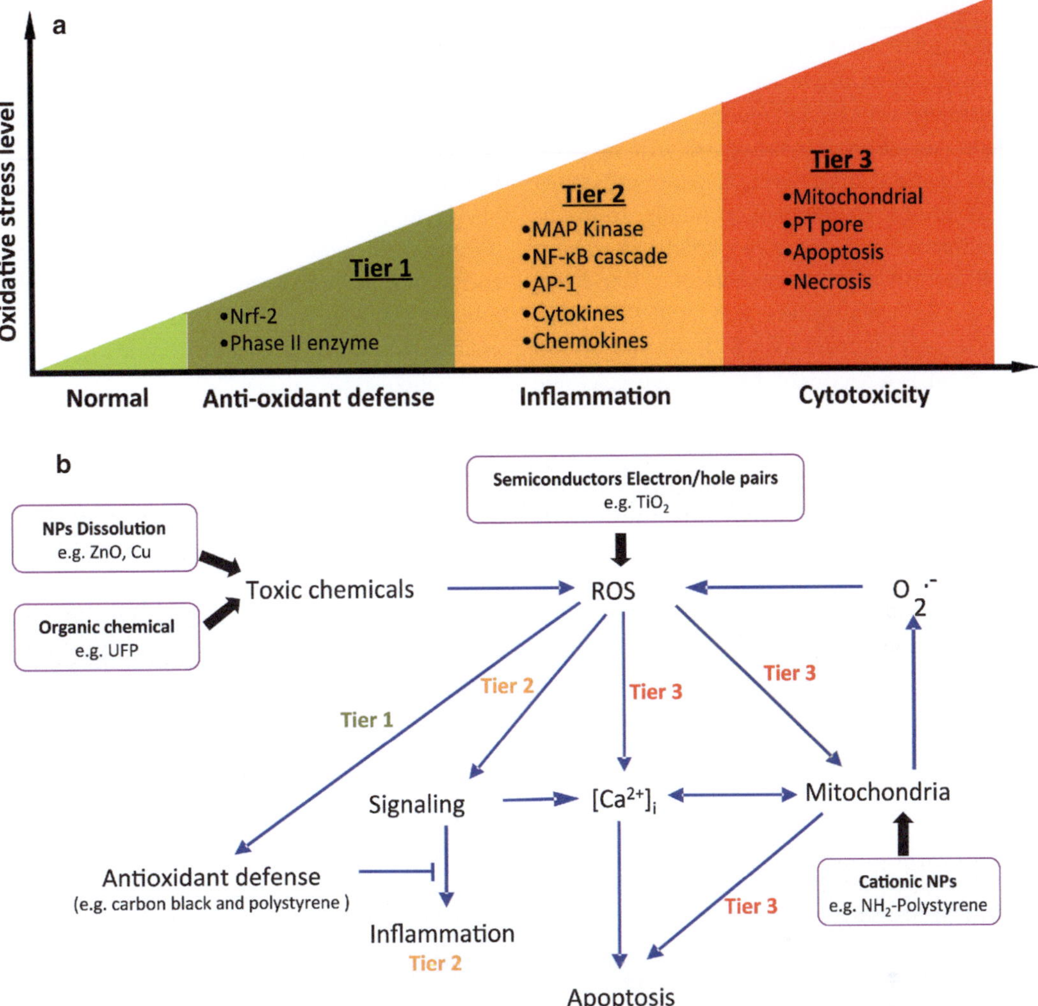

Fig. 4.5 Cellular responses encapsulated in different tiers of hierarchical oxidative stress paradigm useful for hazard assessment of nanomaterials. (**a**) Incremental levels of oxidative stress could be manifested as series of cellular responses that are encapsulated in the hierarchical oxidative stress paradigm. Lower level of oxidative stress, induces antioxidant defence (Tier 1), medium level induces pro-inflammatory effects (Tier 2) and oxidative stress at the higher level induces cytotoxic responses leading to cell death (Tier 3). Each of these response tiers are initiated by specific biological sensors and activation mechanisms. In Tier 1, the transcription factor Nrf2 is activated to enhance the expression of phase II enzymes, which attempts to restore redox equilibrium. If the level of oxidant injury increases (Tier 2), cells express proinflam-matory cytokines by activating signaling pathways such as the mitogen-activated protein kinase (MAPK) and nuclear factor-kappa B (NF-kB) cascades. At the highest level of oxidative stress (Tier 3), interference in mitochondrial inner membrane electron transfer or changing open/closed status of permeability transition pore could lead to effects on ATP synthesis and release of pro-apoptotic factors. (**b**) Induction of hierarchical oxidative stress paradigm through different physico-chemical properties of nanomaterials. Thus, hierarchical oxidative stress paradigm can be used to elucidate hazard potential of nanomaterials that dissolve and release heavy metal ions, nanomaterials with cationic surface charge, redox potential *etc*.

cal oxidant injury paradigm since it incorporates both immediate and downstream responses by nanomaterials. For the *in vitro* screening assay to be a truly predictive toxicological tool, the assay result has to be linked to *in vivo* toxicological analysis [84]. Studies in animal models have demonstrated that the oxidant potential of ambient ultrafine particles correlates well with their propensity to generate atherosclerotic plaques or allergic airway inflammation *in vivo* [92–95]. Similarly, the oxidative stress profiling of ZnO nanoparticles in cellular models has elucidated features that are analogous to the clinical manifestations of metal fume fever, an acute inflammatory condition of the lung resulting from the inhalation of metal oxide nanoparticles in welders [96, 97]. Thus, the development of a Tier 2-like response in macrophages and epithelial cells, characterized by interleukin 8 (IL-8) and TNF-α production, is reflective of the exact cytokines and chemokines in the bronchoalveolar lavage fluid of exposed workers [88]. In addition, metal fume fever is characterized by an adaptive response where prior exposure to welding fumes can avert metal fume fever on subsequent exposure [98]. This is akin to a Tier 1-like adaptive response. Genomic profiling of mononuclear cells from the blood of boilermakers prior to and after exposure to welding fumes further confirms a genetic footprint that includes clustering of pro-inflammatory pathways and oxidative stress genes [99]. In short, cellular events envisaged at different tiers of a hierarchical oxidative stress paradigm can be implemented to screen nanomaterials in order to account for their potential biological outcomes on interaction with cells.

Measuring Inflammatory Responses to Nanomaterials

Cytokines and chemokines play important roles in inflammation that is a major form of lung injury induced by incidental and ENMs. The most commonly tested human and murine inflammatory markers are Interleukin-8 (IL-8), followed by TNF-α and IL-6. In some cases, IL-1b as well as a few other cytokine and stimulating factor levels are measured. The chemokine MIP-2 is usually quantified in rat model systems together with TNF-α and/or IL-6. Several kits based on the enzyme-linked immunosorbent assay (ELISA) principle are available to measure pro-inflammatory responses of cells during interaction with nanomaterials using cell culture supernatants. Cytokines can be determined by cytokine antibody arrays (e.g. RayBio® mouse/rat/human cytokine antibody arrays, from Raybiotech, Inc. GA). The screening platform will look at the detection of pro-inflammatory responses that will be assessed by quantitative ELISA that can measure cytokine and chemokine production in the cellular supernatant in 96 well plate formats. While it is easy to perform these assays in the standard manner, having to analyze a large number of samples in different cell types could be labour intensive, and therefore, it is necessary to automate this process by using robotics and multiplex bead assays that can measure several cytokines simultaneously.

Measuring Cytotoxic Responses to Nanomaterials

Nowadays, a number of methods are available for the detection of cytotoxicity-related parameters under *in vitro* conditions, where the read-out can be colorimteric, fluorimetric or based on luminescence intensity [100]. Majority of the reported cytotoxicity studies employ dyes such as tetrazolium derivative 3-[4,5-dimethylthiazol-2-yl]-2,5-diphenyl tetrazolium bromide (MTT) that is reduced by mitochondrial dehydrogenases to water-insolubleformazanor3-(4,5-dimethylthiazol-2-yl)-5-(3-carboxymethoxyphenyl)-2-(4-sulfophenyl)-2H-tetrazolium (MTS) and 4-[3-(4-iodophenyl)-2-(4-nitrophenyl)-2H-5-tetrazolio]-1,3-benzene disulfonate (WST-1)), which are water-soluble cell-permeant products. Plasma membrane leakage can be detected by assessing culture supernatants for the activity of intracellular enzyme lactate dehydrogenase (LDH) that mediate the oxidation of yellow tetrazolium salt INT to a red formazan. Although cytotoxicity assessments using these dyes are relatively easy and require no expensive equipment, the possible interference of nanomaterials with the catalytic conversion of colourless compounds to coloured formazan products and/or ability of nanomaterials to absorb the reading wavelength could limit their usefulness [101, 102]. Monteiro-Riviere et al. reported that cell viability assays such as MTT and neutral red (NR) produced invalid results with carbon-based nanomaterials (carbon black, Multiwalled carbon nanotubes, C60) and QDs due to Nanomaterial/dye interactions and/or nanomaterial adsorption of the dye/dye products [103]. Several unique nanomaterial properties such as high adsorption capacity, optical properties, catalytic activity, *etc.* are thought to generate inaccurate results [104]. While the chance of inaccurate information from a single dye-based assay may be unavoidable when nanomaterials are tested, the reliability of *in vitro* assays can be improved substantially by simultaneous evaluation of multiple cytotoxicity markers. Fluorescent probes that have been used successfully to study cellular processes due to their supreme sensitivity and selectivity could circumvent some of the disadvantages of UV–VIS spectroscopic methods.

4.5 Faster Cytotoxicity Screening Methods

As discussed in the previous section, the reliability of extrapolation of *in vitro* observations to *in vivo* is dependent on many factors. The cytotoxicity profile of a nanomaterial may vary from one cell type to another because of the possible difference in the cellular uptake and processing of nanomaterials. A rational approach to overcome these shortcomings of *in vitro* evaluations is to incorporate more than one cell type (often different cell lineage) and to conduct multiple cytotoxicity assays simultaneously for a range of reasonable doses and durations of exposures. Such a multidimensional cytotoxicity screening effort will constitute a large number of variables, testing of which is a very cost- and time-consuming process. Automating the key process involved in the cytotoxicity evaluation by using a high-*throughput screening/high-content screening* approach is warranted to overcome these bottlenecks.

4.5.1 Overview of High-Throughput Screening

HTS is the brainchild of the drug industry where its versatile functionalities have been utilized for drug screening and lead compound discovery. An HTS platform is constituted by interlinked modular units that undertake one or more operations in the workflow of compound testing, which include cell seeding to multiwell plates, preparation of working concentrations of compounds, plate washing, addition of compounds and reagents to multiwell plates, reading the plate, data archiving and data analysis. Since all the key operations are automated and their timing is frequently controlled by a central computer for 24/7 operation, the time taken for entire process of lead compound identification is drastically lowered. The reagent consumption in an HTS is substantially lowered since the assays are run with microlitre quantities of assay reagents. Furthermore, the possibility of multiplexing – combining more than a single parameter in one assay – may also reduce the resource requirements in terms of consumables, reagents and machine usage. Because of these reasons, HTS has also been adopted more recently as a screening tool in toxicology in the place of more labour-intensive and descriptive toxicological approaches.

4.5.2 Tiered Oxidative Stress Induced by Nanomaterials as End Points of HTS Analysis

As mentioned previously, nanomaterial-induced oxidative stress involves an incremental series of cellular responses that is explained by '*hierarchical oxidative stress paradigm*'. The different levels of cellular response can be classified as antioxidant defence (Tier 1), pro-inflammatory effects (Tier 2) and cytotoxicity (Tier 3). Each of these response tiers is initiated by specific biological sensors and activation mechanisms. In Tier 1, the transcription factor Nrf2 is activated to enhance the expression of phase II enzymes, which attempts to restore redox equilibrium. If the level of oxidant injury increases (Tier 2), cells express pro-inflammatory cytokines by activating signalling pathways such as the mitogen-activated protein kinase (MAPK) and nuclear factor-kappa B (NF-κB) cascades. These inflammatory effects contribute to disease processes such as asthma and atherosclerosis. At the highest level of oxidative stress (Tier 3), interference in mitochondrial inner-membrane electron transfer or changing open/closed status of permeability transition pore could lead to a drop in ATP synthesis and release of pro-apoptotic factors, ultimately leading to cell death. Cytotoxicity (Tier 3 response) is an example of a downstream injury response to particle effects involving ROS production, surface shedding of toxic metal ions that trigger cellular ROS generation (e.g. interference in mitochondrial electron transduction by cationic nanoparticles or dissolved Zn^{2+}), electronically active surfaces (e.g. semiconductors) and presence of redox active ions (e.g. transition metal impurities in carbon nanotubes) [3, 84]. Therefore, often it is necessary to select multiple biological end points which capture cellular events in Tier 1–3 during an *in vitro* study for reliably extrapolating the observations to an *in vivo* scenario. This aspect was exemplified by the toxicity studies conducted on ZnO nanoparticles by Xia et al [88]. They showed that ZnO nanoparticles could elicit cytotoxicity and pro-inflammatory responses in cellular models that are analogous to the molecular profile evident in metal fume fever. The expression of a Tier 2-like response in macrophages and epithelial cells, characterized by interleukin 8 (IL-8) and TNF-α production, was reflective of similar cytokine production in the lungs and bronchoalveolar lavage fluid of exposed workers [88]. In short, cellular events envisaged at different tiers of a hierarchical oxidative stress paradigm can be implemented to screen nanomaterials in order to account for their potential biological outcomes on interaction with cells.

High-Content Screening

High-content screening (HCS) refers to the use of an automated microscope (e.g. epifluorescence microscope) that captures images of cells stained with reporter fluorescent probes. Generally, images can be acquired in different wavelengths allowing for reading more than one parameter at a time. One of the reasons why HCS is extremely well suited for toxicological screens is that it allows for the resolution of data on a single-cell level and thus allows for monitoring of cell populations which undergo a toxicity reaction. The other advantage is that it allows for unexpected phenotypes to be recorded (pictures can be analyzed repeatedly) and it does not rely on an assumed knowledge of the mode of action of a target. Since the same fluorescent probes can be excited many times and detected, it is possible to monitor the toxicokinetics of nanomaterials in real time. One of the important characteristics of fluorescent image–based assay is the possibility of multiplexing to observe multiple cellular responses in a single assay. The functional linking of cellular events using a multi-parametric assay could generate meaningful cytotoxicity data that goes beyond the simple live/dead scoring. Figure 4.6 represents the major steps involved in the use of HCS for elucidating cytotoxicity of nanomaterials.

George et al. developed a multi-parametric analysis of cellular responses to nanomaterials by combining wavelength-compatible dyes that provide a read-out of Tier 3 oxidative stress effects [22]. The multi-parametric assay is an automated assay that is conducted in 384 well plates containing target cells to which the nanomaterials are added over a wide range of different doses. The Nanomaterial-treated cells are stained with reporter dyes which include the DNA interactive agent, Hoechst 33342, to assess cell number and

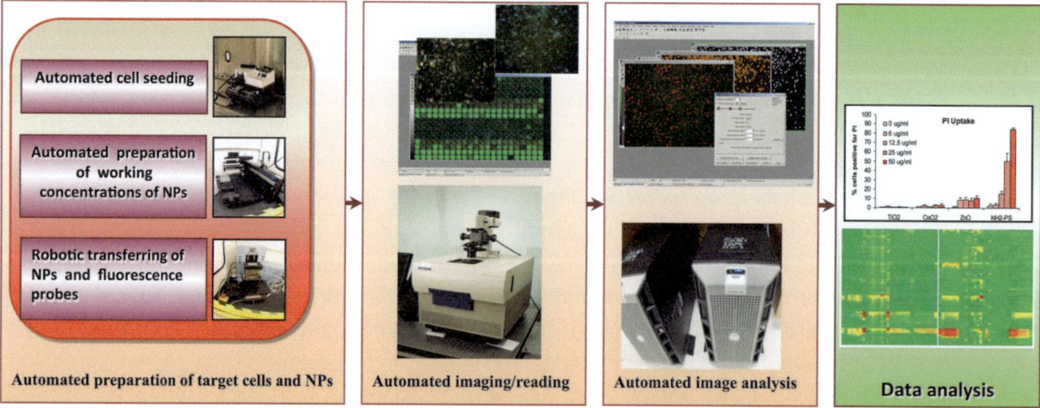

Fig. 4.6 Flow diagram showing sequential operations involved in screening of nanomaterials for toxicity in cells using high content screening approach

nuclear size, JC1 (5,5′,6,6′-tetrachloro-1,1′,3,3′-tetraethylbenzimidazolylcarbocyanine iodide) to assess mitochondrial membrane potential, Fluo 4 to report intracellular calcium flux and propidium iodide (PI) to report increased membrane permeability of dead cells [22]. Following the addition of reporter dyes, the plates are loaded for image acquisition using an automated epifluorescence microscope. The images in three different wavelengths (blue, green and red) are captured from each well of 384 well plates using 10× objective lens. The images captured in different wavelengths can be combined to generate composite images. These microscopic images are automatically analyzed by software. First, the images are 'segmented' into sub-cellular structures (nucleus and cytoplasm) based on morphological features such as object size and fluorescence distribution. Images to assess total cell number and PI uptake can be scored from the nuclear segment, while assessment of mitochondrial responses and Ca^{2+} levels were taken from the cytoplasmic segment of the cell. The possibility of evaluating more than one cytotoxic event from the same image allows for calculating relative occurrence of a cytotoxic event. For example, based on the total number of cells (counted from Hoechst 33342 stained region) and the total number of cells with PI signal, it is possible to calculate the % of cells positive for PI signal. Figure 4.7 gives the utility of fluorescent probes for scoring cytotoxicity of nanomaterials.

Response marker	Fluorescent Probe	Excitation/ Emission	Response	Utility
Cell number/cell location	Hoechst 33342	355/465	Stain nucleus	Locating and counting of cells
Mitochondrial superoxide	MitoSox	480/580	Increased red fluorescence	Oxidative stress
Mitochondrial membrane potential	JC1	480/530-590	Shift of red to green fluorescence	Loss of MMP
Intracellular calcium	Fluo-4	480/510	Increased green fluorescence	Increased Intra-cellular calcium level
PI Uptake	Propidium Iodide	540/620	Increased red fluorescence	Compromised cell membrane integrity

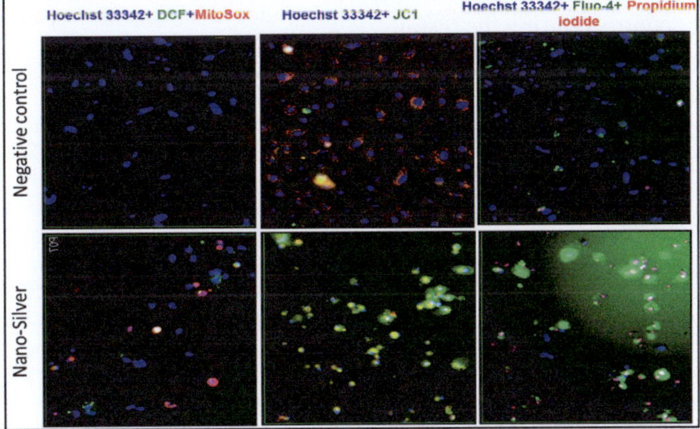

Fig. 4.7 The *table* showing the utility of fluorescence markers for assessing cytotoxicity of nanomaterials *Lower panel*: Cellular images showing differential staining of RAW 264.7 cells with or without silver nanomaterials. Staining of silver nanomaterial treated cells with Hoechst 33342, DCF and MitoSox shows *blue*, *green* and *red* fluorescence indicating cell nucleus, intracellular hydrogen peroxide and mitochondrial superoxide, respectively. In the case of Hoechst 33342 and mitochondrial dye JC1, healthy cells shows blue nuclei and red mitochondria (negative control) while the mitochondrial depolarization causes the cytoplasm to fluoresce green (silver nanomaterial treated cells). Cells treated with silver nanomaterials and stained with a dye cocktail of Hoechst 33342, Fluo-4 and Propidium iodide shows blue nuclei, green fluorescence because of intracellular calcium and *red* fluorescence because of cell membrane damage. Depending upon the color profile, the percentage of cells affected can be assayed and nanoparticles can be ranked for their cytotoxic potential

4.6 Challenges of Cellular Studies and Future Perspective

Majority of the reported cytotoxicity studies employ dyes such as MTT, MTS and alamar blue for colorimetric evaluation of cell metabolism or evaluating cell membrane integrity using neutral red or propidium iodide uptake. While these cytotoxicity markers can be successfully employed for toxic chemical agents, the unique material properties of the nanomaterials may interfere with assay reagents. Monteiro-Riviere et al. reported that cell viability assays such as MTT and neutral red (NR) produced invalid results with carbon-based nanomaterials (carbon black, Multiwalled carbon nanotubes, C60) and QDs due to Nanomaterial/dye interactions and/or nanomaterial adsorption of the dye/dye products [103]. Several unique nanomaterial properties such as high adsorption capacity, optical properties, catalytic activity, *etc.* are thought to generate inaccurate results [104]. While the chance of inaccurate information from a single dye-based assay may be unavoidable when nanomaterials are tested, the reliability of *in vitro* assays can be improved substantially by simultaneous evaluation of multiple cytotoxicity markers and by incorporating appropriate controls during the testing.

One of the major concerns using the *in vitro* cytotoxicity assessment is that it can never represent the whole event happening in an in vivo situation. The particle behaviour in terms of aggregation and settling could differ under *in vitro* and *in vivo* test conditions. The interaction of nanomaterials with biomolecules present in circulatory and interstitial fluids could complicate the reliable extrapolation of cytotoxicity data to *in vivo* situations. Under *in vivo* conditions, nanomaterials may get conditioned with other body fluids such as lymph, saliva, lung lavage fluid, *etc.* depending on the route and course of exposure. In order to improve the reliability of extrapolation from *in vitro* studies, one has to identify the profile of biomolecules likely to interact with nanomaterials and cell models and responses of relevance to routes of exposure.

4.7 Concluding Remarks

Identification of the toxic potential at an early stage of nanomaterial generation allows for physical and/or chemical modifications to circumvent their toxicity. The mechanisms of toxicity mediated by nanomaterials can be tested according to the known paradigms of toxicity. The known paradigms of toxicity include (1) induction of reactive oxygen, (2) cationic charge and (3) dissolution of heavy metals. Once the mechanism of nanomaterials has been established, it may be possible to purposefully reduce the toxicity by chemical and/or physical modification. The process of identifying mechanisms of toxicity and purposeful amelioration could be a rational approach for the safe design of nanomaterials, as it has been shown with nanomaterials of ZnO [22], MSNPs [105] and rare earth materials [106]. We can expect to find more ENM toxicity paradigms and pathways which can be screened for, and the generation of a nano-QSAR should help us in our quest for safe nanomaterials.

References

1. Meyer DE, Curran MA, Gonzalez MA. An examination of existing data for the industrial manufacture and use of nanocomponents and their role in the life cycle impact of nanoproducts. Environ Sci Technol. 2009;43(5):1256–63.
2. The National Nanotechnology Initiative Strategic Plan, 2004, Nanoscale Science, Engineering, and Technology Subcommittee. National Science and Technology Council, Executive Office of the President; 2004.
3. Nel A, et al. Toxic potential of materials at the nanolevel. Science. 2006;311(5761):622–7.
4. Oberdörster G, Oberdörster E, Oberdörster J. Nanotoxicology: an emerging discipline evolving from studies of ultrafine particles. Environ Health Perspect. 2005;113.
5. Xia T, Li N, Nel AE. Potential health impact of nanoparticles. Annu Rev Public Health. 2009;30(1):137–50.
6. ICRP. Human respiratory tract model for radiological protection. A report of a Task Group of the International Commission on Radiological Protection. Ann ICRP. 1994;24:1–482.
7. Gao H, Shi W, Freund LB. Mechanics of receptor-mediated endocytosis. Proc Natl Acad Sci U S A. 2005;102(27):9469–74.

8. Li X. Size and shape effects on receptor-mediated endocytosis of nanoparticles. J Appl Phys. 2012;111(2) doi:10.1063/1.3676448.

9. Garnett MC, Kallinteri P. Nanomedicines and nanotoxicology: some physiological principles. Occup Med. 2006;56(5):307–11.

10. Pan Y, et al. Size-dependent cytotoxicity of gold nanoparticles. Small. 2007;3(11):1941–9.

11. Allouni ZE, et al. Agglomeration and sedimentation of TiO2 nanoparticles in cell culture medium. Colloids Surf B Biointerfaces. 2009;68(1):83–7.

12. Derjaguin BV, Landau LD. Theory of the stability of strongly charged lyophobic sols and of the adhesion of strongly charged particles in solutions of electrolytes. Acta Physicochim URSS. 1941;14:733–62.

13. Rezwan K, et al. Change of xi potential of biocompatible colloidal oxide particles upon adsorption of bovine serum albumin and lysozyme. J Phys Chem B. 2005;109(30):14469–74.

14. Jiang JK, Oberdorster G, Biswas P. Characterization of size, surface charge, and agglomeration state of nanoparticle dispersions for toxicological studies. J Nanopart Res. 2009;11(1):77–89.

15. Gupta AK, Gupta M. Synthesis and surface engineering of iron oxide nanoparticles for biomedical applications. Biomaterials. 2005;26(18):3995–4021.

16. Hajdú A, et al. Surface charging, polyanionic coating and colloid stability of magnetite nanoparticles. Colloids Surf A Physicochem Eng Aspects. 2009;347(1–3):104–8.

17. Schulze C, et al. Not ready to use – overcoming pitfalls when dispersing nanoparticles in physiological media. Nanotoxicology. 2008;2(2):51–U17.

18. Foucaud L, et al. Measurement of reactive species production by nanoparticles prepared in biologically relevant media. Toxicol Lett. 2007;174(1–3):1–9.

19. Bihari P, et al. Optimized dispersion of nanoparticles for biological in vitro and in vivo studies. Part Fibre Toxicol. 2008;5(1):14.

20. Porter D, et al. A biocompatible medium for nanoparticle dispersion. Nanotoxicology. 2008;2(3):144–54.

21. Ji Z, et al. Dispersion and stability optimization of TiO2 nanoparticles in cell culture media. Environ Sci Technol. 2010;44:7309–14.

22. George S, et al. Use of a rapid cytotoxicity screening approach to engineer a safer zinc oxide nanoparticle through iron doping. ACS Nano. 2010;4(1):15–29.

23. George S, et al. Surface defects on plate-shaped silver nanoparticles contribute to its hazard potential in a fish gill cell line and zebrafish embryos. ACS Nano. 2012;6(5):3745–59.

24. Wang X, Xia T, Ntim SA, Ji ZX, George S, Meng H, Zhang HY, Castranova V, Mitra S, Nel AE. Quantitative techniques for assessing and controlling the dispersion and biological effects of multiwalled carbon nanotubes in mammalian tissue culture cells. ACS Nano. 2010;4:7241–52.

25. Wang X, Xian T, Ntim SA, Ji Z, Lin S, Meng H, Chung C, George S, Zhang H, Wang M, Li N, Yang Y, Castranova V, Mitra S, Bonner J, Nel AE. Dispersal state of multiwalled carbon nanotubes elicits profibrogenic cellular responses that correlate with fibrogenesis biomarkers and fibrosis in the murine lung. ACS Nano. 2011;5:9772–87.

26. Decuzzi P, et al. Size and shape effects in the biodistribution of intravascularly injected particles. J Control Release. 2010;141(3):320–7.

27. Meng H, et al. Aspect ratio determines the quantity of mesoporous silica nanoparticle uptake by a small GTPase-dependent macropinocytosis mechanism. ACS Nano. 2011;5(6):4434–47.

28. Stanton MF, Layard M, Tegeris A, Miller E, May M, Morgan E, Smith A. Relation of particle dimension to carcinogenicity in Amphibole Asbestoses and other fibrous minerals. J Natl Cancer Inst. 1981;67:965–75.

29. Donaldson K, et al. Pulmonary toxicity of carbon nanotubes and asbestos — similarities and differences. Adv Drug Deliv Rev. 2013;65(15):2078–86.

30. Wagner JC, Berry G. Mesothdiomas in rats following inoculation with asbestos. Br J Cancer. 1969;23:567–81.

31. Davis JMG. A review of experimental evidence for the carcinogenicity of man-made vitreous fibres. Scand J Work Environ Health. 1986;12:12–7.

32. Lippmann M. Effects of fiber characteristics on lung deposition, retention, and disease. Environ Health Perspect. 1990;88:311–7.

33. Tsuda A, et al. Alveolar cell stretching in the presence of fibrous particles induces interleukin-8 responses. Am J Respir Cell Mol Biol. 1999;21(4):455–62.

34. Brown SC, et al. Influence of shape, adhesion and simulated lung mechanics on amorphous silica nanoparticle toxicity. Adv Powder Technol. 2007;18(1):69–79.

35. Ji Z, et al. Designed synthesis of CeO_2 nanorods and nanowires for studying toxicological effects of high aspect ratio nanomaterials. ACS Nano. 2012;6(6):5366–80.

36. Perrault SD, et al. Mediating tumor targeting efficiency of nanoparticles through design. Nano Lett. 2009;9(5):1909–15.

37. Meng H, et al. Codelivery of an optimal drug/siRNA combination using mesoporous silica nanoparticles to overcome drug resistance in breast cancer in vitro and in vivo. ACS Nano. 2013;7(2):994–1005.

38. Ryman-Rasmussen JP, Riviere JE, Monteiro-Riviere NA. Surface coatings determine cytotoxicity and irritation potential of quantum dot nanoparticles in epidermal keratinocytes. J Invest Dermatol. 2007;127(1):143–53.

39. Lovric J, et al. Differences in subcellular distribution and toxicity of green and red emitting CdTe quantum dots. J Mol Med. 2005;83(5):377–85.

40. Verma A, et al. Surface-structure-regulated cell-membrane penetration by monolayer-protected nanoparticles. Nat Mater. 2008;7(7):588–95.

41. Stasko NA, et al. Cytotoxicity of polypropylenimine dendrimer conjugates on cultured endothelial cells. Biomacromolecules. 2007;8(12):3853–9.

42. Sayes CM, et al. The differential cytotoxicity of water-soluble fullerenes. Nano Lett. 2004;4(10):1881–7.

43. Sayes CM, et al. Functionalization density dependence of single-walled carbon nanotubes cytotoxicity in vitro. Toxicol Lett. 2006;161(2):135–42.

44. Aydin Sevinç B, Hanley L, Antibacterial activity of dental composites containing zinc oxide nanoparticles. J Biomed Mater Res B Appl Biomater. 2010;94B(1):22–31.

45. George S, et al. Use of a high-throughput screening approach coupled with in vivo zebrafish embryo screening to develop hazard ranking for engineered nanomaterials. ACS Nano. 2011;5(3):1805–17.

46. Xia T, et al. Decreased dissolution of ZnO by iron doping yields nanoparticles with reduced toxicity in the rodent lung and zebrafish embryos. ACS Nano. 2011;5(2):1223–35.

47. Saha N, Dubey AK, Basu B. Cellular proliferation, cellular viability, and biocompatibility of HA-ZnO composites. J Biomed Mater Res B Appl Biomater. 2012;100B(1):256–64.

48. Sousa CJA, et al. Synthesis and characterization of zinc oxide nanocrystals and histologic evaluation of their biocompatibility by means of intraosseous implants. Int Endod J. 2014;47:416–24.

49. Lührs AK, Geurtsen W. The application of silicon and silicates in dentistry: a review. In: Müller WG, Grachev M, editors. Biosilica in evolution, morphogenesis, and nanobiotechnology. Berlin/Heidelberg: Springer; 2009. p. 359–80.

50. Castranova V, Vallyathan V. Silicosis and coal workers' pneumoconiosis. Environ Health Perspect. 2000;108:675–84.

51. Elias Z, et al. Cytotoxic and transforming effects of silica particles with different surface properties in Syrian hamster embryo (SHE) cells. Toxicol In Vitro. 2000;14(5):409–22.

52. Donaldson K, Tran CL. Inflammation caused by particles and fibers. Inhal Toxicol. 2002;14(1):5–27.

53. Rimal B, Greenberg AK, Rom WN. Basic pathogenetic mechanisms in sificosis: current understanding. Curr Opin Pulm Med. 2005;11(2):169–73.

54. Dutta D, Moudgil BM. Crystalline silica particles mediated lung injury. Kona Powder Part. 2007;25:76–87.

55. Zhang H, et al. Processing pathway dependence of amorphous silica nanoparticle toxicity: colloidal vs pyrolytic. J Am Chem Soc. 2012;134(38):15790–804.

56. Melo MAS, et al. Novel dental adhesives containing nanoparticles of silver and amorphous calcium phosphate. Dent Mater. 2013;29(2):199–210.

57. Ahn S-J, et al. Experimental antimicrobial orthodontic adhesives using nanofillers and silver nanoparticles. Dent Mater. 2009;25(2):206–13.

58. Navarro E, et al. Toxicity of silver nanoparticles to Chlamydomonas reinhardtii. Environ Sci Technol. 2008;42(23):8959–64.

59. Zhao C-M, Wang W-X. Comparison of acute and chronic toxicity of silver nanoparticles and silver nitrate to Daphnia magna. Environ Toxicol Chem. 2011;30(4):885–92.

60. Glover RD, Miller JM, Hutchison JE. Nanoparticle dynamics on surfaces and potential sources of nanoparticles in the environment. ACS Nano. 2011;22:8950–7.

61. Pal S, Tak YK, Song JM. Does the antibacterial activity of silver nanoparticles depend on the shape of the nanoparticle? A study of the gram-negative bacterium escherichia coli. Appl Environ Microbiol. 2007;73(6):1712–20.

62. Morones JR, et al. The bactericidal effect of silver nanoparticles. Nanotechnology. 2005;16(10):2346.

63. Xinling TM, Tsuji, Syntheses of silver nanowires in liquid phase. Nanowires Science and Technology, Nicoleta Lupu (Ed.), ISBN: 978-953-7619-89-3, InTech, Available from: http://www.intechopen.com/books/nanowires-science-and-technology/syntheses-of-silver-nanowires-in-liquid-phase. 2010. Last accessed on Oct 29 2014.

64. Cui F-Z, Ge J. New observations of the hierarchical structure of human enamel, from nanoscale to microscale. J Tissue Eng Regen Med. 2007;1(3):185–91.

65. Imbeni V, et al. The dentin-enamel junction and the fracture of human teeth. Nat Mater. 2005;4(3):229–32.

66. Catledge SA, et al. Nanostructured ceramics for biomedical implants. J Nanosci Nanotechnol. 2002;2(3–1):293–312.

67. Reynolds EC, et al. Fluoride and casein phosphopeptide-amorphous calcium phosphate. J Dent Res. 2008;87(4):344–8.

68. Huang S, et al. Remineralization potential of nano-hydroxyapatite on initial enamel lesions: an in vitro study. Caries Res. 2011;45(5):460–8.

69. Chen L, et al. The role of surface charge on the uptake and biocompatibility of hydroxyapatite nanoparticles with osteoblast cells. Nanotechnology. 2011;22(10):105708.

70. Yuan Y, et al. Size-mediated cytotoxicity and apoptosis of hydroxyapatite nanoparticles in human hepatoma HepG2 cells. Biomaterials. 2010;31(4):730–40.

71. Shi Z, et al. Size effect of hydroxyapatite nanoparticles on proliferation and apoptosis of osteoblast-like cells. Acta Biomater. 2009;5(1):338–45.

72. Cho MJ, et al. The impact of size on tissue distribution and elimination by single intravenous injection of silica nanoparticles. Toxicol Lett. 2009;189(3):177–83.

73. Fitzpatrick JAJ, et al. Long-term persistence and spectral blue shifting of quantum dots in vivo. Nano Lett. 2009;9(7):2736–41.

74. Chen Z, Chen H, Meng H, Xing G, Gao X, Sun B, Shi X, Yuan H, Zhang C, Liu R, Zhao F, Zhao Y, Fang X. Bio-distribution and metabolic paths of sil-

ica coated CdSeS quantum dots. Toxicol Appl Pharmacol. 2008;230:364–71.

75. Chen YS, et al. Assessment of the in vivo toxicity of gold nanoparticles. Nanoscale Res Lett. 2009; 4(8):858–64.

76. Hartung T. Toxicology for the twenty-first century. Nature. 2009;460(7252):208–12.

77. Choi JY, Ramachandran G, Kandlikar M. The impact of toxicity testing costs on nanomaterial regulation. Environ Sci Technol. 2009;43(9):3030–4.

78. Gilbert N. Chemical-safety costs uncertain. Nature. 2009;460:1065.

79. Abbott A. Toxicity testing gets a makeover. Nature. 2009;461:158.

80. Damoiseaux R, et al. No time to lose-high throughput screening to assess nanomaterial safety. Nanoscale. 2011;3(4):1345–60.

81. Gordon RJ, Lowy FD. Pathogenesis of methicillin-resistant staphylococcus aureus infection. Clin Infect Dis. 2008;46(Supplement 5):350–9.

82. Olmedo DG, et al. Effect of titanium dioxide on the oxidative metabolism of alveolar macrophages: an experimental study in rats. J Biomed Mater Res A. 2005;73A(2):142–9.

83. Wang J, et al. Acute toxicity and biodistribution of different sized titanium dioxide particles in mice after oral administration. Toxicol Lett. 2007;168(2): 176–85.

84. Meng H, et al. A predictive toxicological paradigm for the safety assessment of nanomaterials. ACS Nano. 2009;3(7):1620–7.

85. Li N, et al. Particulate air pollutants and asthma. A paradigm for the role of oxidative stress in PM-induced adverse health effects. Clin Immunol. 2003;109(3):250–65.

86. Nel AE, et al. Understanding biophysicochemical interactions at the nano-bio interface. Nat Mater. 2009;8:543–57.

87. Xia T, et al. Comparison of the abilities of ambient and manufactured nanoparticles to induce cellular toxicity according to an oxidative stress paradigm. Nano Lett. 2006;6(8):1794–807.

88. Xia T, et al. Comparison of the mechanism of toxicity of zinc oxide and cerium oxide nanoparticles based on dissolution and oxidative stress properties. ACS Nano. 2008;2(10):2121–34.

89. Xia T, et al. Cationic polystyrene nanosphere toxicity depends on cell-specific endocytic and mitochondrial injury pathways. ACS Nano. 2008;2(1):85–96.

90. Jung EJ, et al. Pro-oxidative DEP chemicals induce heat shock proteins and an unfolding protein response in a bronchial epithelial cell line as determined by DIGE analysis. Proteomics. 2007;7(21): 3906–18.

91. Xiao GG, et al. Use of proteomics to demonstrate a hierarchical oxidative stress response to diesel exhaust particle chemicals in a macrophage cell line. J Biol Chem. 2003;278(50):50781–90.

92. Araujo JA, et al. Ambient particulate pollutants in the ultrafine range promote early atherosclerosis and systemic oxidative stress. Circ Res. 2008;102(5): 589–96.

93. Li N, Xia T, Nel AE. The role of oxidative stress in ambient particulate matter-induced lung diseases and its implications in the toxicity of engineered nanoparticles. Free Radic Biol Med. 2008;44(9): 1689–99.

94. Li N, et al. The adjuvant effect of ambient particulate matter is closely reflected by the particulate oxidant potential. Environ Health Perspect. 2009;117(7): 1116–23.

95. Araujo J, Nel A. Particulate matter and atherosclerosis: role of particle size, composition and oxidative stress. Part Fibre Toxicol. 2009;6(1):24.

96. Martin CJ, et al. Zinc exposure in Chinese foundry workers. Am J Ind Med. 1999;35(6):574–80.

97. Rohrs LC. Metal-fume fever from inhaling zinc oxide. Arch Intern Med. 1957;100(1):44–9.

98. Gordon T, Fine JM. Metal fume fever. Occup Med State Art Rev. 1993;8(3):505–17.

99. Wang ZX, et al. Global gene expression profiling in whole-blood samples from individuals exposed to metal fumes. Environ Health Perspect. 2005;113(2): 233–41.

100. Galluzzi L, et al. Guidelines for the use and interpretation of assays for monitoring cell death in higher eukaryotes. Cell Death Differ. 2009;16(8): 1093–107.

101. Casey A, et al. Spectroscopic analysis confirms the interactions between single walled carbon nanotubes and various dyes commonly used to assess cytotoxicity. Carbon. 2007;45(7):1425–32.

102. Worle-Knirsch JM, Pulskamp K, Krug HF. Oops they did it again! Carbon nanotubes hoax scientists in viability assays. Nano Lett. 2006;6(6):1261–8.

103. Monteiro-Riviere NA, Inman AO, Zhang LW. Limitations and relative utility of screening assays to assess engineered nanoparticle toxicity in a human cell line. Toxicol Appl Pharmacol. 2009; 234(2):222–35.

104. Kroll A, et al. Current in vitro methods in nanoparticle risk assessment: limitations and challenges. Eur J Pharm Biopharm. 2009;72(2):370–7.

105. Zhang H, et al. Differential expression of syndecan-1 mediates cationic nanoparticle toxicity in undifferentiated versus differentiated normal human bronchial epithelial cells. ACS Nano. 2011;5(4): 2756–69.

106. Li R, et al. Surface interactions with compartmentalized cellular phosphates explain rare earth oxide nanoparticle hazard and provide opportunities for safer design. ACS Nano. 2014;8:1771–83.

Nanomedicine: Size-Related Drug Delivery Applications, Including Periodontics and Endodontics

5

Xu Wen Ng, Raghavendra C. Mundargi, and Subbu S. Venkatraman

Abstract

In this chapter, we discuss polymer- and liposome-based nanocarriers used in the delivery of bioactive molecules, from drugs to proteins. The focus is on the enhancements in efficacy of bioactive molecules when nanotechnology is used for delivering them. The perspective centres around commercial and clinical successes and a rationalization of these successes. Microparticulate systems are also discussed in relation to their nano-counterparts, and the advantages of nano size are emphasized in relevant applications. In general, the main application of nanocarriers is in cancer therapy; however, with the ability to programme sustained release of bioactive molecules from certain types of nanoparticles, other applications in ocular, cardiovascular and periodontic/endodontic therapy may be possible.

Abbreviations

CMC	Critical micellar concentration
DNA	Deoxyribonucleic acid
DOX	Doxorubicin
DPPC	Dipalmitoylphosphatidylcholine
DSPE	Distearoyl-phosphatidyl-ethanolamine
DXY	Doxycycline
EPM	Extracellular polymeric matrix
EPR	Enhanced Permeation and Retention
FDA	Food and Drug Administration
GI	Gingival index
HLE	*Harungana madagascariensis Lam. Ex Poir.*
IOP	Intraocular pressure
IV	Intravenous
MIC	Minocycline
NC	NanoCarriers
PCL	Poly (caprolactone)
PD	Probing depth
PDT	Photodynamic therapy
PEG	Poly(ethylene glycol)
PI	Phosphatidylinositol
PLA	(Poly L-lactide)
PLGA	(Poly (D,L-lactic acid and Glycolic acid copolymer)
PS	Photosensitizers
PVA	(Poly vinyl Alcohol)

X.W. Ng, PhD • R.C. Mundargi, MSc, PhD
S.S. Venkatraman, PhD (✉)
School of Materials Science & Engineering,
Nanyang Technological University,
N4.1 Nanyang Dr, Singapore 63978, Singapore
e-mail: ASSUBU@NTU.EDU.SG

A. Kishen (ed.), *Nanotechnology in Endodontics: Current and Potential Clinical Applications*,
DOI 10.1007/978-3-319-13575-5_5, © Springer International Publishing Switzerland 2015

PVP	Poly(vinyl pyrrolidone)
RES	Reticulo-endothelial system
RPE	Retinal pigmented epithelium
S. oralis	*Streptococcus oralis*
S. sanguis	*Streptococuss sanguis*
SA	Steraylamine
SESD	Spontaneous emulsification solvent diffusion
TCL	Tetracycline
TCS	Triclosan
TEM	Transmission electron microscopy
T_g	Glass transition temperatures
T_m	Melting points
TPP	Tripolyphosphate
TSA	Tissue-specific antigen
ULV	Uni lamellar vesicle

5.1 Introduction

In this review, we will focus on the status of research and development of polymer- and liposome-based nanoparticles as carriers for therapeutic drug delivery and more specifically for targeted drug delivery. For the purposes of this review, a nanoparticle is defined as having dimensions below 1 µm; consequently, we will not discuss particles much larger than a micron. In addition, we will not discuss the use of ceramic or metallic particles, which have their own niche in this field. The review thus covers polymer-based nanoparticles including self-assembling systems such as micelles and liposomes.

Liposomal carriers dominate the field of approved products in nanomedicine. Other particles in the nano range have been approved, but these are nanocrystalline drug molecules [1] or drug molecules conjugated to proteins [2], which are mostly aimed at better solubilization than the control of drug release. Several liposomal carriers have in fact been approved [3], although it is not clear that all these are carriers in the nanometer range. A survey of the literature from 2002 onwards shows that approximately 2,600 articles feature nanocarriers, with 1,060 falling under drug delivery systems. Similarly, a total of 4,382 patents have been filed for nanoparticulate drug delivery since 1987. So this is an extremely fertile field of research and will continue to be fertile for the foreseeable future.

Translation of the research into commercial products appears to be slow but steady. As of 2006, the number of approved nanotherapeutic products stood at 23 in total [4] and has increased to about 33 in 2013 [5]. The same review also categorizes the nanotherapeutic products by disease: cancer therapeutics dominated the field of approved products, with zero products approved in either ocular or dental fields.

Microparticulate systems have been around longer than nanopartiulate systems, with the first product approved in 1989: the biodegradable Lupron-Depot® formulation of PLGA (oly(lactic-co-glycolic acid) microspheres containing leupro-lide acetate for prostate cancer therapy. These are 'solid' microparticles, in which the drug is dispersed in a matrix of a biodegradable polymer: the drug releases through a diffusional mechanism. In fact, such formulations are now commonplace, with the main advantage being the ability to carry reasonable amounts of drug and being easily formulated into injectable (i.e. of acceptable viscosity) formulations.

In general, nanoparticles are touted to have advantages over microparticles. The most important advantage appears to be that of extravasation and of cellular penetration. There are two approaches to producing nanoparticles: commu-nition of larger particles by milling or other grinding techniques and by the conventional 'synthetic' processes of emulsification/dialysis/lyophilization. Generally, processes such as spray-drying cannot produce particles in the nano range. For liposomal particles, ultrafiltration is frequently employed to weed out larger particles.

This review will focus on nanoparticulate systems in various stages of development: approved, in clinicals, in pre-clinicals and in an early research phase.

5.2 Nanoparticulates: Structure, Preparation and Characterization

The different particle types are usually made by different processes, and these processes have been reviewed [6]. We summarize briefly those processes here.

5.2.1 Polymeric Nanoparticles

Milling

Milling (which is basically a grinding process) in conjunction with sieving/filtration has been used to prepare polymeric nanoparticles. The starting point may be spray-dried particles produced in the micron range provided the drugs are stable to spray-drying; otherwise, emulsion or precipitation processes may be employed to generate micron-sized particles incorporating drugs. Milling generally leads to non-spherical particles, but this is acceptable from a drug delivery standpoint.

Emulsion Methods

This is by far the preferred approach to obtain drug-incorporated polymeric nanocarriers. When the drug is hydrophobic, it is co-dissolved with polymers in a common solvent and then emulsified into water, usually containing surfactants. Evaporation of solvent followed by lyophilization is typically used to obtain drug-containing polymeric particles. Size ranges tend to be in the micron range, so further size reduction is obtained via milling or filtration methods.

To incorporate a hydrophilic drug, it is dissolved in water first, followed by emulsification into an organic polymer solution. This mixture is then emulsified into water containing surfactant to produce a water/oil/water (w/o/w) emulsion. Solvent drying followed by lyophilization leads again to micron-sized particles in general.

Size control may be further exerted by homogenization [7] and by surfactants [8]. Speeds of approximately 10,000 rpm were required to bring sizes of PLGA (poly (D,L-lactic acid and glycolic acid copolymer) particles down to below 1,000 nm, while PVA (polyvinyl alcohol) concentrations of greater than 5 % did not further decrease PLA (poly L-lactic acid) particle sizes below 120–150 nm.

Variations on the emulsion method have been proposed. For example, Quintanar-Guerrero et al. [8] have prepared both nanospheres and nanocapsules by the use of an oil in the organic phase. In this scheme, to prepare nanocapsules, a polymer and oil (miglyol) are first dissolved in ethyl acetate (EtAc)/water and then emulsified into a PVA solution in water/EtAc mixture. When water is added to this emulsion, EtAc diffuses out into the water, forming nano-sized structures, which are then solidified by solvent removal followed by ultracentrifugation. Nanospheres of about 400 nm can be made by eliminating miglyol from the mixture, while nanocapsules of sizes 180–300 nm are made by addition of varying amounts of miglyol. In these cases, the 'skins' of the capsules is made of the polymer. This method enables greater loading of hydrophobic drugs in the core of the nanocapsule.

5.2.2 Polymeric Nanomicelles

Polymeric nanomicelles are produced using a different approach altogether. The starting point is an amphoteric molecule (typically, PLA-poly(ethylene glycol) (PEG) copolymer if biodegradability is desired), which is dissolved in water at increasing concentrations until its critical micellar concentration (CMC) is reached. Once the CMC is known, the polymer and drug are dissolved first in an organic solvent (if the drug is hydrophobic), then the mixture is dialyzed against water. The concentration of the polymer in the final amount of water is adjusted to be well above the CMC. Following dialysis, the 'particles' may be isolated by centrifugation and/or freeze-drying. Reconstitution in water or saline should regenerate the micellar structure as long as the micelles are stable in an ionic medium. Usually, the CMC is increased by the presence of ionic species.

Hydrophilic drugs cannot be easily incorporated into simple spherical micellar structures. For this, we need to use another type of self-assembling system, the liposomes. Refer to Fig. 5.1 for schematics of liposomes, micelles, dendrimers and solid nanoparticles.

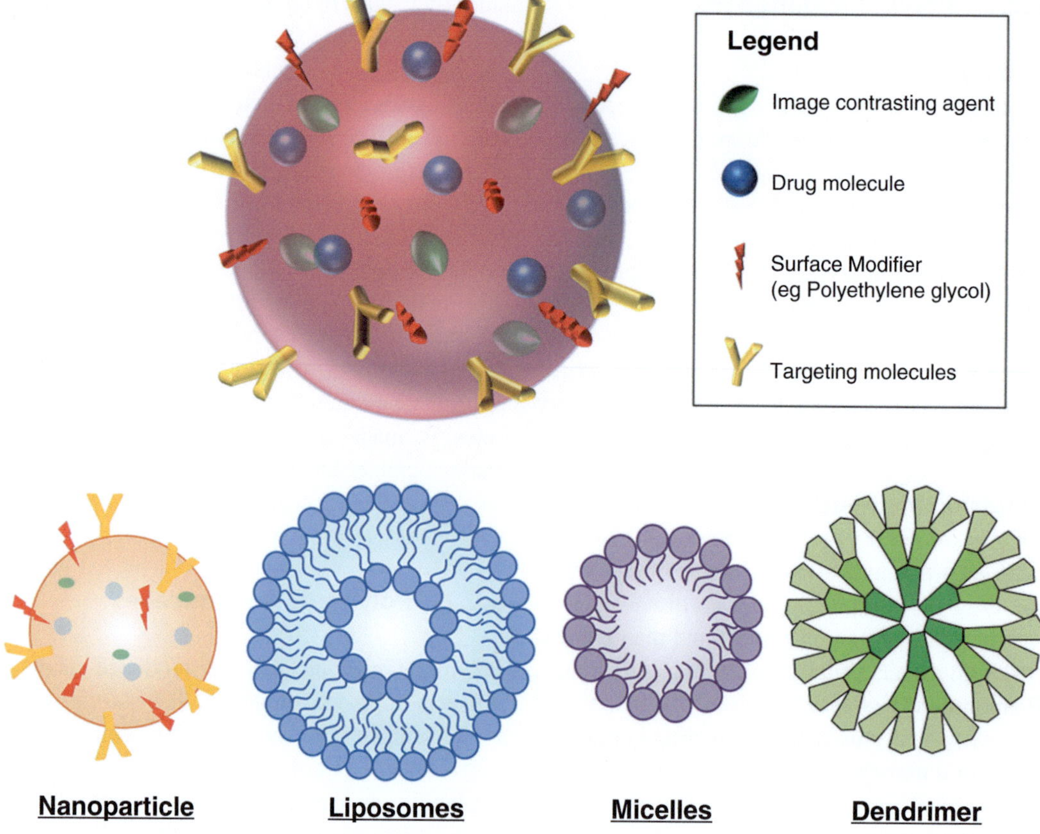

Fig. 5.1 Pictorial descriptions of different types of polymeric nanoparticles and their respective functionalities. Surface modifiers, targeting molecules (ligands) and incorporated agents are shown

5.2.3 Liposomes

Nano-sized liposomes can be prepared using the rehydration technique, followed by extrusion. Generally, known amounts of lipids, with and without cholesterol, are dissolved in a low-boiling organic solvent such as ethanol at 60 °C. Using a rotary evaporator, the organic solvent is evaporated off to form a thin film. This is then hydrated with an aqueous solution containing the drug or protein. Rehydration leads usually to the formation of multilamellar vesicles (MLV), which have sizes in the range of 0.5–5 μm. These are downsized to about 70–500 nm (unilamellar vesicle, or ULV) by extrusion through membrane filters at high pressures [9] to produce liposomes with incorporated drugs. In the above example, the drug is incorporated by the 'passive' loading method. For more hydrophilic drugs, an 'active'

loading method is required to incorporate therapeutically meaningful quantities. For passive loading, a lipophilic drug is dissolved in the organic phase, while a hydrophilic drug is dissolved in the aqueous phase. An example of an active method is the use of a pH gradient [10] or an electrochemical gradient [11, 12] for different drug types.

5.2.4 Dendrimers

Dendrimers are essentially hyperbranched polymer structures that can potentially 'encapsulate' drugs or proteins. The control of release is exerted primarily through diffusion and in some cases by degradation. These hyperbranched structures (polymers) are prepared by a very specific reaction sequence, usually starting with an amine-terminated molecule.

Such a molecule is reacted with an acrylate ester and then subsequently with ethylene diamine to yield a 'full-generation' dendrimer [13]. Repetition of the above reactions yields a highly branched structure with internal 'cavities' that may hold metal atoms or other guest molecules by virtue of the presence of amine groups. Molecules may be conjugated to the 'interior' groups as well as the 'surface' groups, and these molecules may be a drug, peptide, antibody or PEG. Conjugation generally opens up possibilities for selective targeting of tissue but seldom for sustained release applications.

5.3 Clinical Applications

5.3.1 Cancer Chemotherapy

We now review the applications of the above classes of nanoparticulate carriers in different therapies. By far, the greatest attention has been paid to targeting tumour tissue as nanoparticles have the ability to traverse easily out of blood vessels into tumour tissue in comparison to their microparticulate cousins. In addition, surface modification of nanoparticles for evasion of the reticuloendothelial system (RES) as well as for penetration of selected tissue is no less feasible for nanoparticles. So in what follows, we discuss the relative successes of the different particle types in targeting tumour tissue.

Active and Passive Targeting

Two concepts have been used for targeting cancer tissue. Passive targeting involves injectable drug carriers that have been surface modified (Fig. 5.1) to evade the RES such that their blood lifetime is relatively long. Long-lived particles have a much greater chance of reaching the blood vessels surrounding solid tumours, and then extravasate by virtue of their size. Once in tumour tissue, the relative lack of lymphatic drainage allows for slow release of the payload into the surrounding tissue. The whole effect goes by the name of enhanced permeability and retention (EPR).

Active targeting relies on conjugating a targeting ligand (usually an antibody) to the surface of the particle such that the ligand targets cancerous tissue only. Most of the work to date has focussed on folate and transferrin receptor targeting; these two receptors are over-expressed in cancerous cells. Others have tried to exploit the presence of a tissue-specific antigen (TSA) whose antibody may be used for the targeting. In this approach, the nanosize is not as critical, although it still helps in facile extravasation.

Liposomes
Approved Products

Liposomal delivery systems have been by far the most successful of the nanoparticulate carriers. No fewer than four pharmaceutical products that use liposomes as a carrier have been approved: of these, Doxil® was approved in 1995 for the treatment of ovarian and other cancers. Other approved liposomal products include DaunoXomeTM, LipoDox and Myocet, while another liposomal formulation (Synergene Therapeutics) for activating the p53 (tumour suppressor) gene has successfully completed Phase 1 trials. In the discussion below, we will focus on the development of Doxil and comment on the shortcomings and advantages of liposomal nanocarrier systems.

Doxil

The story of successful development of the very first nanoliposmal carrier is both long and intriguing [14]. The story includes many discoveries wrapped into one product: for example, the idea of 'pegylation' of the carrier to increase blood circulation lifetimes and the concept of 'active drug loading' into the core of the liposome were both crucial to successful development. The important features of this so-called stealth liposome are depicted in Fig. 5.2. The lipid molecule, which is distearoyl-phosphatidyl-ethanolamine (DSPE), is conjugated at its amine end to a 1,900-MW mPEG molecule [15]. Doxorubicin (Dox) is partially crystallized in the aqueous core (Fig. 5.2), and the control of its release through the membrane is accomplished mostly by the crystallization and somewhat by incorporation of cholesterol in the membrane: the cholesterol tends to make the membrane more rigid and thus slow down diffusion.

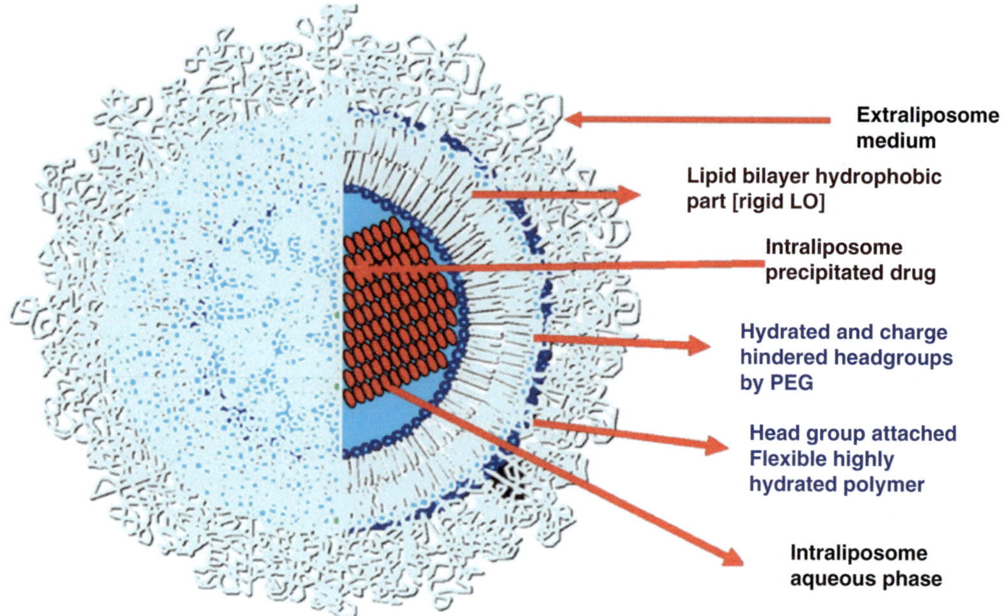

Fig. 5.2 Pictorial description of "stealth" liposome: Doxil (Reprinted from Barenholz [14]. With permission from Elsevier). *LO* 'liquid ordered' and *PEG* 'polyethylene glycol' represents different domains of Doxil

Most of the drug (~12 % by weight of lipid) is claimed to be inside the aqueous core of the liposomes. Such high loadings are generally only possible with what are called active loading methods. This uses a salt such as ammonium sulfate inside the core of the liposome to bind to the drug molecule as drug sulphate and precipitate which then drives the influx of drug to the liposome core [12].

The liposomal drug thus formulated works on the principle of 'passive targeting'. Upon intravenous (IV) administration, the carrier (with incorporated drug, Dox) circulates for about 30–45 h without being phagocytosed. During this period, very little of the Dox is released into the bloodstream: hence, controlling (minimizing) the release for at least 45 h is crucial to the success of this mode of drug administration. The long blood circulation time of the pegylated carrier is partly responsible for accumulation at the tumour site: statistically, the chance of particles accumulating near the tumour tissue is high because of the number of blood vessels feeding it. The size of the liposomal particles (80–100 nm) [16] is the second reason for accumulation into tumour tissue by virtue of 'extravasation' or leakage through the highly permeable capillaries surrounding tumour tissue. It is now generally accepted that extravasation is possible only for particles less than 600 nm in size. Clinically, this is highly significant because accumulation in liver and spleen is considerably reduced for the stealth particles, and this translates directly to tumour size reduction and its maintenance over 100 days compared to free drug injections and thus to a substantial improvement in the survival rate of mice with colon carcinoma.

Doxil® was a pioneer in the field of passive targeting. Its success prompted several imitations, each of which touted advantages over Doxil®. One example is Myocet®, which claimed an even higher loading of Dox (~25 %) in particles of about 190 nm. Interestingly, this product consists of non-pegylated egg PC/cholesterol lipids and appears to minimize one side effect of Doxil®, which is the so-called hand-foot syndrome [17], that is exacerbated with long-circulating carriers. Myocet® is approved in Europe for treating metastatic breast cancer (co-administration with Herceptin) and is being considered for approval in the USA. Unlike Doxil®, which is stored as a suspension, Myocet® is a lyophilized powder that is reconstituted prior to IV administration.

Liposomal Systems in Pre-clinical and Clinical Phases

Other cancer targets (besides simply killing tumour cells) have been addressed. One such is activation of the tumour suppressor gene p53, especially in lung cancer. Early attempts to do so were reported by researchers at the M. D. Anderson Cancer Center in Houston, with successful suppression of primary and metastatic lung tumour growth [18] in animal studies. Co-authors included employees of Introgen, which subsequently seemed to have abandoned liposomal delivery in favour of adenoviral carriers for delivering the gene [19]. Since then, however, Introgen appears to have folded due to lack of funding.

As mentioned above, another company that has reported clinical trials with a liposomal vector is SynerGene Therapeutics based in Washington D.C., USA. From various company reports, this appears to be a liposomal vehicle with targeting ligands, presumably a folate receptor–targeting ligand. The payload is likely to be a p53 gene also. A patent [20] describes an 'immunoliposome' which is based on the liposome component DOPE (dioleyl phosphatidyl ethanolamine) incorporating a transferrin receptor–targeting ligand. The immunoliposome is complexed with p53 wild-type genes and targets the transferrin receptor, which is over-expressed in many tumours.

Polymeric Nanomicelles

Liposomes are relatively narrow distribution in their sizes following extrusion. The sizes are generally not very stable against both aggregation and fusion or disassembly over time; these are self-assembled structures that are in a state of thermodynamic equilibrium at the temperature of interest and with the requisite concentration of lipids. The self-assembling nature confers shape, stability and uniformity, however, unlike solid nanoparticles. Similarly, spherical polymeric nanomicelles in fact can spontaneously form under the appropriate conditions in aqueous or non-aqueous media depending on the molecular structure and extent of amphiphilic character.

In terms of papers and patents, nanomicellar drug delivery has been a fertile field. For polymeric molecules to self-assemble to spherical micelles or lamellar structures, block co-polymers are ideal, with one block being hydrophilic and the other hydrophobic. Triblocks, diblocks and even four-armed blocks have been used [21, 22]. Relative to liposomes, success in the clinic has been harder to come by, but there are some notable exceptions, as we will see in the next section.

The key to the success of a polymeric nanomicelle carrier is CMC [23]. Although micelles are prepared at concentrations that ensure they are the predominant species, upon injection (IV) into the body, there is immediate and substantial dilution, which may lead to disassembly of the micelles. Hence, an effective micellar carrier has a very low CMC to ensure that the self-assembled structure is retained following administration. In general, the CMC is influenced by these factors:

1. The hydrophobic–hydrophilic 'mismatch': the greater the difference between the two groups forming the micellar molecule, the lower the CMC
2. The lengths of the two segments: in general, the longer these segments, the lower the CMC

Usually above the CMC, spherical micelles are generated, although other structures are possible and have been observed. For drug delivery, spherical carriers that maintain their shape over time of storage are preferred for stability of release of drugs. Drug partitioning is also enhanced when the drug hydrophobicity is matched with that of the hydrophobic segment. In general, the lower the CMC, the lower the amount of free molecules in the 'suspension', as illustrated in Fig. 5.3.

Clinical Candidates

The use of nanomicelles for passive targeting to certain tumours has been spearheaded by Japanese researchers at the National Cancer Center /Research Institute East in Chiba, Japan,

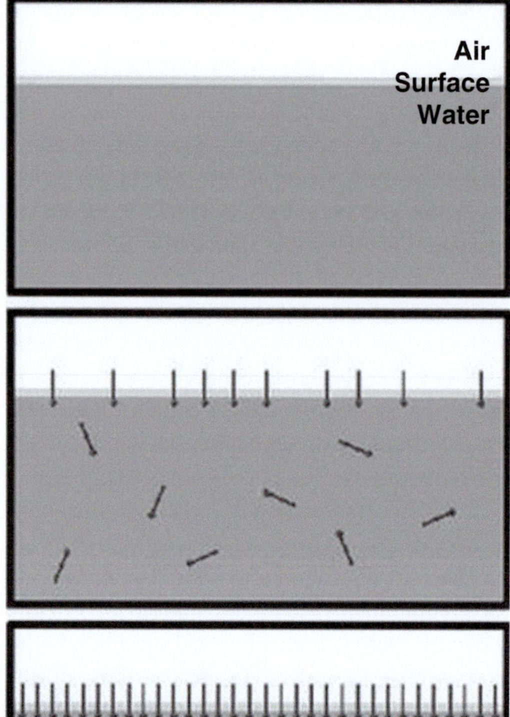

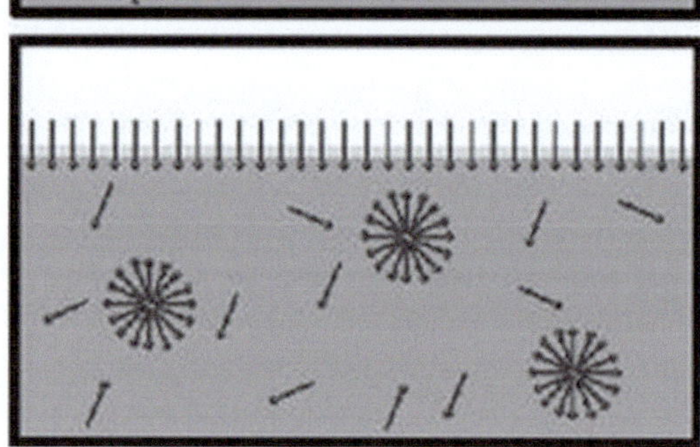

Fig. 5.3 Pictorial depiction of the formation of micelles by addition of amphiphilic molecules to water (Reprinted from Critical micelle concentration [23].with permission from Creative Commons Attribution)

in conjunction with the University of Tokyo. The approach is to encapsulate the generally hydrophobic drug candidates in the micellar core and manipulate release predominantly through degradation of the hydrophilic segment. Equilibration of drugs between the core and the hydrophilic segment is discouraged via low partitioning.

Years of exploratory work have led to the development of three promising nanomicellar formulations, paclitaxel (NK105), doxorobucin (NK911) and cisplatin (NC-6004). The statuses of these products are briefly covered below.

NK105

The NK in the name of this experimental product is due to 'Nippon Koyaku", the Japanese company that is developing and financing the clinical trials. The polymer is a diblock copolymer of PEG and modified polyaspartate [24] where half of the aspartate groups are converted to the more hydrophobic derivative, 4-phenyl 1-butanolate. As explained above, this derivatization enhances the hydrophobicity of the segment and leads to a lower CMC. The overall MW of the polymer used was 20,000, of which the PEG block was 12,000 and the aspartate block was 8,000. Approximately 20 % of the polymer weight can be loaded with paclitaxel held in the micellar core by hydrophobic self-association. The micelles obtained in water or aqueous media were lyophilized to obtain particles which upon reconstitution yielded micelles of average diameter about 85 nm, with size ranging from 20 to 430 nm.

Paclitaxel is a potent anti-cancer drug that suffers from poor bioavailability due to its low solubility. In addition, systemic injection with a solubilizer such as Cremaphore EL induces hypersensitivity reactions. Moreover, peripheral neuropathic reactions as well as neutropenia have been reported with repeated use of paclitaxel [25]. For all these reasons, a better delivery system that can at least partially target tumours is highly desirable for paclitaxel. Following promising animal data, a Phase 1 trial was conducted in 2007 on 19 patients [26] who had solid tumours refractory to conventional chemotherapy. NK105 showed slow clearance, with a half-life of about 5–6 h (compared to about 1 h for injected paclitaxel), thus enabling passive targeting to tumours. Neuropathy, which is a common side effect of paclitaxel, was not observed. Hypersensitivity was also negligible, even without co-administration of steroids. Since then, a Phase II (efficacy) study has been completed, and a Phase III study in breast cancer patients [27] is under way. While the timeline has been a long one, it appears that NK105 is likely to be approved for use in certain cancers.

NC-6004

Based on early work at the University of Tokyo [28], the company NanoCarriers (NC) has developed a micellar delivery system for another cancer drug, Cisplatin. Cisplatin (a Pt (II) complex with *cis*-dichlorodiamine) is an anti-cancer drug that suffers from various limitations, including nephrotoxicity and neuropathy, with repeated use [29]. The micellar system using PEG and poly(glutamate) has been developed to incorporate cisplatin via complexation of the glutamate unit to Pt(II) [30, 31]. This is a novel way to improve cisplatin loading and control its release. In fact, the *in vitro* release of cisplatin into saline is triggered by chloride substitution for glutamate in the complex and shows no burst but sustained release over 150 h as shown in Fig. 5.4. The size of the particles remains stable, at about 30 nm over 72 h, and presumably decreases beyond that due to micellar dissociation into unimers.

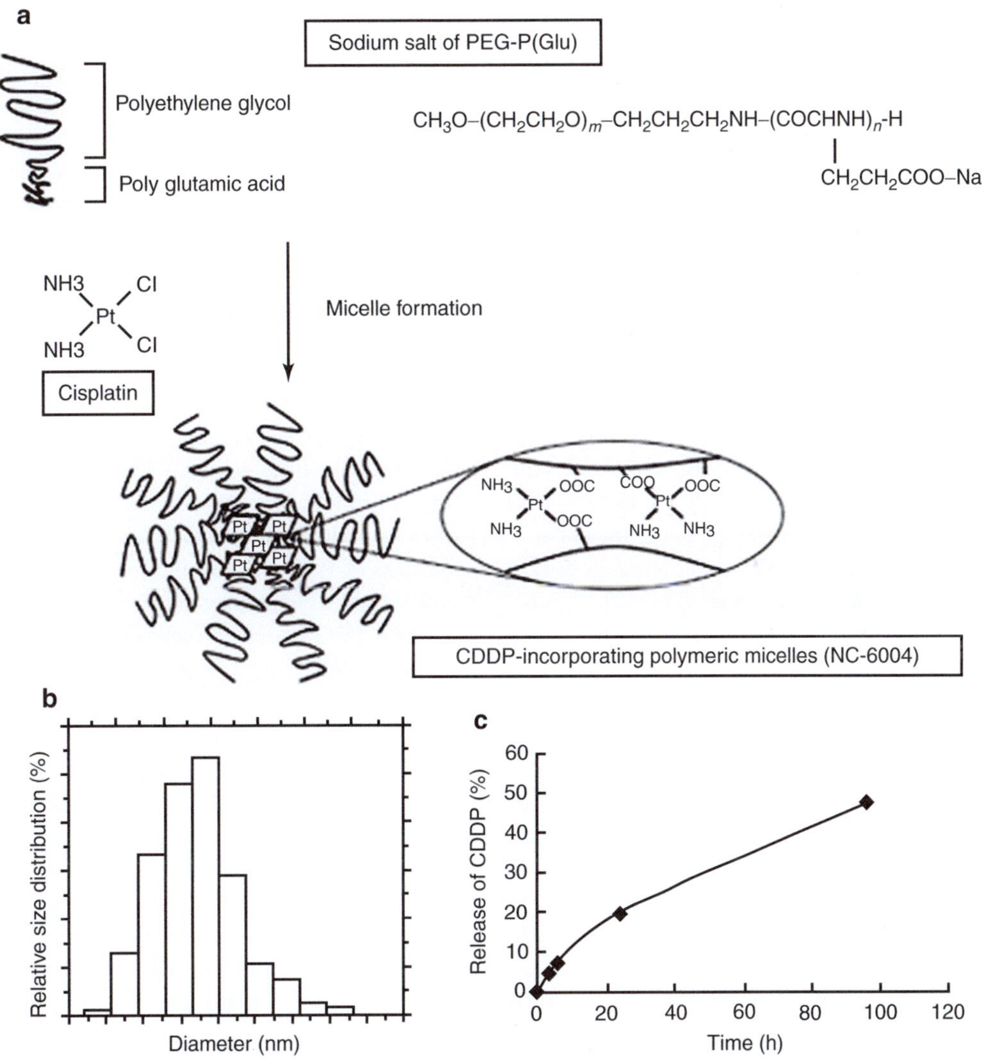

Fig. 5.4 Schematic of cisplatin structure and its incorporation into a micelle via complexation with a polyamino acid (polyglutamate). The hydrophilic polymer, PEG, also acts as anti-phagocytic layer, similar to pegylated liposomes (Reprinted from Uchino et al. [93]. With permission from Nature Publishing Group). (**a**) Chemical structures of CDDP and PEG–P(Glu) block copolymers, and the micellar structures of CDDP-incorporating polymeric micelles (NC-6004). (**b**) The particle size distribution of NC-6004 measured by the dynamic light-scattering method. The mean particle size of NC-6004 was approximately 30 nm. (**c**) Release of CDDP from NC-6004 in saline at 37°C.

Clearance and tissue distribution are superior for the micellar formulation of cisplatin compared to free cisplatin. In the mouse study [30], tumour accumulation over 24 h is roughly 5–10 fold higher than free cisplatin depending on tumour location. Tumour inhibition was demonstrably superior with the micellar formulation for mice bearing colon adenocarcinoma. As of now, this compound is in Phase III trials in Asia and in Phase II in the USA. It is expected to win approval in a couple of years.

NK911

A micellar formulation for Dox has been developed as well, and it is interesting to compare micellar Dox against liposomal Dox. The micellar system for Dox is similar to that for paclitaxel (NK105), utilizing the same PEG–aspartate copolymer (PEG MW=5,000; polyaspartate has 30 repeats) [32]. The Dox in this case is conjugated to the aspartate moiety for greater loading efficiency. The conjugated drug confers sufficient hydrophobicity to enable formation of stable micelles. This enhanced hydrophobicity allows more packing of 'incorporated' Dox (as distinct from 'conjugated' Dox). The conjugated Dox is not active and is not released from the micelle, while the incorporated Dox is released slowly over 24 h. The mean diameter of the micellar particles is 42 nm after freeze-drying of the micellar solution.

An interesting comparison with Doxil (the liposomal Dox) shows the micellar Dox to be cleared 400 times faster than Doxil [33]. However, it is not clear that this lower blood life-time necessarily translates into lowered accumulation in tumour tissue as the volume of distribution for Doxil appears higher than that for micellar Dox. Another study [34] showed that *in vitro*, the Doxil formulation was more stable to leakage of drugs than the NK911, which also points to potentially greater side effects caused by free Dox in systemic administration. It is likely that for these reasons, development of NK911 has been discontinued.

Other Nanomicelle Systems and Studies

There is an excellent review of the large amount of research in the field of fully degradable micellar nanocarriers [35]. The major polymeric systems studied appear to be block copolymers of hydrolyzable polyesters with PEG. The hydrolyzable polyesters include poly (caprolactone) (PCL) [36, 37], PLA [21, 22] and PLGA [38]. In addition, 'stealth' entities other than PEG have been studied as micelles, including Poly(vinyl pyrrolidone) or PVP and PVA. However, no other entity has been shown to work as well as PEG.

Previous studies have reported the optimum PEG length and configuration needed for stealthiness [39]. In our study, we synthesized triblock PLA-PEG-PLA copolymers (ABA type) as well as PEG-PLA-PEG copolymers (BAB type) with varying lengths of PEG and PLA and measured uptake by blood cells *in vitro*. We also compared the uptake behaviour of triblocks with diblocks of PEG-PLA. The lower the uptake, the greater is the 'stealthiness' or longer the blood lifetime. Confocal microscopy shows that as the size of the nanomicelle increases, the phagocytic uptake increases compared to the smaller particle. In general, uptake was greater for ABA triblocks compared to diblocks and BAB triblocks of similar PEG and PLA lengths; in addition, it was found that the PEG segment length had to be above 5,000 MW in order for any effect to be seen for uptake reduction by blood cells. Surface concentration of PEG was not the only determinant factor for reducing uptake; the surface conformation of the PEG was important as well. More free PEG molecules (i.e. with conformational freedom) exerted a bigger effect compared to more constrained molecules.

Loading of hydrophobic drugs into the micelle is generally easier than the loading of hydrophilic drugs. As noted above, paclitaxel has been loaded to about 20 % in a typical micelle. However, control of release in the clinically tested formulations is still inadequate. In our own work, we find that this control of release can be improved substantially [21, 22] by the use of the relatively hydrophobic polyesters, which have the added advantage of full biodegradability into harmless products.

Loading of more hydrophilic drugs and control of its release requires special techniques, including drug conjugation. Fortunately, most of the anti-cancer drugs tend to be fairly hydrophobic.

This is in fact the main reason that these aqueous micelles have been predominantly studied for anti-cancer therapy. Loading of antibiotics, for example, will not be high, and loaded drugs may release quicker than other drug types. One example of hydrophilic drug loading is in the example shown above for cisplatin using poly glutamate as a conjugating agent (NC6004). Cisplatin is moderately hydrophilic, and the loading can be increased to 35–40 % by conjugation. Although other approaches have been proposed including the so-called core surface-cross-linked micelles [40], it is not clear that the increased loading is without an accompanying increase in release rate.

Solid Nanoparticles

These are nanoparticles that are composed mainly of polymers that are not fluid at 37 °C but have glass transition temperatures (T_g) or melting points (T_m) above 37 °C. Of the polymers evaluated to date, PLGA is the most versatile in that the properties can be fine tuned by varying the ratio of lactide to glycolide in the copolymer: for example, the T_g may be varied between 45 and 70 °C in changing the ratio from 50 % of L-LACTIDE to 85 % of L-lactide. The copolymers are fully biodegradable, and the degradation products are relatively harmless. The overall degradation rate may also be manipulated by changing the lactide-to-glycolide ratio. In this section, we will focus on PLGA nanoparticles as applied to cancer therapy.

Solid nanoparticles are defined as matrix formulations whereby the drug is either dissolved or dispersed in a polymer that is processed into a nanoparticulate form (other than via self-assembly), whereas nanocapsules are core–shell particles made with two different types of polymers or with lipids as cores and polymeric shells. Of the two, it appears that nanocapsules are difficult to fabricate and most core–shell particles tend to have micron dimensions. Thus, most of the discussion below relates to matrix-type nanoparticles. The main advantages of using these particles in preference to micelles and liposomes are simplicity of manufacture, greater stability and better control over release of drugs. Disadvantages are relative difficulty in attaching PEG or ligands to the particle. As mentioned above, matrix microparticulate drug delivery systems have been around for a long time, but nanoparticulate ones are rare. As far as we are aware, none of the Food and Drug Administration (FDA)–approved nanosystems is based on solid nanoparticles.

PLGA nanoparticles can be made using a variety of ways, from emulsion techniques to spray-drying. The emulsion technique is preferred with the use of ultrasound for size reduction. In a variation of the emulsion technique called the spontaneous emulsification solvent diffusion (SESD) method, a mixed solvent system is used for the polymer solution, which is then emulsified into an aqueous medium containing a surfactant. To minimize agglomeration of the particles, a low-MW PVA surfactant is used, and two completely water-miscible solvents can be used for the polymer phase [41]. This technique produces non-aggregating particles in the size range 100–200 nm that remains stable even after freeze-drying.

Such particles can incorporate a wide range of drug molecules, from low-MW hydrophobic molecules [42] to proteins [43] and even plasmid deoxyribonucleic acid (DNA) [44]. As mentioned earlier, such nanoparticles may again be surface modified: PLGA has also been coated with PEG and PEG-like molecules including the surfactant molecule PEG-PLA-PEG [45]. Grafting of PEG molecules to surfaces of already formed nanoparticles is seldom reported, most of the 'chemically bound' PEG nanoparticles relying instead on copolymerization with PLA [46] or PLGA [47] or even polycyanoacrylate [48].

Passive Targeting

In spite of several years of active research following the approval of the first PLGA microsphere formulation (Zoladex) in 1989, nanoparticulate PLGA, decorated with surface-bound PEG, has not been shown to be superior to either liposomes or micelles for passive targeting to tumours. The reasons for this are twofold: (a) the PLGA nanoparticles tend to aggregate over time, and lyophilization and reconstitution does not always produce particles of the same size range, and (b) although drug loading is increased with these solid NPs, release is retarded due to diffusional barriers. Too much retardation is not useful in passive targeting as slow release at the tumour leads to sub-cytotoxic levels of drugs.

Active Targeting

Given the lack of advantage in passive targeting, most of the PLGA nanoparticle work has concentrated on the so-called active targeting approach. In this approach, a ligand is attached to the particle that binds to receptors or molecules that are unique to cancerous cells while carrying their cargo of cytotoxic drugs. The most studied ligand is the folate ligand that binds to folate receptors that are usually over-expressed in cancerous cells, particularly in epithelial malignancies such as breast cancer and endometrial cancer. The concept is to attach the folate ligand covalently to the PEG molecule usually with a spacer such as PEG. An example of this approach is found in Liang et al. [49]. In this example, a new terpolymer had to be synthesized which contains blocks of PLGA, PEG and folate. Incomplete conjugation of folate at one end leads to some level of 'stealthiness' (Fig. 5.5) due to the 'free' PEGs on the surface.

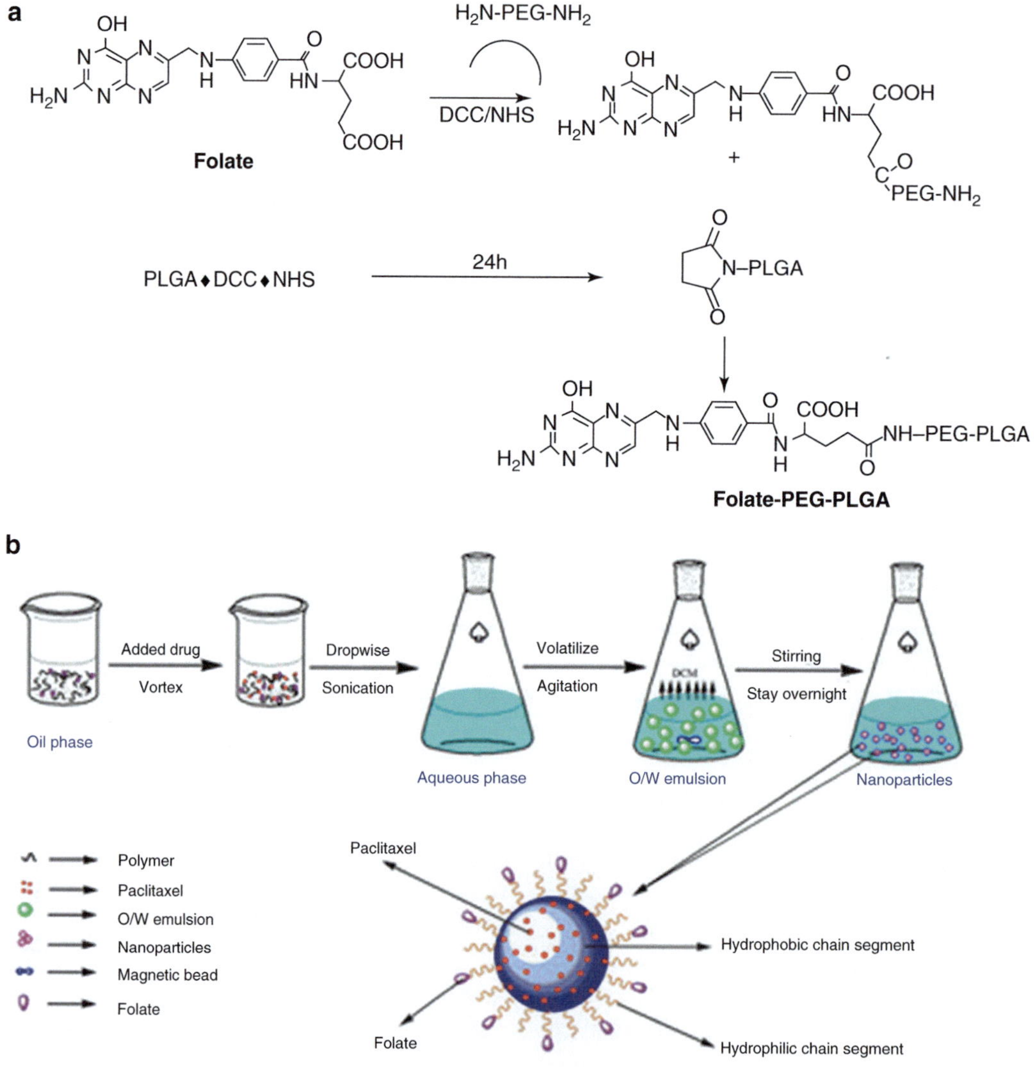

Fig. 5.5 Schematic of synthesis of folate-decorated PLGA-PEG nanoparticles (Reprinted from Liang et al. [49]. With permission from Elsevier). (**a**) Structure of FOL–PEG–PLGA. (**b**) Schematic representing formulation of FOL–PEG–PLGA nanoparticles and the process for drug loading

The nanoparticles generated by volatilization of the organic solvent from the emulsion were about 220 nm in average diameter and a surface charge of −8 mV with the folate added. Using *in vitro* studies, it was found qualitatively that the folate-conjugated particles were internalized to much greater extent by human carcinoma cell lines when compared to PLGA-PEG particles. Incorporating paclitaxel into these particles and then injecting (via the tail vein) into mice bearing endometrial cancer cells showed greater tumour volume reduction for the folate-conjugated nanoparticles. Many more such examples can be found in the literature. In spite of encouraging animal data, none of these ligand-bearing nanoparticles have been tested in humans. It does appear that 'stealthiness' and targeting are mutually incompatible features of a nanoparticle. If sufficient stealthiness is achieved by PEGylation, generally the targeting capability is compromised, and vice versa. The best solution therefore appears to be local administration of the NPs rather than intravenous administration.

Comparing Nanoliposomes, Nanomicelles and Solid Nanoparticles for Drug Delivery

Stability Issues

In general, nanoliposomal carriers appear to be more stable than nanomicelles, particularly to drug leakage. For cancer applications, it is critical that substantial leakage of drugs must not occur during the period of carrier circulation in blood (i.e. within the half-life in blood). Doxil, for example, sustains the release of drugs in ambient conditions to much beyond 45–50 h so that the drug remains encapsulated until its encounter with tumour cells.

One explanation for this greater stability to leakage is given by Lasic [50]. Lasic argues that liposomes do not form unilamellar or multilamellar structures spontaneously, unlike micelle formation from block copolymers. This means that the lamellar structures are not thermodynamically favoured. Yet, paradoxically, once they are formed (by an extrusion process, i.e. external energy input), these structures tend to be more stable due to what Lasic calls 'kinetic trapping'. This kinetic trapping is essentially caused by

hydration of the polar heads, and thus it results in lack of penetration of the lipid cores by water. This stability is crucial to clinical success especially if intravenous injection is used to administer the formulation.

As noted above, micelles are thermodynamically favoured structures for block copolymers in solvent (i.e. good solvent for one of the constituents). The stability of these micelles, however, is dependent on the actual magnitude of the CMC: the lower the CMC, the greater is the tendency for these micelles to form and, of course, greater the resistance to structure breakdown upon dilution. Achieving sufficiently low CMCs is a challenge. Since micellar formulations are naturally diluted upon injection into the bloodstream, it is speculated that they tend to become monomeric entities upon injection; this leads to immediate drug release. Another complication is that added salts generally increase CMCs and hence break down existing micelle structures. Liposomes are not susceptible to this as the kinetic entrapment ensures that dilution has very little effect on the liposome. In other words, water penetration into the core of a liposome is not enhanced by adding more water, *hence the apparent stability of liposomes following injection.* Incorporation of cholesterol in liposomal formulations is also believed to enhance this kinetic stability.

Solid nanoparticles do not have leakage issues when stored at ambient temperature. However, nanoparticles aggregation over time is a significant issue. Additionally, heterogeneity of shape (and size) appears to be more of a problem with nanoparticles, with consequent variability in drug release rates. These have precluded their widespread use in cancer therapy to date.

Blood Lifetimes

In general, whichever entity has greater stability in blood will probably have enhanced blood lifetimes, greater tumour tissue accumulation (although this one can be influenced by other factors such as size and distribution in specific tissues), better control of drug release and enhanced bioactivity. Stability in this context also refers to stability against phagocytosis by elements of the RES. There is widespread agreement that the nano-sized particles have greater blood lifetimes than micron-sized ones; furthermore, that attachment of PEG molecules confers enhanced resistance to phagocytosis for these nanocarriers. There is experimental evidence that nanoliposomes exhibit longer blood lifetimes than nanomicelles of comparable size. Thus, it appears that the 'presentation' of the PEG corona to the phagocytic cells is not the same for liposomal PEG and micellar PEG. Exactly why this is so has not been clarified yet.

For solid nanoparticles, the PEG must be covalently attached to the matrix polymer via an extra synthesis step. In principle, this should work just as well for solid NPs as for liposomes and micelles. However, the heterogeneity in shape/size perhaps contributes to a more rapid clearance of PLGA particles as compared to liposomes. For non-cancer applications, however, blood lifetimes are not critical, but control of drug release is, as we will see in the examples that follow.

5.3.2 Ocular Applications

Sustained delivery of therapeutic agents is desirable for many ocular diseases, including glaucoma and macular degeneration, both chronic conditions without any known cures. However, the impermeability of the corneal epithelium and rapid clearance by tears preclude the use of topical eye drops for sustained delivery in spite of several years of research. Drugs applied via eye drops penetrate the barriers in the front of the eye at the best by about 5 % of dose; the dosing lasts for about 20–30 min on average. For the diseases to be discussed below, longer-duration drug action is highly desirable both from an efficacy as well as a patient compliance standpoint.

Glaucoma

Glaucoma is a chronic, progressive optic neuropathy that causes irreversible blindness. It is the major cause of irreversible blindness worldwide. The global burden of glaucoma is estimated to rise to affect 80 million worldwide by 2020 primarily due to an increasing ageing population in the world. Elevated intraocular pressure (IOP) is the only known effective modifiable risk factor. The use of daily eye drops containing hypotensive agents to lower the IOP remains the first-line treatment for glaucoma. Like any chronic therapy, ocular hypotensive agents require patient adherence. In addition, patients have trouble using eye drops and applying them correctly in the eye. A consequence to poor patient adherence is treatment failure and a poor outcome from disease progression. It is estimated that at least 10 % of blindness is directly attributed to poor patient adherence to prescribed medications.

Clearly, eye drops cannot sustain IOP lowering for more than 24 h at best. To achieve sustained duration of IOP lowering, other modes of administration are needed. One approach is to use a 'punctual plug' which effectively plugs the tear duct; if an IOP-lowering drug is incorporated into the body of the plug, it may be released slowly over time. The company Ocular Therapeutics has reported the development and testing in humans of such a plug containing the drug travoprost [51]. Although such a mode of administration [52] (Fig. 5.6) offers promise for longer-duration therapy, it is more invasive than a sub-conjunctival injection that can also sustain the duration of IOP lowering for at least 3 months following a single injection. For this, nanoliposomes have been found to be successful [53]. The challenge is to incorporate sufficient amounts of the drug into the nanoliposomes such that a single 150 mL injection contains sufficient drug to last over 3 months; secondly, the release of drug must be slow enough to sustain IOP lowering over the same time period. The nanosize helps to reduce opacity in the anterior chamber as well as not to be irritating to the patient.

Work in our group showed that the prostaglandin analogue, latanoprost, may be loaded up to 10 % by weight of lipids in liposomes [54] and the release sustained over 2 months *in vitro*. *In vivo* studies in diseased monkeys showed duration of IOP lowering for up to 4 months. There were no deleterious local effects of the sub-conjunctival injection: the bleb that is formed initially resolves in 24 h. This approach opens up possibilities for sustained therapy of anterior chamber diseases that are chronic in nature. The required dosing over 3 months is usually high for

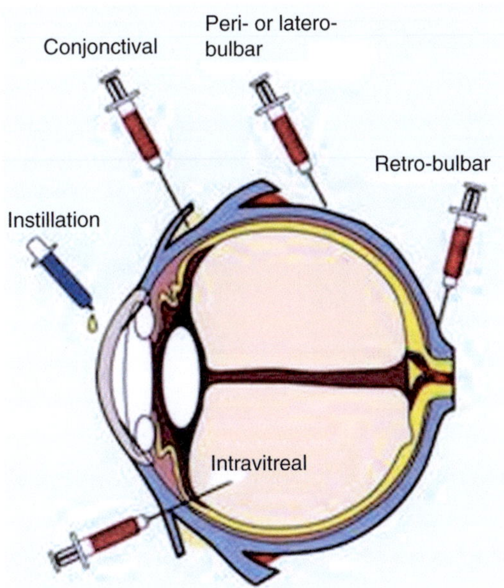

Fig. 5.6 Routes of administration of nanocarriers in the eye (Courtesy of Claudia Di Tommaso)

most drugs, and so a suitable nanocarrier system must be identified that can incorporate such large doses without losing shape or aggregating.

Posterior Segment Diseases

One site of action where sustained delivery capability is needed is the posterior eye segment. There are several conditions that require sustained therapy: diabetic oedema, macular degeneration (the so-called wet kind) and infections, including cytomegaloviral infections. Topical administration is not a feasible option due to the impermeability of the various barriers. Currently, all the three conditions are treated with either intra-vitreal implants or intra-vitreal injections. Both procedures have associated risk factors, such as retinal detachment and vitreous haemorrhage. Furthermore, such procedures also suffer from poor patient acceptance. Thus, sustained delivery options for the back of the eye are an important area of research and have been for some years.

Using PLA nanoparticles incorporating rhodamine, Bourges et al. [55] found that intra-vitreally injected particles migrate slowly to the retinal pigmented epithelium (RPE) cells and stay in the back of the eye for up to 4 months. As no other controls were used, it is not clear if any similar particle would also stay in the vitreous for that duration or whether PLA particles are unique in this regard. Since the PLA particles were not surface treated in any particular way, it is reasonable to assume that similar polymeric particles in a similar size range would also be retained without clearance for 4 months.

The fate of the liposomes in the vitreous humour is less clear, with discrepancies reported between animal models [56]. Residence time in rabbit eyes is longer than in rat eyes: half-life in rabbit eyes is at least 2–3 weeks. Blurring of vision due to aggregation has also been reported with intra-vitreally injected liposomes [56]; hence, for this reason, intra-vitreally injected liposomes have not been translated to the clinic. We suspect that any nanoparticle formulation injected intra-vitreally will suffer from poor dispersion in the viscous vitreous humour and hence tend to aggregate over time. Stabilization against aggregation is a minimum requirement for intra-vitreal administration of nanoparticles.

5.3.3 Periodontic Disease and Endodontic Failure

According to a dental health care report published by the World Health Organization [57], dental caries affects nearly 100 % of the adult population in the world. Periodontic and endodontic diseases are the conditions resulting from inflammatory responses from teeth and its supporting structures [58], as illustrated in Fig. 5.7. Endodontic infection is an infection of the *tooth canal system*, which is mediated by the biofilm contagion (also known as dental plaque). The current and primary treatment option for endodontic infections is root canal surgery. This involves decontamination and disinfection of the pulp of the tooth, and then the canal is plugged with inert fillings to prevent future microbial invasions.

Periodontitis arises as a resultant of chronic inflammation of the periodontium, initiated or associated with the presence of biofilms. Prolonged inflammation results in loosening of tooth or tooth loss. The current treatment involves

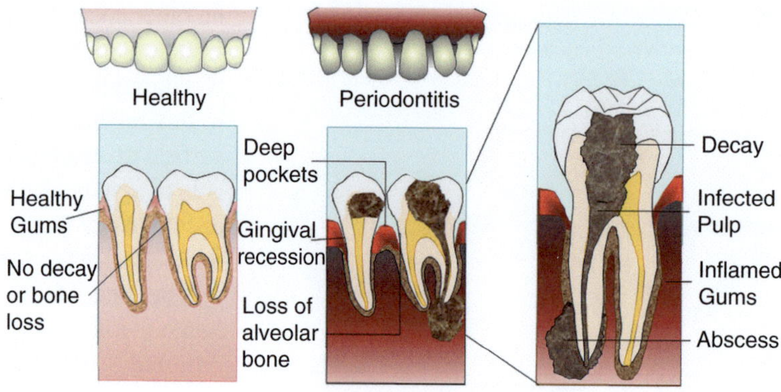

Fig. 5.7 Schematic representation of healthy teeth and diseased teeth: periodontic condition on the *left* and endodontic on the *right*

Table 4.1 List of approved periodontal products

Product	Technology	Indication
ACTISITE®	EVA copolymer containing dispersed tetracycline; controlled release for 7 days	Periodontitis
ARESTIN®	Minocycline incorporated in PLGA microspheres	Periodontitis (used in conjunction with SRP); also periimplant lesions
PERIOCHIP®	Chlorhexidine gluconate in a gelatin matrix delivered as a 4×5 mm chip; sustained release expected	Periodontitis (used in conjunction with scaling & root planning (SRP))
ATRIDOX®	In-situ gelling system consisting of PLGA and doxycycline in NMP solution; forms gel upon injection; controlled release over 7 days	Periodontitis (used in conjunction with SRP)
ELYZOL® dental gel	Bioresorbable gel made from sesame oil and glycerol mono-oleate, containing metronidazole; sustained release characteristics not reported	Periodontitis

mechanical removal of the biofilms, sometimes with the additional use of antimicrobial agents.

There are some common challenges for drug delivery strategies to treat marginal and apical periodontitis. These may be summed up as follows:

(a) Generally, both conditions are caused by bacterial species; systemic treatment is ineffective and may lead to antibiotic resistance.
(b) Localized delivery is critical.
(c) Sustained release over weeks with a single application; the dosing must be above the minimum effective (or inhibitory) concentration.
(d) Access to bacteria population (especially as a biofilm) residing in various parts of the gums, and root canal is crucial.

From the preceding discussions, it is clear that (a), (b) and (c) may be adequately addressed by microparticles incorporating antibiotics. However, the fourth requirement is better addressed by nanoparticles; in fact, nanomedicine may open new treatment options for highly

resistant bacterial biofilm causing persistent periodontic and endodontic diseases.

Many options for localized delivery have been considered for dental drug delivery. However, the approved products (Table 4.1) are discussed below. This list is representative rather than exhaustive. Clearly, many technologies have been tried with varying degrees of success. It would appear that gels and deformable devices have a better chance of making intimate contact with infected regions and tissues. The in situ gelation allows for anchoring. In such systems, the effective mode of action is sustained release of the bioactive molecule (usually an antibiotic) from the 'ge' followed by its penetration into bacterial cells or its neutralization of the bacteria depending on its mechanism of activity. When bacterial biofilms are present, the antimicrobial efficacy decreases, presumably due to the difficulty of drug penetration offered by EPS. Nanotechnology may play a role in these instances.

Endodontic and periodontal pathologies are both bacterial biofilm–mediated diseases. Thus, the suppression of biofilms/microorganisms is challenging [59], and the occurrence of reinfection or persistent infection is not uncommon due to:

(a) Adaptive nature of microorganisms even in nutrient-deprived environments

(b) The protective barrier of extracellular polymeric matrix (EPM) self-generated by the biofilm bacteria

(c) Development of resistance towards antimicrobial agents

(d) Anatomical complexities and apical of root canal

(e) Rapid clearance and salivary flow in the oral environment.

To circumvent these challenges, nanomaterials in the form of nanofibres and nanoparticles have been considered for dental therapy. The electrospun nanofibres have been used as scaffolds for regenerating dental tissue in the case of endodontic failure, but since this concept does not have any drug delivery aspects, it is beyond the scope of this chapter. We will focus on micro- and nanoparticles for localized delivery of anti-infective agents for periodontic and endodontic diseases. In the following sections, we focus on reviewing the applications of microparticles to date and the current research on nano-based delivery systems; in particular, we will discuss the possible advantage of nanomedicine for the management of endodontal and periodontal diseases.

Microparticles

The introduction of micro-based drug delivery systems has great potential to improve the management of endodontal and periodontal disease [60] and form the mainstay of pharmacological intervention. Microparticles have clear benefits over traditional treatment that focusses on irrigation and mechanical means to eradicate dental biofilms because they can provide prolonged antibacterial effects and are extremely effective when directed at sites such as periodontal pockets and root canals.

Mundargi et al. [61] and Álvarez et al. [60] reviewed encapsulation techniques for drugs and proteins into biodegradable polymers by several techniques: water-in-oil-in-water (W/O/W) emulsion, phase separation and spray drying (Fig. 5.8). Among the various natural

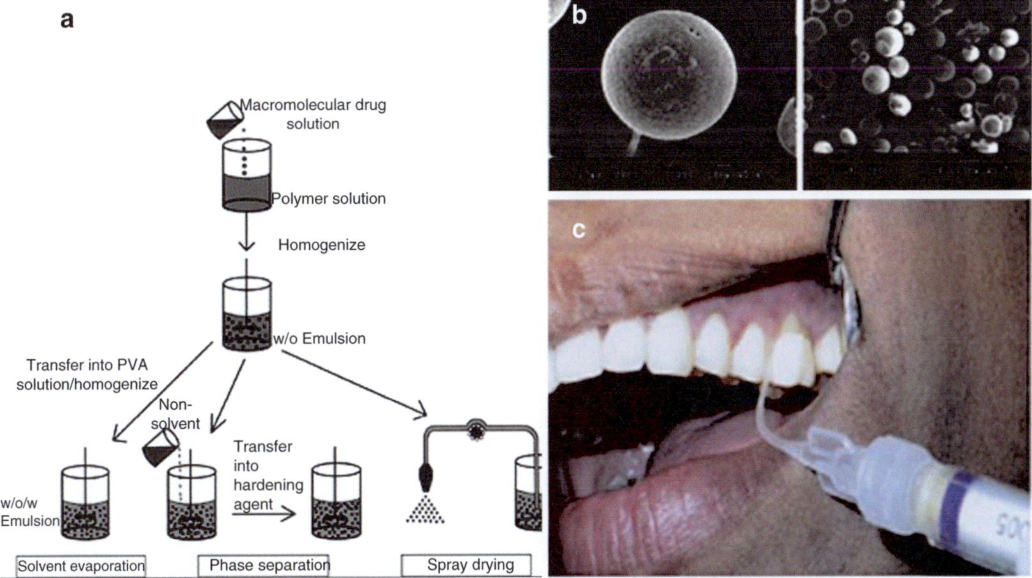

Fig. 5.8 (**a**) Comparison of microencapsulation methods: (1) solvent evaporation, (2) polymer phase separation and (3) spray drying. Aqueous solution is dispersed in the organic polymer solution by ultrasonication (w/o) emulsion; the w/o emulsion is processed further by specific methods to prepare the drug-loaded microparticles; (**b**). Surface morphology of antibiotic loaded PLGA-PCL microspheres; (**c**) Periodontal delivery of microspheres using unit dose cartridge (**a**: Reprinted from Mundargi et al. [61]. With permission from Elsevier; **c, b**: Reprinted from Mundargi et al. [92]. With permission from Elsevier)

and synthetic biopolymers [60, 61], copolymers of the lactide and glycolide family (poly(D,L-lactide-co-glycolide; PLGA) have been the most studied due to their versatile nature with reference to commercial availability, release profile, biodegradation time, biocompatibility and regulatory approval.

To control the periodontal pathogenesis, antibiotics such as doxycycline (DXY), tetracycline (TCL), minocycline (MIN) and metronidazole have been widely used in dentistry. Mundargi et al. encapsulated DXY into PLGA-PCL blend microspheres with size range 90–200 µm, and encapsulation efficiency ranged from 10 to 25 %. The low encapsulation is attributed to the high aqueous solubility of DXY in the continuous phase before solidification of microspheres in the W/O/W technique and the increment in size; the encapsulation is increased with increase in PLGA component in the blend formulation. Different encapsulation approaches are desirable to efficiently encapsulate highly aqueous soluble drugs such as DXY, TCL and MIN by preventing migration of DXY during the W/O/W process by increasing polymer concentration, drug loading and addition of salts (NaCl) to the external water phase. The *in vitro* sustained DXY release upto 1 week and drug concentration in the gingival crevicular fluid is higher than the minimum inhibitory concentration against most of the periodontal pathogens. A clinical study on DXY microspheres by split-mouth design in 30 Asian populations by Srirangarajan et al. [62] was followed for up to 9 months, and the improvement in the clinical parameters, plaque index, gingival index (GI), probing depth (PD) and relative attachment levels is attributed to the localized sustained release of DXY in the periodontal pocket resulting in the significant reduction of PD scores to 0.4–1 mm.

Although spray-drying is an alternative strategy to achieve high encapsulation efficiency in PLGA microspheres with hydrophilic drugs [63], generally the release from spray-dried particles is faster, perhaps due to the larger surface-to-volume ratio [63]. Nevertheless, the clinical and commercial success of minocycline-PLGA microspheres as Arestin® in periodontal pocket therapy has shown the potential of PLGA

particulate delivery systems in the management of periodontal diseases.

Endodontic treatments utilize calcium hydroxide (CH) as an intracanal medicament for its antibacterial properties and ability to form a calcific barrier. This can be effectively achieved by sustained release of Ca^{2+} ions from core–shell calcium hydroxide microspheres (75–150 µm) to the root canals there by retaining higher calcium concentration compared to non-microsphere formulation (UltraCal®XS) for 6 months [64]. In spite of the widespread use of microspheres in dental drug delivery, there are some additional advantages to the use of nanocarriers over microcarriers, and these will be explained in the next section.

Potential of Nanomedicine

According to a survey [65] using the PubMed database, research articles on 'nanoparticles and dental' had only a mere 153 hits as of 2012, while separate searches on 'nanoparticles' and 'dental materials' yielded 27,251 hits and 81,895 hits respectively. These figures suggest that nanotherapeutics has not been explored sufficiently or in depth for treatment of endodontic and periodontic diseases. Furthermore, most of the applications in dental materials were pertaining to the application of nanoparticulates as nanofillers to improve the physico-mechanical properties of restorative materials [66]. In this section, we focus on examining the status of nano-based delivery systems in dental applications and uncover other potential benefits of nanomedicine in endodontics and periodontics.

Polymeric Nanoparticles

Nanoparticles serve as prophylactic approaches to prevention of biofilm formation as well as opening up the possibility of penetration of already-formed biofilms. For these reasons, nanoparticulate drug delivery has been explored for dental diseases. For nanoparticulate carriers, chitosan appears to be a logical choice given its inherent anti-bacterial properties. Chitosan is the name given to partially de-acetylated chitin, a polysaccharide that occurs in the shells of crabs and other crustaceans (Fig. 5.9).

The degree of de-acetylation controls the solubility of chitosan. The higher the de-acetylation,

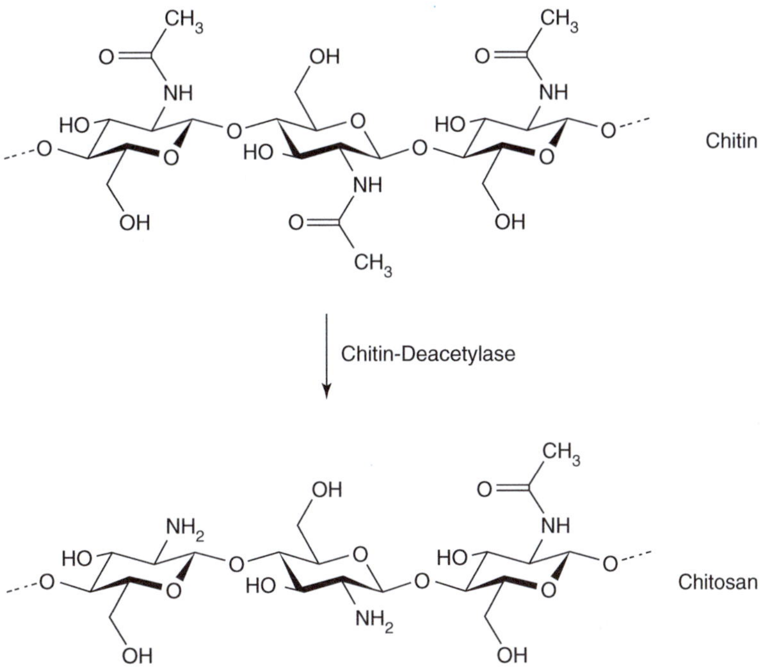

Fig. 5.9 Schematic of Chitosan formation from deacetylation of chitin

the easier it is to dissolve in solvents such as dilute acids or weak acids, where it exists as a protonated species (through the amine groups). The degree of protonation is also related to the degree of de-acetylation; it is this protonation coupled with its structure that enables chitosan to interact/inactivate microbial species. Although the mechanistic details are not very clear, it is generally agreed that at lower pH, with lower–molar mass chitosan and in nano-form chitosan, there is bacterial/cellular entry followed by inhibition of mRNA synthesis. Higher–molar mass chitosan is not expected to enter cells but exerts its action by surface 'complexation', which leads to impermeability of the bacterial wall to nutrients. At higher pH, it is likely that chelation and hydrophobicity of the chitosan play a significant role and alter cell wall permeability. In view of these modes of action, it is clear that soluble chitosan is more effective than solid chitosan unless the chitosan is in nanoparticle form. For more details, please refer to a comprehensive review [67].

Polymeric nanoparticles loaded with various therapeutic agents have shown a commendable degree of success. In general, the drug capacity of nanoparticles is inferior to that of microspheres. This is compensated by a greater ability to penetrate spaces inaccessible to microspheres including bacterial cells and biofilms. Additionally, certain charged nanoparticles have intrinsic antimicrobial activity and hence confer an added advantage over similarly charged microspheres. Researchers from Chosun University reported the successful synthesis of antisense oligonucleotide–loaded chitosan-based nanoparticles [68]. The synthesis of this nanoparticle was based on the complex formation between the oligonucleotides with chitosan followed by addition of tripolyphosphate (TPP). In this work, the authors evaluated the release of oligonucleotides at varying pH conditions and the stability of the nanoparticles in saliva solution. The authors concluded that the Chitosan-TPP nanoparticles showed sustained release of oligonucleotides and are suitable for local therapeutic applications in periodontal management.

In addition to chitosan, PLGA and cellulose acetate phthalate nanoparticles [68–70] have been used to provide the optimum supply of antibacterial agents for endodontal and periodontal disease management. Moulari and co-workers had reported the development of *Harungana madagascariensis Lam. Ex Poir.* (HLE; an ethyl acetate leaf extract)–loaded PLGA nanoparticles [69] on the *in vitro* bacterial activities of oral bacteria strains that caused dental caries and gingivitis infection. The 48 h bacterial activity studies showed that HLE-loaded nanoparticles had better performance compared to direct HLE solution. The key to this successful outcome could be attributed to two distinct advantages of these nanoparticles. The continuous release of HLE at maintained drug concentrations from nanocarriers enhanced the antibacterial activities. The second key lay in the ability of theses polymeric nanoparticles to stay on in the bacterial cells for prolonged periods of time. This is attributed to the hydrophobic nature of the polymer supporting the attachments of these nanoparticles to bacteria.

In an *in vivo* study lasting 2 weeks [70], Pinon-Segindo and co-workers reported the preparation of triclosan(TCS)-loaded nanoparticles specifically for periodontal treatment. Using PLGA, poly(D,L-lactide) and cellulose acetate phthalate, they produced nanoparticles that were approximately 200–450 nm in size. From the *in vitro* release study, 85 % of drugs diffused out of most of the nanoparticle formulations within 2 h. These nanoparticles were tested in a preliminary *in vivo* study in dogs. When the severity of inflammation between days 1, 8 and 15 after the administration of TCS-loaded nanoparticles were compared, the authors reported a decrease in gingival inflammation. However, more investigations need to be carried out to understand the true value of these carriers for reducing inflammation.

Although there are currently only a limited number of endodontal and periodontal nanodelivery systems, the idea of using controlled release nanoparticles for future application in endodontics and periodontics is conceivable and prospective. However, the environmental barriers should be taken into account when developing controlled release nanoparticles for such applications, which includes the continuous loss of therapeutic agents due to turnover of salivary flow and the pharmacokinetics in gingival crevicular fluid flow. Additionally, for these applications, the exploration of charged particles may yield benefits over neutral particles.

Charged and Surface-Charged Nanoparticles

They can find significant application to eliminate endodontic and periodontic pathogens. The use of charged nanoparticles is extremely useful since many of the pathogens are themselves charged. The induction of charge on nanoparticles could be via loading of charged materials or coating the surface with charged moieties. The charged nanoparticles could be used to (1) directly affect the microorganism or (2) alter the microenvironment of the microorganism.

1. Direct interaction with the microorganism: 'Selective eradication'

   'Selective eradication' is a method to directly affect or eradicate the microorganism, where nanoparticles' charge could bind to opposite-charged bacteria species or pathogens. This concept was demonstrated by combining the use of positively charged nanoparticles and photodynamic therapy (PDT) [71, 72] to eradicate negatively charged bacteria and pathogens [72]. Instead of encapsulating therapeutic agents for controlled release, photosensitizers (PSs) are also encapsulated in the nanoparticles. PDT inactivates microorganisms based on the principle that PSs are preferentially taken by bacteria and activation by light generates free radicals and singlet oxygen which kills the microorganism [71–73]. In this work, the transmission electron microscopy (TEM) results showed that the nanoparticles were not internalized by the microorganism but were concentrated on the cell wall. The authors believe that this has rendered the cell wall permeable to the PS that was released from the nanoparticles.

   In the management of endodontal and periodontal disease, nanoparticles encapsulating PS molecules are particularly advantageous over PS molecules alone due to (1) production of more reactive oxygen species associated with its larger 'critical mass' [72], (2) selectivity for charged moieties and (3) reduction in immunogenicity and side effects. However, from the practical point of view, long-term PDT may not be very viable unless treatment could be made in a more portable fashion. Another method would be co-delivery of PS along with anti-microbial agents, where the PS could increase the permeability of the cell wall for more effective entry of continuously released antimicrobial agents.

2. Affect the environment that it resides in: *Biofilm disruption*

Biofilm-producing bacteria generate a protective barrier consisting of extracellular polymeric matrix (EPM). This polysaccharide-rich layer prevents effective penetration of therapeutic agents and contributes to persistence of infection. The best way to circumvent this layer would be to disrupt the EPM, penetrate into the biofilm structure and significantly eliminate the residing bacteria [74]. Silver, chitosan and zinc oxide nanoparticles have been reported to exert cell wall permeability alteration due to electrostatic attractions [75, 76] and/or damage cell membranes via induction of reactive oxygen species [77, 78].

Apart from the aforementioned antibacterial materials, the use of liposomal systems to mimic and target periodontal biofilms has been intensively investigated in recent years. Jones and Kaszuba [79] incorporated phosphatidylinositol (PI) with dipalmitoylphosphatidylcholine (DPPC) liposomes. They reported that the addition of PI facilitated the adsorption of liposome to strains of *Streptococuss sanguis* (*S. sanguis*), which they later ascribed to the interaction between the polyhydroxy groups of liposomes and the bacterial glycol-calyx. In another study [80] carried out by the same group, the researchers developed steraylamine (SA)-containing liposomes to target oral bacteria *S. sanguis* and mutants using the same concept they established in their previous study. They found that increasing the SA content increased the targeting ability of the liposome. Taking it a step further, they investigated the delivery of triclosan and chlorhexidine from these liposomes. From their optimization studies, they found that liposomal delivery of triclosan was most efficient when combinations consisting of DPPC, cholesterol and SA were employed.

Specific Targeting Nanoparticles

Almost 700 bacterial species reside in our oral cavity [81]. Extensive investigations in molecular microbiology have generated a complete picture of the endodontic pathogens. In particular, the identification of the 'more' resistant microorganisms has been established [82]. Biofilm bacteria are highly adaptive and can survive even in nutrient-depleted environments such as root canal treatment [83, 84]. Endodontic biofilm is a particularly common cause of persistent and recurrent infection due to its resistance to antimicrobial agents [85]. Hence, it is important for treatment options to possess a 'nip-in-the-bud' approach where key pathogens are specifically suppressed.

Specific targeting of microorganisms is possible via conjugation of antibodies to the respective nanoparticles. In a work by Robinson et al. [86], they conjugated anti-*oralis* antibodies to liposomes which were targeted to bind to the *Streptococcus oralis* (*S. oralis*) biofilms. They found that targeting efficiency of these anti-*oralis* immunoliposomes were dependent on the number of antibodies per liposome. When the immunoliposomes were compared with the cationic liposomes, the immunoliposomes exhibited lower affinity to the *S. oralis*. This is because immunoliposome adsorption is a result of specific interaction with the strain, whilst charged liposome adsorption is due to non-specific electrostatic forces, and they were found to absorb strongly to other bacterial strains. Hence, the use of such immunoliposomes would be most constructive for precise delivery of anti-microbial agents for specific plaque control.

Future of Nano-Based Delivery Systems

The antimicrobial effects of conventional topical antimicrobials are usually temporal in nature. Their antibacterial activity decreases significantly with time [87]. Endodontic and periodontic diseases are chronic pathologies that require long-term sustained supply of antimicrobial drugs for effective treatment [60]. Furthermore, due to the complex anatomy of the root canal system, 35 % of root canal surfaces are not dealt with regardless of the different antimicrobial methods [88]. Thus, newer treatment approaches are warranted.

Micro-based delivery systems such as Arestin® can provide long-term sustained release of antibacterial agents directly to the periodontium. However, due to the size limitations, microparticles may face difficulties in penetrating into deeper lesions, which is paramount in the cases of severe periodontitis. Nanodelivery systems are promising approaches to address some issues of microparticles including dispersibility, stability and penetration to inaccessible regions owing to its small size. It is conceivable that in the near future, the development of drugs incorporated into nanoparticles similar to Arestin® may bring about greater success.

5.4 Miscellaneous Applications and Future Perspectives

The other application where nanosize becomes important is in delivering bioactive agents into eukaryotic cells. These include delivery of low-MW drugs (for anti-restenosis or anti-thrombotic effects) as well as genes for gene therapy and small interfering ribonucleic acid (siRNA) for anti-sense therapy. Here, the blood lifetime is no longer a predominant requirement as the mode of administration may be via infusors at the site of delivery. An important application in this line is the treatment of thrombus formation and restenosis following angioplasty. This approach involves catheter-based drug delivery. The Dispatch® device, for example, was given a 501 K approval in 1996 for delivery of anti-thrombotic agents such as heparin and urokinase following angioplasty. Since then, the device has been used in a variety of preclinical and clinical studies.

The interesting feature of this indwelling catheter is that it is designed to deliver drug intra-arterially for up to 4 h without blocking blood flow. This makes it an ideal delivery mechanism for various bioactive molecules from drugs to proteins to plasmid DNA. Since the infused drug usually is in a liquid (either suspended or dissolved), the delivery of nanocrystalline drugs and particles becomes easy without having to worry about agglomeration effects or rapid clearance that usually accompanies intravenous administration. Sustained and localized delivery is possible, but surface modification of particles appears to be critical for cellular uptake. Song et al. have developed an *ex vivo* model for measuring arterial uptake using an explanted canine femoral artery prefused with the drug solution for 30 s at 37 °C [89].

In a study of PLGA nanoparticles [90], cationically modified PLGA nanoparticles of size 100 nm were found to be readily taken up by arterial cells. These findings were translated into positive results in a rat study to confirm cellular delivery of dexamethasone incorporated in PLGA particles [91]. The study involved an artery injury model, with 'local' infusion of the drug-containing nanoparticles, over a 3-min period in the carotid artery. The local infusion is accomplished by closing off the arterial segment, creating a closed arterial space into which the drug suspension is infused. The study hints at the tissue delivery of nanoparticles (in this case containing dexamethasone) that may have the potential to reach diseased tissue (minimally/non-invasive). Thus, the future successes for nanoparticle-based delivery systems may be considered in the following areas:

(a) Passive and active targeting of solid tumours
(b) SiRNA delivery
(c) Charge-mediated delivery of anti-infectives and low-MW drugs for periodontic and endodontic conditions
(d) Localized infusion therapy involving intracellular delivery

Acknowledgements We acknowledge support from the School of Materials Science and Engineering, Nanyang Technological University for part of this work.

References

1. Rapamune®, an immunosuppressant, approved in 1999; Emend®, an anti-emetic, approved in 2003.

2. Abraxane®, albumin-bound paclitaxel, approved 2005; IT-101, a camptothecin bound cyclodextrin polymer, in clinical trials currently.

3. Ambisome®, approved in 1997, for fungal infections; Diprivan®, an anaesthetic approved in 1989; and Doxil® approved in 1995 for ovarian cancer.

4. Wagner V, et al. The emerging nanomedicine landscape. Nat Biotech. 2006;24(10):1211–7.

5. Etheridge ML, et al. The big picture on nanomedicine: the state of investigational and approved nanomedicine products. Nanomedicine: Nanotechnol Biology Med. 2013;9(1):1–14.

6. Venkatraman SS, et al. Polymer- and liposome-based nanoparticles in targeted drug delivery. Front Biosci (Schol Ed). 2010;2:801–14.

7. Cegnar M, Kos J, Kristl J. Cystatin incorporated in poly(lactide-co-glycolide) nanoparticles: development and fundamental studies on preservation of its activity. Eur J Pharm Sci. 2004;22(5):357–64.

8. Quintanar-Guerrero D, et al. Preparation and characterization of nanocapsules from preformed polymers by a new process based on emulsification-diffusion technique. Pharm Res. 1998;15(7):1056–62.

9. Olson F, et al. Preparation of liposomes of defined size distribution by extrusion through polycarbonate membranes. Biochim Biophys Acta Biomembr. 1979; 557(1):9–23.

10. Mayer LD, Bally MB, Cullis PR. Uptake of adriamycin into large unilamellar vesicles in response to a pH gradient. Biochim Biophys Acta Biomembr. 1986; 857(1):123–6.

11. Clerc S, Barenholz Y. Loading of amphipathic weak acids into liposomes in response to transmembrane calcium acetate gradients. Biochim Biophys Acta Biomembr. 1995;1240(2):257–65.

12. Haran G, et al. Transmembrane ammonium sulfate gradients in liposomes produce efficient and stable entrapment of amphipathic weak bases. Biochim Biophys Acta Biomembr. 1993;1151(2):201–15.

13. Patri AK, Majoros IJ, Baker Jr JR. Dendritic polymer macromolecular carriers for drug delivery. Curr Opin Chem Biol. 2002;6(4):466–71.

14. Barenholz Y. Doxil® — the first FDA-approved nanodrug: lessons learned. J Control Release. 2012; 160(2):117–34.

15. Vaage J, et al. Therapy of human ovarian carcinoma xenografts using doxorubicin encapsulated in sterically stabilized liposomes. Cancer. 1993;72(12):3671–5.

16. Papahadjopoulos D, et al. Sterically stabilized liposomes: improvements in pharmacokinetics and antitumor therapeutic efficacy. Proc Natl Acad Sci. 1991; 88(24):11460–4.

17. Myocet. Wikipedia; 2014. http://en.wikipedia.org/wiki/Myocet. [cited 2014 Jan 30].

18. Ramesh R, et al. Successful treatment of primary and disseminated human lung cancers by systemic delivery of tumor suppressor genes using an improved liposome vector. Mol Ther. 2001;3(3):337–50.

19. 10Q Detective. Introgen therapeutics: empty promises for cancer patients and shareholders. Seeking alpha April 2007. http://seekingalpha.com/article/ 33114-introgen-therapeutics-empty-promises-for-cancer-patients-&-shareholders. [cited 2014 Jan 30].

20. Xu L, et al. Fragment-targeted immunoliposomes for systemic gene delivery, U.S. Patent, Editor; 2009.

21. Venkatraman SS, et al. Micelle-like nanoparticles of PLA–PEG–PLA triblock copolymer as chemotherapeutic carrier. Int J Pharm. 2005;298(1):219–32.

22. Jie P, et al. Micelle-like nanoparticles of star-branched PEO–PLA copolymers as chemotherapeutic carrier. J Control Release. 2005;110(1):20–33.

23. Critical micelle concentration. Wikipedia; 2014. http://en.wikipedia.org/wiki/Critical_micelle_concentration. [cited 2014 30 January].

24. Hamaguchi T, et al. NK105, a paclitaxel-incorporating micellar nanoparticle formulation, can extend in vivo antitumour activity and reduce the neurotoxicity of paclitaxel. Br J Cancer. 2005;92(7):1240–6.

25. Rowinsky EK, Donehower RC. Paclitaxel (Taxol). N Engl J Med. 1995;332(15):1004–14.

26. Hamaguchi T, et al. A phase I and pharmacokinetic study of NK105, a paclitaxel-incorporating micellar nanoparticle formulation. Br J Cancer. 2007;97(2): 170–6.

27. A phase III study of NK105 in patients with breast cancer. ClinicalTrials.gov; 2012 http://clinicaltrials.gov/ct2/show/study/NCT01644890. [cited 2014 Jan 30].

28. Kataoka K, et al. Polymeric micelle containing cisplatin enclosed therein and use thereof. 2003; US 2003/0170201 A1.

29. Pinzani V, et al. Cisplatin-induced renal toxicity and toxicity-modulating strategies: a review. Cancer Chemother Pharmacol. 1994;35(1):1–9.

30. Nishiyama N, et al. Novel cisplatin-incorporated polymeric micelles can eradicate solid tumors in mice. Cancer Res. 2003;63(24):8977–83.

31. NC-6004 Nanoplatin™. NanoCarrier; 2013. http://www.nanocarrier.co.jp/en/research/pipeline/02.html. [cited 2014 Jan 30].

32. Nakanishi T, et al. Development of the polymer micelle carrier system for doxorubicin. J Control Release. 2001;74(1–3):295–302.

33. Matsumura Y, et al. Phase I clinical trial and pharmacokinetic evaluation of NK911, a micelle-encapsulated doxorubicin. Br J Cancer. 2004;91(10):1775–81.

34. Tsukioka Y, et al. Pharmaceutical and biomedical differences between micellar doxorubicin (NK911) and liposomal doxorubicin (Doxil). Jpn J Cancer Res. 2002;93(10):1145–53.

35. Torchilin VP. Micellar nanocarriers: pharmaceutical perspectives. Pharm Res. 2007;24(1):1–16.

36. Allen C, et al. Polycaprolactone-b-poly(ethylene Oxide) block copolymer micelles as a novel drug delivery vehicle for neurotrophic agents FK506 and L-685,818. Bioconjug Chem. 1998;9(5):564–72.

37. Kim SY, et al. Methoxy poly(ethylene glycol) and ε-caprolactone amphiphilic block copolymeric micelle containing indomethacin: II. Micelle formation and drug release behaviours. J Control Release. 1998;51(1):13–22.

38. Gref R, et al. 'Stealth' corona-core nanoparticles surface modified by polyethylene glycol (PEG):

influences of the corona (PEG chain length and surface density) and of the core composition on phagocytic uptake and plasma protein adsorption. Colloids Surf B Biointerfaces. 2000;18(3–4):301–13.

39. Ma LL, Jie P, Venkatraman SS. Block copolymer 'stealth' nanoparticles for chemotherapy: interactions with blood cells in vitro. Adv Funct Mater. 2008; 18(5):716–25.

40. Xu P, et al. Highly stable core-surface-crosslinked nanoparticles as cisplatin carriers for cancer chemotherapy. Colloids Surf B Biointerfaces. 2006;48(1): 50–7.

41. Murakami H, et al. Preparation of poly(dl-lactide-co-glycolide) nanoparticles by modified spontaneous emulsification solvent diffusion method. Int J Pharm. 1999;187(2):143–52.

42. Beletsi A, et al. Simultaneous optimization of cisplatin-loaded PLGA-mPEG nanoparticles with regard to their size and drug encapsulation. Curr Nanosci. 2008;4(2):173–8.

43. Desai MP, et al. Immune response with biodegradable nanospheres and alum: studies in rabbits using staphylococcal enterotoxin B-toxoid. J Microencapsul. 2000;17(2):215–25.

44. Cohen H, et al. Sustained delivery and expression of DNA encapsulated in polymeric nanoparticles. Gene Ther. 2000;7(22):1896–905.

45. Dunn SE, et al. In vitro cell interaction and in vivo biodistribution of poly(lactide-co-glycolide) nanospheres surface modified by poloxamer and poloxamine copolymers. J Control Release. 1997;44(1): 65–76.

46. De Jaeghere F, et al. Formulation and lyoprotection of poly(lactic acid-co-ethylene oxide) nanoparticles: influence on physical stability and in vitro cell uptake. Pharm Res. 1999;16(6):859–66.

47. Gref R, et al. Biodegradable long-circulating polymeric nanospheres. Science. 1994;263(5153): 1600–3.

48. Peracchia MT, et al. Complement consumption by poly(ethylene glycol) in different conformations chemically coupled to poly(isobutyl 2-cyanoacrylate) nanoparticles. Life Sci. 1997;61(7):749–61.

49. Liang C, et al. Improved therapeutic effect of folate-decorated PLGA-PEG nanoparticles for endometrial carcinoma. Bioorg Med Chem. 2011;19(13): 4057–66.

50. Lasic DD. On the thermodynamic stability of liposomes. J Colloid Interface Sci. 1990;140(1):302–4.

51. QLT shows positive efficacy trends from data in plug combinations in phase II studies for glaucoma using latanoprost punctual plug delivery system. QLT, Inc. http://www.qltinc.com/newsCenter/2012/121025. htm. [cited 2014 Jan 30].

52. Michael Möller. In vitro and in vivo studies of polymeric micelles for ophthalmic applications. Universite de Geneve. http://www.unige.ch/sciences/pharm/f/ la_section/docu/compet/51.pdf. [cited 2014 Jan 30].

53. Natarajan JV, et al. Nanomedicine for glaucoma: liposomes provide sustained release of latanoprost in the eye. Int J Nanomedicine. 2012;7:123–31.

54. Natarajan JV, et al. Sustained drug release in nanomedicine: a long-acting nanocarrier-based formulation for glaucoma. ACS Nano. 2014;8(1):419–29.

55. Bourges JL, et al. Ocular drug delivery targeting the retina and retinal pigment epithelium using polylactide nanoparticles. Invest Ophthalmol Vis Sci. 2003; 44(8):3562–9.

56. Bochot A, Fattal E. Liposomes for intravitreal drug delivery: a state of the art. J Control Release. 2012; 161(2):628–34.

57. Petersen PE, et al. The global burden of oral diseases and risks to oral health. Bull World Health Organ. 2005;83(9):661–9.

58. Difference Between Periodontist & Endodontist. Intelligent Dental. http://www.intelligentdental. com/2011/10/24/difference-between-periodontist-endodontist/. [cited 2014 Jan 30].

59. del Pozo JL, Patel R. The challenge of treating biofilm-associated bacterial infections. Clin Pharmacol Ther. 2007;82(2):204–9.

60. Álvarez AL, Espinar FO, Méndez JB. The application of microencapsulation techniques in the treatment of endodontic and periodontal diseases. Pharmaceutics. 2011;3(3):538–71.

61. Mundargi RC, et al. Nano/micro technologies for delivering macromolecular therapeutics using poly(D, L-lactide-co-glycolide) and its derivatives. J Control Release. 2008;125(3):193–209.

62. Srirangarajan S, et al. Randomized, controlled, single-masked, clinical study to compare and evaluate the efficacy of microspheres and gel in periodontal pocket therapy. J Periodontol. 2011;82(1):114–21.

63. Patel P, et al. Microencapsulation of doxycycline into poly(lactide-co-glycolide) by spray drying technique: Effect of polymer molecular weight on process parameters. J Appl Polym Sci. 2008;108(6):4038–46.

64. Strom TA, et al. Endodontic release system for apexification with calcium hydroxide microspheres. J Dent Res. 2012;91(11):1055–9.

65. Socialstyrelsen, Editor. Release of nanoparticles from dental materials. Ministry of Health and Social Affair (Sweden): Stockholm; 2013. p. 14–5.

66. Chogle SM, et al. Preliminary evaluation of a novel polymer nanocomposite as a root-end filling material. Int Endod J. 2011;44(11):1055–60.

67. Kong M, et al. Antimicrobial properties of chitosan and mode of action: a state of the art review. Int J Food Microbiol. 2010;144(1):51–63.

68. Dung TH, et al. Chitosan-TPP nanoparticle as a release system of antisense oligonucleotide in the oral environment. J Nanosci Nanotechnol. 2007;7(11): 3695–9.

69. Moulari B, et al. Potentiation of the bactericidal activity of Harungana madagascariensis Lam. ex Poir. (Hypericaceae) leaf extract against oral bacteria using poly (D, L-lactide-co-glycolide) nanoparticles: in vitro study. Acta Odontol Scand. 2006;64(3): 153–8.

70. Pinon-Segundo E, et al. Preparation and characterization of triclosan nanoparticles for periodontal treatment. Int J Pharm. 2005;294(1–2):217–32.

71. Wilson M. Lethal photosensitisation of oral bacteria and its potential application in the photodynamic therapy of oral infections. Photochem Photobiol Sci. 2004;3(5):412–8.
72. Patel NB. Targeted methylene blue-containing polymeric nanoparticle formulations for oral antimicrobial photodynamic therapy. In: Bouvé College of Health Sciences. Department of Pharmaceutical Sciences. Northeastern University; 2009. p. 11.
73. Wilson M, Burns T, Pratten J. Killing of Streptococcus sanguis in biofilms using a light-activated antimicrobial agent. J Antimicrob Chemother. 1996;37(2):377–81.
74. Dunne Jr WM, Mason Jr EO, Kaplan SL. Diffusion of rifampin and vancomycin through a Staphylococcus epidermidis biofilm. Antimicrob Agents Chemother. 1993;37(12):2522–6.
75. Jung B-O, et al. Preparation of amphiphilic chitosan and their antimicrobial activities. J Appl Polym Sci. 1999;72(13):1713–9.
76. Rabea EI, et al. Chitosan as antimicrobial agent: applications and mode of action. Biomacromolecules. 2003;4(6):1457–65.
77. Lovric J, et al. Unmodified cadmium telluride quantum dots induce reactive oxygen species formation leading to multiple organelle damage and cell death. Chem Biol. 2005;12(11):1227–34.
78. Shrestha A, et al. Nanoparticulates for antibiofilm treatment and effect of aging on its antibacterial activity. J Endod. 2010;36(6):1030–5.
79. Jones MN, Kaszuba M. Polyhydroxy-mediated interactions between liposomes and bacterial biofilms. Biochim Biophys Acta. 1994;1193(1):48–54.
80. Jones MN, et al. The interaction of phospholipid liposomes with bacteria and their use in the delivery of bactericides. J Drug Target. 1997;5(1):25–34.
81. Paster BJ, et al. The breadth of bacterial diversity in the human periodontal pocket and other oral sites. Periodontol. 2000;2006(42):80–7.
82. Narayanan LL, Vaishnavi C. Endodontic microbiology. J Conserv Dent. 2010;13(4):233–9.
83. Costerton JW, et al. Bacterial biofilms in nature and disease. Annu Rev Microbiol. 1987;41:435–64.
84. George S, Kishen A, Song KP. The role of environmental changes on monospecies biofilm formation on root canal wall by Enterococcus faecalis. J Endod. 2005;31(12):867–72.
85. Distel JW, Hatton JF, Gillespie MJ. Biofilm formation in medicated root canals. J Endod. 2002;28(10):689–93.
86. Robinson AM, Creeth JE, Jones MN. The specificity and affinity of immunoliposome targeting to oral bacteria. Biochim Biophys Acta. 1998;1369(2):278–86.
87. Neelakantan P, Subbarao CV. An analysis of the antimicrobial activity of ten root canal sealers–a duration based in vitro evaluation. J Clin Pediatr Dent. 2008;33(2):117–22.
88. Peters OA, Schonenberger K, Laib A. Effects of four Ni-Ti preparation techniques on root canal geometry assessed by micro computed tomography. Int Endod J. 2001;34(3):221–30.
89. Song CX, et al. Formulation and characterization of biodegradable nanoparticles for intravascular local drug delivery. J Control Release. 1997;43(2–3):197–212.
90. Labhasetwar V, et al. Arterial uptake of biodegradable nanoparticles: effect of surface modifications. J Pharm Sci. 1998;87(10):1229–34.
91. Guzman LA, et al. Local intraluminal infusion of biodegradable polymeric nanoparticles. A novel approach for prolonged drug delivery after balloon angioplasty. Circulation. 1996;94(6):1441–8.
92. Mundargi RC, et al. Development and evaluation of novel biodegradable microspheres based on poly(d, l-lactide-co-glycolide) and poly(epsilon-caprolactone) for controlled delivery of doxycycline in the treatment of human periodontal pocket: in vitro and in vivo studies. J Control Release. 2007;119(1):59–68.
93. Uchino H, Matsumara Y, Negishi T, et al. Cisplatin-incorporating polymeric micelles (NC-6004) can reduce nephrotoxicity and neurotoxicity of cisplastin in rats. Br J Cancer. 2005;93(6):678–87.

Nanoparticles for Endodontic Disinfection

Anil Kishen and Annie Shrestha

Abstract

The widespread recognition of microbial biofilm as the contributory factor for human infection and constant increase in antimicrobial resistance warrants the discovery of a reliable and effective antimicrobial strategy to combat infectious diseases. Treatment of infected root canals presents with a major challenge of bacterial persistence after treatment. Use of various antibacterial nanoparticles presents as a potential treatment strategy to improve the elimination of biofilm bacteria from the root canal system. Nanoparticles have been developed to improve the root canal disinfection as well as to seal the canal space during root canal treatment. For effective therapeutic effect with nanoparticles, information about the effectiveness of treatment, location of infection, and delivery efficiency of nanoparticles should be well understood. This chapter discusses the current limitations in achieving effective root canal disinfection followed by examples of different nanoparticles that are being developed and tested for this purpose.

6.1 Introduction

In the ongoing war against antibiotics, the bacteria seem to be winning, and the drug pipeline is verging on empty. [1]

Antimicrobial resistance is constantly on the rise, leading to a major hindrance in the treatment of many infectious diseases [2–4]. The conservative management of such infections involving topical or systemic antibiotics has been shown to be ineffective owing to several factors (Table 6.1) [4]. Approximately 60 % of the current human infections have been associated with the presence of bacterial biofilms, which includes both implant-related infections and non-implant-related infections [5]. Additionally, widespread systemic use of antibiotics is the cause of multidrug resistance and superinfection due to their untoward effects on commensal microbial flora (nonpathogenic) [6]. Currently the search for newer antibiotics is on the decline, which is inversely proportional to

A. Kishen, PhD, MDS, BDS (✉)
A. Shrestha, PhD, MSc, BDS
Department of Endodontics,
Faculty of Dentistry, University of Toronto,
124 Edward Street, Room 348, Toronto, ON
M5G1H4, Canada
e-mail: anil.kishen@utoronto.ca;
annie.shrestha@mail.utoronto.ca

A. Kishen (ed.), *Nanotechnology in Endodontics: Current and Potential Clinical Applications*,
DOI 10.1007/978-3-319-13575-5_6, © Springer International Publishing Switzerland 2015

Table 6.1 Limitations of current topical or systemic antimicrobial treatment strategies to manage infectious diseases

1. Microbial factors	(a) Antibiotic resistant mutant strains	MRSA
		Vancomycin resistant enterococci
	(b) Antibiotic resistant mechanisms	Exchange of genetic materials
		Deficiency of specific porin channels
		Promotion of active drug efflux
		Thickening of the peptidoglycan layer of the outer wall
	(c) Structure and organization of microbes	Variation in the outer wall in different classes of bacteria and fungi
		Formation of biofilms
		Protozoal existence as tropic feeding stage of resting cystic stage
2. Antibiotic misuse		Excessive or inappropriate prescription
		Failure to complete treatment regimen
		Widespread use of antibiotics in livestock feedstuff

Adapted from Cunha [4]. With permission from Elsevier

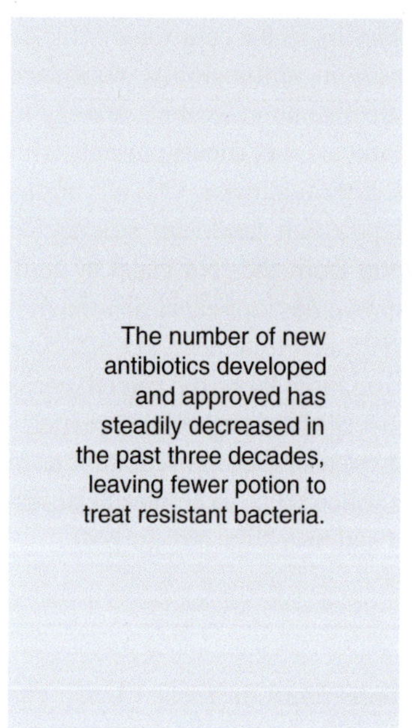

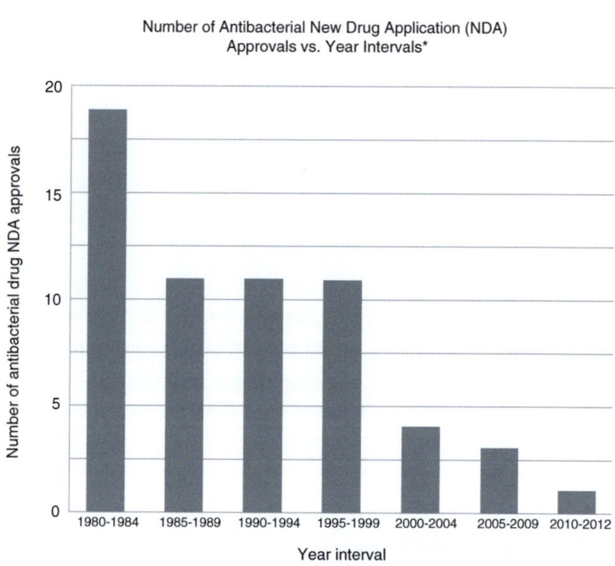

Fig. 6.1 Tomorrow's antibiotics: the drug pipeline (Adapted from Prevention CfDCa [7]. With permission from CDC.org)

the rise in antibiotic-resistant bugs (Fig. 6.1) [7]. The widespread recognition of microbial biofilm as the contributory factor for human infection and drying pipeline of antibiotics has led to a dire situation, which warrants the identification of a reliable and effective antimicrobial strategy to combat infectious diseases, which in the past have been easily treatable [2, 8, 9].

6.2 Antibacterial Nanoparticles

Nanoparticles (NPs) are microscopic particles with one or more dimensions in the range of 1–100 nm [10]. Nanotechnology, literally, is derived from the words "nano" and "technology" [11]. Nano is defined as one billionth of a quantity, represented mathematically as 10^{-9}.

NPs possess properties that are very unique from their bulk counterparts. A wide range of applications has been explored with NPs. Some of their applications are drug/gene delivery, fluorescent labeling for imaging, detection of pathogens, probing DNA structure, tissue engineering, tumor destruction, etc. [12]. NPs that display antibacterial properties may be termed antibacterial NPs. These NPs have been found to have a broad spectrum of antimicrobial activity and far lesser propensity to induce microbial resistance when compared to antibiotics.

Metallic nanoparticles of copper, gold, titanium, and zinc have attracted particular attention, with each of them having different physical properties and spectra of antimicrobial activity [13, 14]. It is known that magnesium oxide (MgO) and calcium oxide (CaO) slurries acted upon both gram-positive and gram-negative bacteria in a bactericidal manner [15], while zinc oxide (ZnO) slurry acted in a bacteriostatic manner and exhibited stronger antibacterial activity against gram-positive than gram-negative bacteria [15]. The antibacterial powders of magnesium oxide (MgO), calcium oxide (CaO), and ZnO generated active oxygen species, such as hydrogen peroxide and superoxide radical, which is responsible for their antibacterial effect. These NPs of metallic oxides with their high surface area and charge density exhibited greater interaction with bacteria and subsequently produced markedly high antibacterial efficacy [16]. Furthermore, it is apparent that bacteria are far less likely to acquire resistance against metallic nanoparticles than other conventional antibiotics.

Heavy metal ions are known to have different cytotoxic effects on bacterial cell functions [13, 17]. Copper ions may induce oxidative stresses [13] and affect the redox cycling, resulting in cell membrane and DNA damages. Zinc ions applied above the essential threshold level inhibit bacterial enzymes including dehydrogenase [18], which in turn impede the metabolic activity [13]. Silver ions inactivate proteins and inhibit the ability of DNA to replicate [14]. NPs synthesized from the powders of Ag, CuO, and ZnO are currently used for their antimicrobial activities. The electrostatic interaction between positively charged NPs and negatively charged bacterial cells and the accumulation of large numbers of NPs on the bacterial cell membrane have been associated with the loss of membrane permeability and cell death [17].

6.3 Bacterial Biofilms as the Therapeutic Target for Antimicrobials

Bacteria have been associated with almost 80 % of human infectious diseases [2, 19, 20]. Biofilm is the preferred mode for bacterial growth in oral disease conditions such as dental caries, periodontal disease, and apical periodontitis (endodontic disease) [21]. Bacteria persisting as biofilms have received major attention due to the difficulty in the clinical management of biofilm-mediated infections. Costerton defined biofilms as "a structured community of bacterial cells enclosed in a self-produced polymeric matrix and adherent to an inert or living surface" [22]. Typically, bacteria occupy 10–20 % of the total volume of the biofilm, and the remainder is the extracellular matrix [23, 24]. The extracellular matrix of a biofilm is polyanionic exopolysaccharides secreted by the bacteriumitself. They consist of polysaccharides, proteins, nucleic acids (extracellular DNA), and salts, totally making up to 95 % of the biofilm volume [24]. This hydrated envelope not only protects the bacteria from noxious threats but also acts as a scavenger to trap and concentrate nutrients for favorable growth [8]. These hydrated structures are known to possess water channels that facilitate efficient interchange of nutrients and waste between bacterial cells and bulk fluid. These channels also act as primitive circulatory system and transfer information between bacterial microcolonies in a biofilm.

The bacterial cells in a biofilm are heterogeneously arranged as microcolonies that are adherent to a solid surface (Fig. 6.2) [21]. These microcolonies are dense aggregates of bacteria assembled based on the physiological and metabolic states within the biofilms. Depending upon

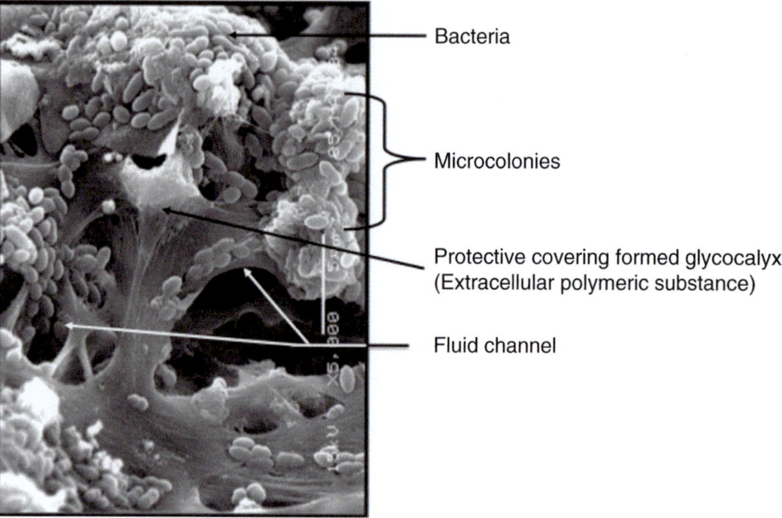

Fig. 6.2 SEM image showing ultrastructure of bacterial biofilm formed on dentin. The major components of biofilm structures are indicated (Courtesy of Anil Kishen)

the environmental conditions and fluid shear forces, mature biofilms can detach microcolonies. This detachment of cells from matured biofilms termed *seeding dispersal*, could be either a continuous process such as erosion or massive loss of biofilm (sloughing). The microcolonies that are shed from the biofilm possess traits of the parent biofilm bacteria; consequently, they may lead to a distant infection [21, 25]. The antimicrobial resistance in a biofilm is a complex phenomenon, wherein one or more mechanisms go hand in hand. The mechanisms that contribute toward antimicrobial resistance in biofilm bacteria are categorized as

1. *Protection by Exopolysaccharide (EPS) matrix*: An EPS matrix is known to impart resistance to antimicrobials by acting as a physical barrier or neutralizing the chemicals applied. The diffusion of antimicrobials is limited to the surface of a biofilm structure enabling bacteria to survive deeper in the biofilms. The anionic EPS also acts as a chemical barrier. The charge and interwoven dense structure reduces the penetration of antimicrobials' by ionic or electrostatic interactions [26, 27]. Mostly the antimicrobials used are positively charged hydrophilic compounds that can bind with EPS. Antimicrobial agents such as iodine, chlorine, and peroxygens have been shown to be neutralized by constituents of a biofilm matrix [28].

2. *Physiological state of bacteria*: Persister cells are a small population of nongrowing cells within a biofilm that contribute to antibiotic tolerance [29]. These phenotypically different bacteria have very low metabolic activity, which allows them to enter a dormant state and survive with minimal nutrient requirement. As most antibiotics target actively growing cells, these persister cells are spared. However, once the antibiotic is stopped and a favorable condition is provided, these persisters are known to repopulate the biofilm, leading to treatment relapse [29].

3. *Altered microenvironment and phenotype*: The mature biofilm consists of multilayered microcolonies of bacteria embedded in an EPS matrix. This creates a gradient of nutrients and redox potential that offers surface bacteria advantage of higher growth rates [30]. The deep-lying bacteria have limited access to nutrients and oxygen and the metabolic rates are reduced, thus allowing bacteria to resist antimicrobials. The enhanced tolerance to antibiotics has been suggested as these chemicals target bacterial cellular processes such as DNA replication or translation as in actively growing cells. The thicker the biofilms with an abundant EPS matrix, the more tenacious they become to be removed by antimicrobials. In addition, exposure to stress or low levels of antimicrobials is known to result

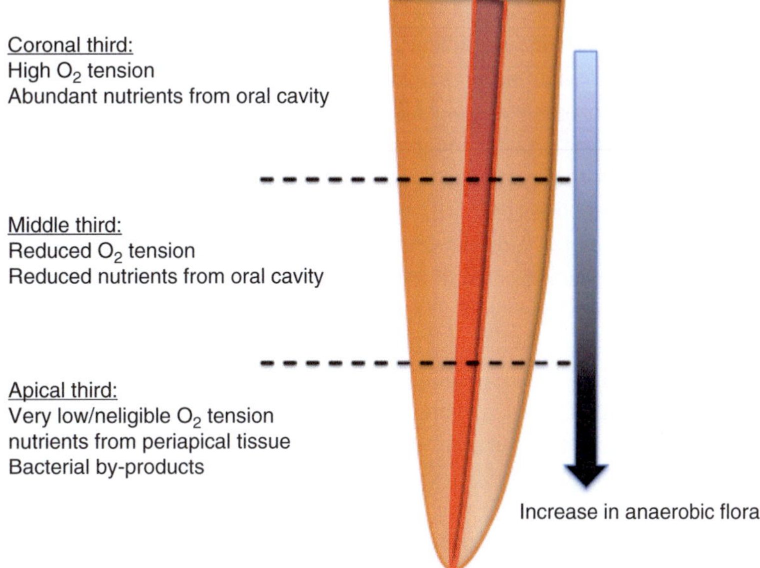

Fig. 6.3 The root canal space presents with selective pressures allowing specific bacteria to survive depending on their nutritional and environmental requirements (Adapted from Chavez de Paz [42]. With permission from Elsevier)

in the expression of certain stress genes, shock proteins, and multidrug efflux pumps in biofilm bacteria [31]. The multidrug efflux pumps are membrane-bound active pumps found in both gram-positive and gram-negative bacteria and have been considered as one of the key factors in case of biofilm resistance [32, 33].

6.3.1 Endodontic Biofilms

Apical periodontitis (endodontic disease) is a bacterial biofilm–mediated infection [34]. In this disease process, the causative bacteria exist within the apical region of the root canals and release toxins/irritants, which induce periapical inflammation and host immune response [35, 36]. Till date, no specific correlation has been established between bacterial species in the root canal and clinical presentation of apical periodontitis [37]. In 1987, Nair and colleagues showed that bacteria in infected root canals exist as clusters of bacterial aggregates, a morphology analogous to biofilm bacteria [38]. Today, based on different histopathological studies, it is stressed that apical periodontitis is a biofilm-mediated disease [39–41].

A root canal system exhibits a tough environmental condition with regard to the availability of nutrients and oxygen [42]. Thus, an infected root canal contains a restricted assortment of bacterial species (approximately 5–10) [43, 44], which is markedly less when compared to more than 500 diverse species of bacteria in the oral cavity. No single microorganism is associated with the occurrence of disease rendering them with pathogenicity. The selection of particular combinations of bacteria could be attributed to the endodontic milieu that provides selective habitat, which supports the development of specific proportions of strict and facultative anaerobic flora [43, 45] (Fig. 6.3). Based on the "community-as-pathogen concept," it is possible at this point to infer that some communities are more related to disease than others [46]. The main treatment goals of endodontic disease is thus to eliminate microbial biofilms from the infected root canals. Unfortunately, complete elimination of bacterial biofilms from the complex anatomy of root canals has been a daunting task for clinicians.

Previous studies showed that the presence of viable microbes at the time of filling was one of the crucial factors for the successful outcome of endodontic treatment [47, 48]. About 79 % of endodontically treated teeth with the presence of bacteria showed periapical lesions as compared to 28 % in those treated without bacteria [47].

Ricucci and Siqueira studied the prevalence of biofilms in primary and persistent infections and reported intraradicular biofilms in almost 74 % of the treated root canals with apical periodontitis at the apical segment [39]. Though extraradicular biofilms have been reported as a causative factor for persistent infections [49, 50], their prevalence has been reported to be on the lower side (6 %). As bacteria persisting within the treated root canals have been considered as the major cause for treatment failure [42, 51, 52], research has focused toward (a) understanding the limitations of the current root canal irrigation methods and (b) development of newer antibiofilm strategies for endodontic application.

6.4 Antimicrobials in Root Canal Therapy

Topical antimicrobials used as irrigants and medications are suggested for effective elimination of bacteria within the root canals [53, 54]. The success rate of root canal treatment is suggested to be 74–95 % [55–58]. The current level of evidence showed that despite the advancements in treatment strategies, the success rates have not increased for the past four to five decades [59]. This could be mainly attributed to the limitations of current technologies to deal with the challenges of the disease process [60]. Clinical outcome studies confirmed that the prognosis of endodontic treatment depend on the stage of disease process and whether the treatment is primary or retreatment [55–58, 61–66]. Follow-up study after 4–6 years of initial endodontic treatment revealed that the presence of periapical infection at the time of treatment reduced the healing rate from 92 to 74 % [61].

Endodontic therapy treats a complex tissue system consisting of the tooth–pulp–periradicular complex. Conventionally, chemical antimicrobials (topical) are used within the root canals in combination with mechanical instrumentation to achieve "microbe-free" root canals prior to filling the root canal with an inert filling material [2, 8, 67]. Ethylenediaminetetraacetic acid (EDTA) and sodium hypochlorite (NaOCl) are commonly used for root canal debridement. NaOCl is a deproteinating agent that results in the heterogeneous removal of organic substrate from dentin, whereas EDTA is a chelating agent that demineralizes the dentin and exposes the surface collagen fibrils [68]. These irrigants alone or in combinations are inadequate in achieving the above treatment goals. Therefore, newer antimicrobials are warranted for root canal disinfection. An effective treatment strategy that possesses marked antibiofilm capability (to eliminate residual endodontic bacteria/biofilm) without producing undue damaging effects on the dentin hard tissue and cytotoxic effect on periapical tissue will overcome the shortcomings associated with the current antimicrobial strategy in endodontic treatment.

Generally, therapeutic strategies for biofilms focus on (1) destroying or inactivating the resident bacteria within the biofilm structure or (2) disrupting the biofilm structure and simultaneously killing the resident microbes. Figure 6.4 shows the schematic representation of different antibiofilm approaches [67]. It includes the application of topical antimicrobials that produce slow destruction of the biofilm structure, antimicrobials that destroy persister cells or quorum-sensing signals in a biofilm, antimicrobial application in combination with strategies that enhance its diffusion into the biofilm structure, antimicrobials that diffuse into the biofilm structure producing biofilm bacterial killing and antimicrobial that destroys both the biofilm matrix and resident bacteria from a biofilm structure [69].

The requirements of an ideal root canal irrigant have been suggested to be (1) broad antimicrobial activity to eliminate biofilms of anaerobic and facultative anaerobic bacteria and yeasts; (2) dissolved necrotic pulp tissue; (3) ability to inactivate the endotoxins; (4) ability to remove smear layer following instrumentation and provide a clean root canal surface; and (5) biocompatibility to vital periapical tissues if extruded beyond the root canal space [70]. Till date, there is no antimicrobial irrigant that could perform all these functions single-handedly. Therefore,

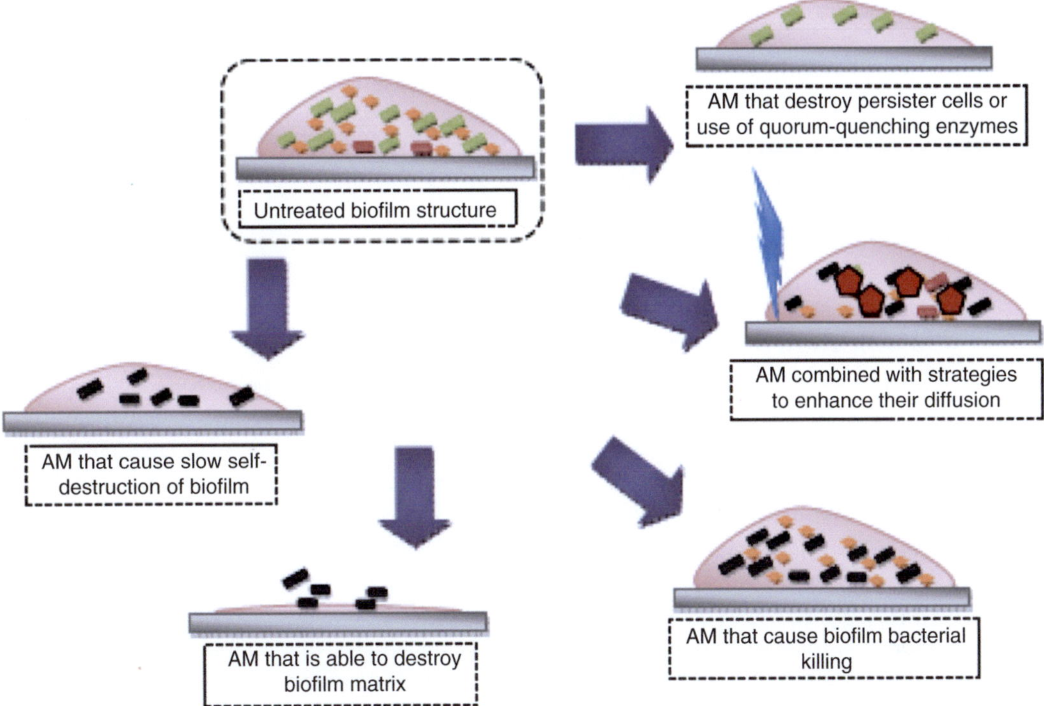

Fig. 6.4 Schematic diagram showing different antibiofilm strategies. *AM* antimicrobial (Adapted from Kishen [67]. With permission from John Wiley & Sons)

combinations of two or more irrigants are used to achieve a bacteria-free root canal space prior to filling [70]. Studies have shown that despite thorough instrumentation and irrigation procedures, bacteria still exist in significantly high numbers within the uninstrumented areas of root canals such as isthmuses and lateral canals [52, 54, 71, 72].

The current limitations in the endodontic disinfection strategies are not only due to the biofilm mode of growth within the root canals but also due to the anatomical complexities of root canal systems, dentin structure/composition, and factors associated with the chemical disinfectants collectively [67]. Microcomputed tomography–based studies revealed that 35–42 % of the root canal surfaces remain untouched by the instruments invariable to the instrumentation methods used [73]. Histological sections of the apical portions of the prepared root canals with instrumentation and irrigation procedures revealed the presence of bacteria in 86 % of cases mainly as

biofilms in the complex anatomical sites [52]. In terms of bacterial reentry/recolonization, studies comparing the efficacy of different obturation techniques and materials have shown that microbial leakage occurs through the filled root canals [74]. Therefore, it may be concluded that even after intracanal medications, complete elimination of bacteria from the root canal system could not be achieved [48, 75]. Due to the shortcomings of the current antibiofilm strategies during root canal treatment, advanced disinfection strategies are being developed and tested. The newer disinfection strategies should aim to circumvent these challenges by eliminating biofilm bacteria not only from the main canals but also from the uninstrumented portions and anatomical complexities of the root canal system without inducing untoward effects on dentin substrate and periradicular tissue. In the following paragraphs, newer nanoparticles that hold significant potential for eliminating endodontic biofilms will be reviewed.

Table 6.2 Nanoparticles available based on the composition

Nanoparticles based on composition				
Inorganic	Metallic	Polymeric	Quantum dots	*Functionalized*
Zinc oxide	Gold	Alginate	Cadmium sulfide	With:
Iron oxide	Silver	Chitosan	Cadmium selenide	Drugs
Titanium dioxide	Iron			Photosensitizers
Cerium oxide	Copper			Antibodies
Aluminum oxide	Magnesium			Proteins

6.5 Classification and Types of Antibacterial Nanoparticles

Nanomaterials could be classified according to composition and various morphologies such as spheres, rods, tubes, and prisms. The nanoparticles based on different composition that are being developed for antibacterial purposes are shown in Table 6.2. Most of these newly developed nanoparticles are in the initial phases of laboratory testing. Table 6.3 shows the list of literature pertinent to the use of various nanoparticles used for improving the root canal disinfection.

6.5.1 Chitosan Nanoparticles

Chitosan (poly (1, 4), β-d glucopyranosamine), a derivative of chitin, the second most abundant natural biopolymer, has received significant interest in biomedicine [76–78]. The industrial extraction of chitin is generally obtained from crustaceans such as crabs, lobsters, and shrimps, making up to 10^{13} kg in the biosphere [79]. The structure of chitin closely resembles that of cellulose, and both act as a structural support and defense material in living organisms. Chitin has two reactive groups: primary (C-6) and secondary (C-3) hydroxyl groups allowing for various chemical modifications. Chitosan has an additional amino (C-2) group on each deacetylated unit.

Chitosan is known as a versatile biopolymer that could be synthesized in various forms such as powder (micro- and nanoparticles), capsules, films, scaffolds, hydrogels, beads, and bandages

[77]. Chitosan has a structure similar to extracellular matrix components and hence is used to reinforce the collagen constructs [80]. This hydrophilic polymer with large numbers of hydroxyl and free amino groups can be subjected to numerous chemical modifications and grafting [81–83]. Nanoparticles of chitosan have been developed mainly for drug/gene delivery applications.

Nanoparticles of chitosan could be synthesized or assembled using different methods depending on the end application or the physical characteristics required in the nanoparticles [77]. Chitosan nanoparticles (CS NPs), by virtue of their charge and size, are expected to possess enhanced antibacterial activity. In addition, chitosan possesses several characteristics such as being nontoxic toward mammalian cells, color compatibility to tooth structure, cost effectiveness, availability, and ease of chemical modification. CS NPs can be delivered within the anatomical complexities and dentinal tubules of an infected root canal to enhance root canal disinfection [84].

Antibacterial Properties

Chitosan and its derivatives such as carboxymethylated chitosan showed a broad range of antimicrobial activity, biocompatibility, and biodegradability [85–88]. The exact mechanisms of antibacterial action of chitosan and its derivatives are still not vivid. However, a more commonly proposed mechanism is contact-mediated killing that involves the electrostatic attraction of positively charged chitosan with the negatively charged bacterial cell membranes (Fig. 6.5). This might lead to the altered cell wall permeability, eventually resulting in rupture of cells and

Table 6.3 List of literature available on the use of nanoparticles for antibacterial purpose to achieve root canal disinfection

Author	Nanoparticles used	Bacteria tested	Observations
Waltimo et al. (2007)	Bioactive glass 45S5	*E. faecalis* in planktonic	Micron to nano size increased the killing
Kishen et al. (2008)	ZnO Chitosan	*E. faecalis* in planktonic	CS -highest antibacterial Combination-highest leaching property
Waltimo et al. (2009)	Bioactive glass- Nano/Micron combination	3 weeks old *E. faecalis* biofilm in root canals	Nano BAG did not show any antibacterial efficacy
Mortazavi et al. (2010)	BAG of different sizes	Planktonic *E. coli, P. aeruginosa, S. typhi,* and *S. aureus*	Antibacterial activity decreased with decrease in size
Pagonis et al. (2010)	MB loaded PLGA	3 days old *E. faecalis* biofilm in root canals	Significant reduction of biofilm
Chogle et al. (2011)	C18 organoclay	*E. faecalis* in planktonic	Reduced apical microleakage when mixed with polymer
Shrestha et al. (2010)	ZnO Chitosan	*E. faecalis* in planktonic and biofilms	Total elimination of planktonic Significant reduction of biofilms Retained antibacterial property after aging
Shrestha and Kishen (2012)	Chitosan in the presence of tissue inhibitors	*E. faecalis* in planktonic	Pulp and BSA significantly inhibited the antibacterial effect
DaSilva et al. (2013)	Sealer incorporated with chitosan nanoparticles with/without canal surface treatment with different formulations of CS	*E. faecalis* 7 days in a chemostat-based biofilm fermentor	Inhibition of biofilm formation within the sealer-dentin interface Surface treatment with chitosan showed higher inhibition of biofilm formation
Abramovitz et al. (2013)	Quaternary ammonium polyethyleneimine nanoparticles (QPEI NP) incorporated into standard TRMs	*S. mutans and E. faecalis* in planktonic form	Incorporation of 2 % wt/wt QPEI NP significantly increased the sealing ability of TRMs
Shrestha et al. (2013)	Rose bengal functionalized chitosan nanoparticles (CSRBnp)	*E. faecalis* in planktonic and biofilms	Photoactivation of CSRBnp resulted in reduced viability of biofilms and disruption of biofilm structure.
Shrestha (Kishen*) et al. (2013)	Rose bengal functionalized chitosan nanoparticles (CSRBnp) in the presence of tissue inhibitors	*E. faecalis* in planktonic form	Pulp and BSA significantly inhibited the antibacterial effect
Shrestha (Kishen*) et al. (2014)	Photoactivated rose bengal functionalized chitosan nanoparticles (CSRBnp)	*E. faecalis* in planktonic and 21 days biofilms	CSRBnp displayed properties of both CS and RB in a photoactivable nano-structure that performed the dual function of targeted elimination of bacterial-biofilms
Shrestha and Kishen (2014)	Rose bengal functionalized chitosan nanoparticles (CSRBnp)	Multispecies biofilms (*Streptococcus oralis, Prevotella intermedia, and Actinomyces naeslundii*) grown on dentin sections for 21 days	CSRBnp showed complete disruption of multispecies biofilm with a reduction in viable bacteria and biofilm thickness

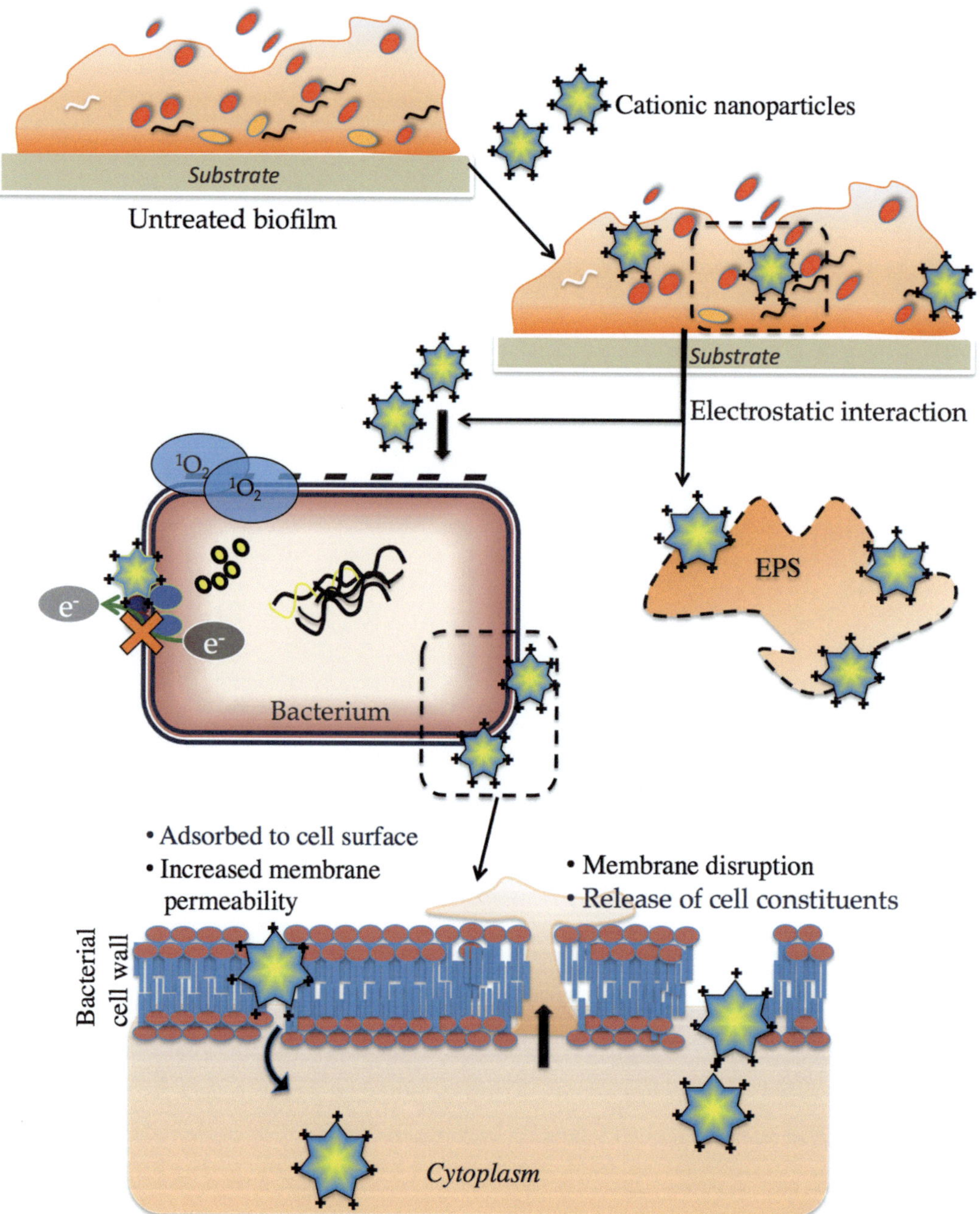

Fig. 6.5 A schematic diagram illustrating the antibacterial mechanism of nanoparticles with positive charge (e.g. Chitosan). Mature bacterial biofilm consisting of abundant EPS and bacteria enclosed. When cationic nanoparticles are introduced for treatment of biofilms, it can interact with both EPS and bacterial cells. The initial electrostatic interaction between positively charged nanoparticles and negatively charged bacterial surface. Bacterial killing occurs upon contact mediated lipid peroxidation via production of reactive oxygen species (ROS). The membrane damage and increased permeability of unstable membrane eventually leads to ingress of nanoparticles into the cytoplasm and release of cytoplasmic constituents. EPS secreted by bacteria in biofilm may interact with the nanoparticles and prevent from interacting with bacteria and thus reducing the antibacterial efficacy

leakage of the proteinaceous and other intracellular components [89, 90]. Rabea et al. proposed that chitosan, due to its chelating property, sequesters trace metals/essential nutrients and inhibits enzyme activities essential for bacterial cell survival [89]. Under transmission electron microscopy, the bacterial cells were seen to be completely enveloped in the chitosan forming an impermeable layer [86]. This could have resulted in the prevention of transport of essential solutes leading to cell death. In case of fungi, chitosan was hypothesized to enter cells and reach nuclei, bind with DNA, and inhibit RNA and protein synthesis.

Chitosan has excellent antibacterial, antiviral, and antifungal properties [89]. In case of bacteria, gram-positive bacteria were more susceptible than gram-negative ones. The minimal inhibitory concentrations ranged from 18 to 5,000 ppm depending upon the organism, pH, degree of deacetylation (DD), molecular weight, chemical modifications, and presence of lipids and proteins [89, 91]. The DD is known to influence the antibacterial activity. With higher DD, the number of amine groups increases per glucosamine unit, and thus, chitosan showed higher antibacterial efficacy [92]. Another study tested the efficacy of CS NPs and ZnO NPs in disinfecting and disrupting biofilm bacteria and the long-term efficacy of these nanoparticulates following aging. *Enterococcus faecalis* (ATCC & OG1RF) in planktonic and biofilm forms were tested in this study. These studies demonstrated that the rate of bacterial killing by NPs depended on the concentration and duration of interaction. Total elimination of planktonic bacteria was observed in contrast to the biofilm bacteria, which survived even after 72 h (Fig. 6.6). Both CS NPs and ZnO NPs were found to retain their antibacterial properties after aging for 90 days [85].

6.5.2 Bioactive Glass Nanoparticles

Bioactive glass (BAG) received considerable interest in root canal disinfection due to its antibacterial properties. BAG consists of SiO_2, Na_2O, CaO_2, and P_2O_5 at different concentrations. Nanometric bioactive glass (BAG) used by Zehnder et al. [93] was amorphous in nature, ranging from 20 to 60 nm in size. They highlighted that the increase in pH is mainly responsible for the antimicrobial activity. Furthermore, the release of Ca^{2+}, Na^+, $PO4^{3-}$, and Si^{4+} could lead to formation of bonds with the mineralized hard tissues.

Antibacterial Properties

The antibacterial mechanism of BAG has been attributed to several factors acting together [94]:

1. High pH: increase in pH due to release of ions in an aqueous environment.
2. Osmotic effects: increase in osmotic pressure above 1 % is inhibitory for many bacteria.
3. Ca/P precipitation: induced mineralization on the bacterial surface.

BAG was used for in vitro root canal disinfection studies [93, 95, 96]. When compared with calcium hydroxide, the latter showed significantly more antibacterial effect than BAG in preventing residual bacterial growth [93]. Waltimo et al. [97] suggested that an ideal preparation of 45S5 bioactive glass suspensions/slurries for root canal disinfection should combine high pH induction with capacity for continuing release of alkaline species. They demonstrated that BAG nanometric slurry had a 12-fold higher specific surface area than the micrometric counterpart. However, the latter had a considerably higher alkaline capacity and disinfected significantly better. The nano-BAG was found to be less effective in eliminating biofilms [97] as compared to the planktonic counterparts [98, 99]. The contradictory result in the antibacterial efficacy of BAG with transition from micron- to nano-sized has been contributed to the tenfold increase in silica release and solution pH elevation by more than three units [98].

6.5.3 Silver Nanoparticles

Use of silver compounds and nanoparticles is widespread mainly owing to its antibacterial

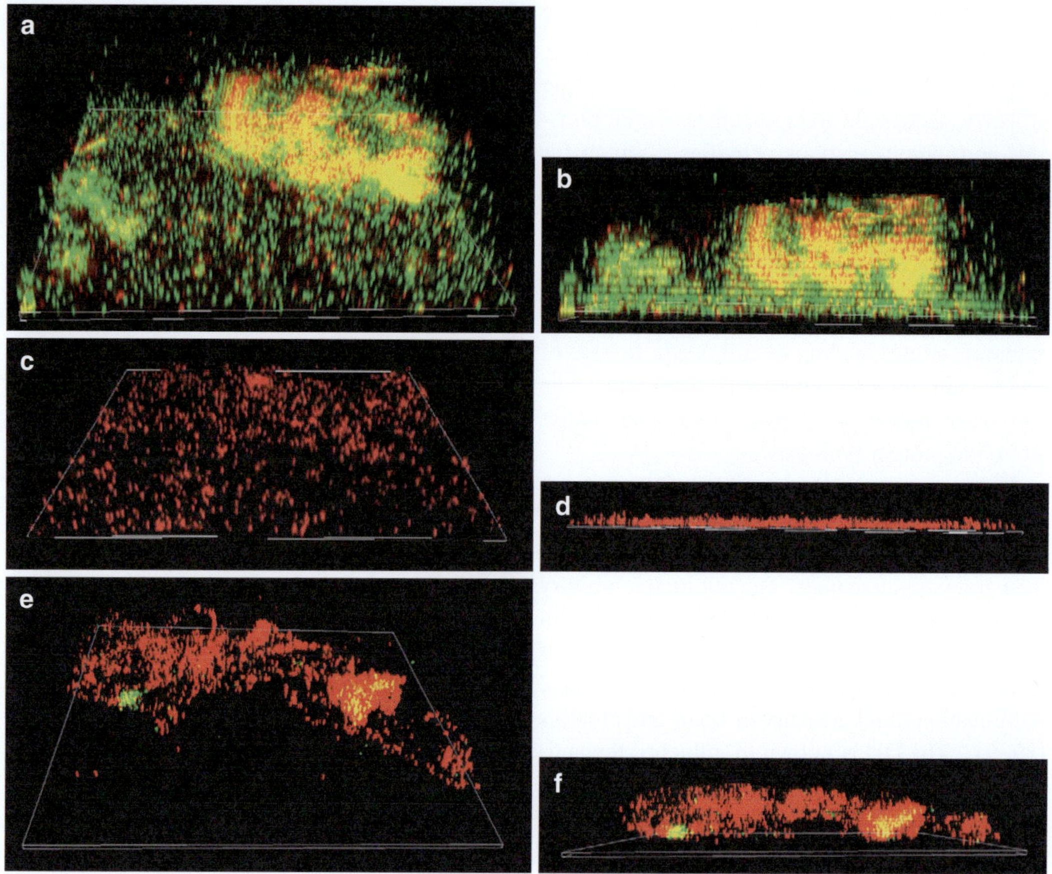

Fig. 6.6 The 3-dimensional CLSM reconstruction of *E. faecalis* (ATCC 29212) biofilm (**a**, **b**) and after treatment with antibacterial ZnO-np (**c**, **d**) and CS-np (**e**, **f**). The number of live bacterial cells was reduced significantly, and the 3-dimensional structure was also disrupted. (**b**, **d**, and **f** show the sagittal sections of the biofilm structure) (original magnification) (Adapted from Shrestha et al. [85]. With permission from John Wiley & Sons)

property. As early as the 1800s, silver has been used for the prevention of infection in burn cases [100]. Prior to the introduction of antibiotics in the early twentieth century, oral intake of aqueous colloidal dispersions of silver for the prevention of infection has been reported [101]. In dentistry, silver and its nanoparticles have been tested for application as dental restorative material, endodontic retrofill material, dental implants, and caries inhibitory solution [102].

Antibacterial Properties

The antibacterial property of silver is mainly due to its interaction with the sulfhydryl groups of proteins and DNA, altering the hydrogen bonding/respiratory chain, unwinding of DNA, and interference with cell wall synthesis/cell division [103, 104]. At the macroscopic level, these silver nanoparticles are known to destabilize the bacterial membrane and increase permeability leading to leakage of cell constituents [105]. However, the toxicity of silver ions has been a major limitation for its widespread use. Research has been directed toward developing silver nanoparticles with specific antibacterial activity and lower cytotoxicity to host cells [106].

Silver in its metallic state is inert but in the presence of moisture is ionized and forms silver ions. These silver ions are highly reactive, bind to tissue proteins, and induce tissue changes. These silver ions bring about structural changes in the bacterial cell wall and nuclear membrane, disrupt the membrane barrier, and result in cell death [107]. The toxicity of silver nanoparticles is proportional to the concentration of free silver ions released. Studies have assessed the potential of silver nanoparticles for root canal disinfection. The antimicrobial activity is the main characteristic for its selection to be used as an irrigant. Hiraishi et al. [108] tested silver diamine fluoride against *E. faecalis* biofilms in vitro. The 48 h biofilms were completely eliminated following 60 min of interaction with 3.8 % silver diamine fluoride and were also found to deposit on the surface of dentin and penetrate up to 40 μm into dentinal tubules. However, the toxicity still remains to be the main concern. Another study showed that the antibiofilm efficacy of AgNPs for root canal disinfection depended on the mode of application, gel being more effective than solution [109]. A 0.02 % AgNP gel as medicament significantly disrupted the structural integrity of the biofilm and resulted in the least number of post-treatment residual viable *E. faecalis* cells compared with 0.01 % AgNP gel, calcium hydroxide groups, and syringe irrigation with 0.1 % AgNP solution. They suggested that the prolonged duration of interaction between positively charged AgNPs and negatively charged resident biofilm bacteriawhen used as medicament for 7 days resulted in marked destruction of biofilm structure and killing of biofilm bacteria. All these in vitro *studies* warrant further experiments in well-characterized in vitro models/in vivo situations to confirm the advantages of using AgNPs for root canal disinfection.

Two main issues of using silver nanoparticles are the potential browning/blackening of dentin and toxicity toward mammalian cells. Filho et al. [110] tested two different concentrations of silver nanoparticles (90 μm) in Wistar rat models. The in vivo tissue reaction was evaluated, which showed that a mild tissue response to Ag NPs was observed especially in a lower concentration (23 ppm). Moderate chronic inflammatory response was found after 15 days in case of 47 ppm silver nanoparticles, which was comparable with 2.5 % sodium hypochlorite. The toxic concentrations of silver ions are approximately 1–10 mg/L and of silver nanoparticles are 10–100 mg/L for eukaryotic cells [101]. Depending on the size, the nonphagocytosing cells take up nanoparticles either by endocytosis or macropinocytosis. Other than the size and structure of AgNPs, the presence of proteins in a biological medium greatly influences the interaction and toxicity of these nanoparticles toward cells.

6.5.4 Nanoparticle-Incorporated Root Canal Sealers

Root canal filling is expected to fill and seal the prepared canal space so as to prevent bacterial recontamination. One of the desired properties of root canal sealers is their antibacterial property that can be utilized to (a) eliminate bacteria remaining in the root canal system and (b) prevent bacterial recolonization in case of leakage. Studies have found that endodontic microorganisms have high affinity for gutta-percha, one of the most common root filling materials [111, 112]. Bacterial adhesion and formation of biofilms on gutta-percha could lead to the persistence of infection in the root canals. The long-term antibacterial property of root canal sealers in such cases would be highly advantageous to enhance the degree of root canal disinfection. However, the common sealers are known to possess antibacterial activity for a maximum period of 1 week, with most of them showing a significant decrease in antibacterial properties immediately after it sets [113–115].

The antimicrobial property of a ZnO-based root canal sealer and resin-based root canal sealer loaded with CS and ZnO NPs was examined recently [87]. It was highlighted that the addition of antibacterial NPs in root canal sealers would improve the direct (based on direct antibacterial assay) and diffusible antibacterial effects (based on a membrane-restricted antibacterial assay) of the root canal sealers. Studies also showed that the application of CS NPs reduced the adherence of *E. faecalis* to root canal dentin. The treatment of root dentin with ZnO NPs, ZnO/CS Mix, CS-layer-ZnO, or CS NPs produced significant reduction in the adherence of *E. faecalis* to dentin.

ZnO NPs, ZnO/CS Mix, and CS-layer-ZnO produced ~95 % reduction in bacterial adherence, while CS NPs produced ~80 % reduction in bacterial adherence. Dentin treated with chlorhexidine and then with nanoparticles showed the maximum reduction (97 %) in bacterial adherence [54].

Previous studies have suggested that the addition of NPs did not deteriorate the flow characteristics of the root canal sealer. ZnO NP–loaded root canal sealer showed a better antibacterial property and ability to diffuse antibacterial components, which is of particular importance in a root canal environment [87]. Another study tested CS NPs incorporated in root canal sealers with and without dentin surface treatment with chitosan modifications [116]. The bovine root canals were surface treated with phosphorylated chitosan or chitosan-conjugated Rose Bengal and photoactivated mainly to stabilize the exposed dentin collagen. The specimens that were filled with CS NP–incorporated sealers without surface treatment revealed significantly lesser biofilms at the filling–dentin interface even after 4 weeks of aging process.

Quaternary ammonium polyethylenimine (QAPEI) nanoparticles were also utilized to improve the antibacterial efficacy of various root canal sealers and temporary restorative materials [117–120]. Barros et al. [117] incorporated QAPEI nanoparticles into AH Plus and pulp canal sealer EWT (PCS). The QAPEI nanoparticles were of 58 ± 18 nm in size with a positive charge of 68.5 ± 1.9 mV. Incorporation of these nanoparticles into the sealers increased the wettability of both AH Plus and PCS. The surface charge of the set sealers also increased significantly when nanoparticles were added. Though fresh PCS showed strongest antibacterial efficacy with and without the QAPEI nanoparticles, both the tested sealers lost their antibacterial efficacy after 7 days of setting. In case of 2 % QAPEI nanoparticles and PCS, the bacterium was still evident after 30–60 min of direct contact. The presence of quaternary ammonium cationic groups on the surface of the

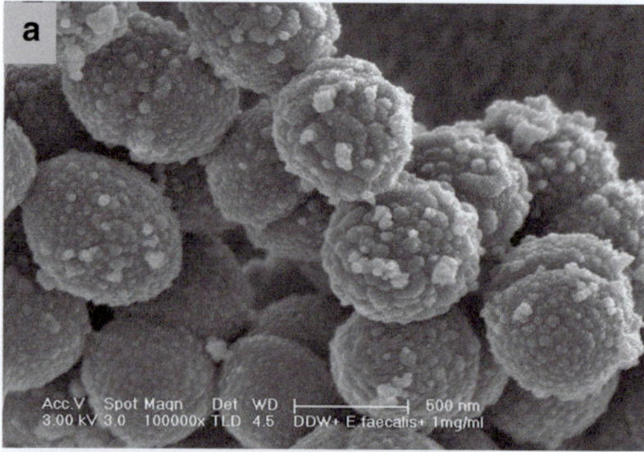

Fig. 6.7 Determination of the antibacterial activity of QPEI nanoparticles in suspension. Scanning electron micrographs of *E. faecalis*. (**a**) In DDW with QPEI nanoparticles- depicting attached nanoparticles on bacterial membranes; (**b**) In DDW- depicting regular bacterial membranes (Adapted from Beyth et al. [120]. With permission from PLoSOne.org)

sealers has been suggested to be responsible for the superior antibacterial efficacy. Although Barros et al. [117] did not show much improvement of AH Plus with incorporation of QAPEI nanoparticles, another study displayed significantly increased antibacterial efficacy of AH Plus and QAPEI nanoparticles [119]. Similarly, Beyth et al. [120] showed that incorporation of QAPEI nanoparticles in RCS (two-paste epoxy–amine resin sealer) resulted in total bacterial growth inhibition. These QAPEI nanoparticles were found to interact with *E. faecalis* cell walls (Fig. 6.7), resulting in completely bacterial elimination after 1 h interaction. At the same time, these nanoparticles, when bound to the sealer surface, impeded nanoparticle penetration into eukaryotic cells that mitigated the toxicity [118, 120]. QAPEI nanoparticles exert bacteri-

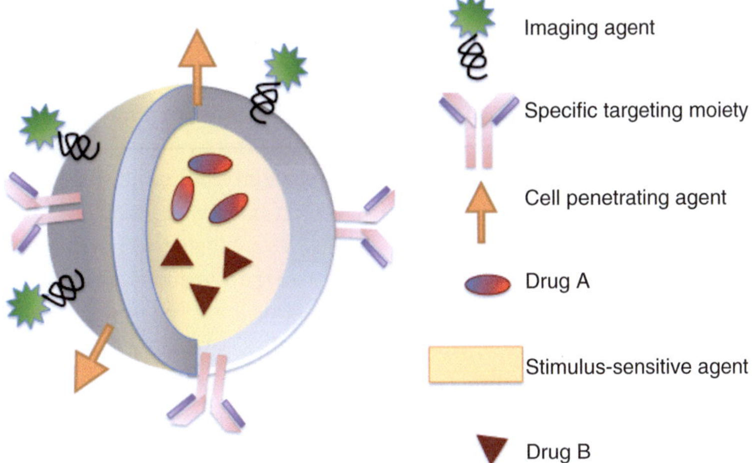

Fig. 6.8 Development of novel strategies for controlled released of drugs will provide nanoparticles with the capability to deliver two or more therapeutic agents. Multifunctional nanoparticles for drug delivery. Multifunctional nanocarriers can combine a specific targeting agent (usually an antibody or peptide) with nanoparticles for imaging (such as quantum dots or magnetic nanoparticles), a cell-penetrating agent (e.g. the polyArg peptide TAT), a stimulus-sensitive element for drug release, a stabilizing polymer to ensure biocompatibility (polyethylene glycol most frequently) and the therapeutic compound (Adapted from Sanvicens and Marco [124]. With permission from Elsevier)

cidal effects mainly by adsorption and penetration through the bacterial cell wall [121], following which they interact with the protein and fat layer in the cell membrane blocking the exchange of essential ions. The destabilization of the cell membrane leads to leakage of intracellular constituents and cell death. The use of the QAPEI nanoparticles is limited to in vitro studies, thus clinical interpretations should be made with caution.

BAG nanoparticles have been recommended to promote closure of the interfacial gap between the root canal walls and core filling materials [122]. The study used a combination of polyisoprene (PI) or polycaprolactone (PCL) and nanometric BAG 45S5 that could create a hydroxyapatite interface and thus ultimately make the use of an endodontic sealer unnecessary. It was concluded that PI and PCL composites with BAG showed promising results as single root canal filling materials. Incorporation of BAG fillers into the polymers under investigation made the resulting composite materials bioactive and improved their immediate sealing ability.

6.5.5 Nanoparticle Functionalization

The word *functionalize* from Merriam-Webster's dictionary means to organize (as work or management) into units performing specialized tasks. Often, the nanoparticles do not possess suitable surface properties for specific applications. Functionalizing the original nanoparticles in a controlled manner with specific molecules could influence the colloidal stability of these nanoparticles and eventually lead to targeted application/delivery effects [123]. When biomolecules are conjugated with the nanoparticles, the term *biofunctionalization* has been used. Functionalization could alter the surface composition, charge, and structure of the material wherein the original bulk material properties are left intact (Fig. 6.8) [124].

In a functionalized nanoparticle, the nanoparticle matrix usually forms the core substrate (Fig. 6.9). It can be used as a convenient surface for molecular assembly and may be composed of inorganic or polymeric materials. The core particle is often protected by several

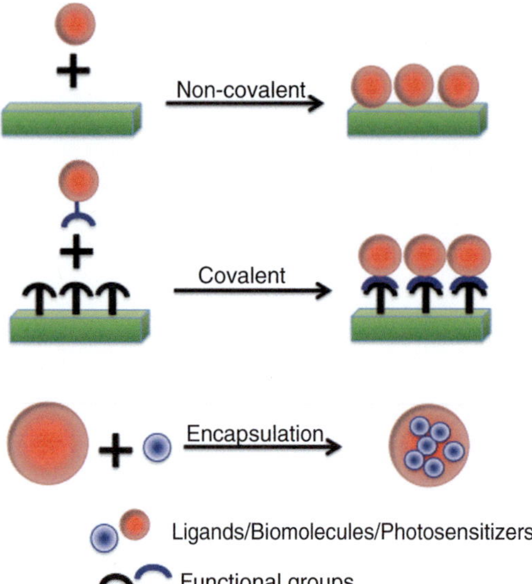

Fig. 6.9 Surface modification or functionalization of nanoparticles through non-covalent, covalent and encapsulation approaches

monolayers of inert material, for example, silica. Organic molecules that are adsorbed or chemisorbed on the surface of the particle are also used for this purpose. The same layer might act as a biocompatible material. However, more often, an additional layer of linker molecules is required to proceed with further functionalization. This linear linker molecule has reactive groups at both ends. One group is aimed at attaching the linker to the nanoparticle surface, and the other is used to bind various moieties like biocompatibles (dextran), antibodies, fluorophores, etc. depending on the function required by the application.

Functionalized nanoparticles containing various reactive molecules or decorated with peptides or other ligands have led to new possibilities of combating antimicrobial resistance [125, 126]. Functionalization of chitosan permits combining or conjugating chitosan, a versatile polymer with different reactive molecules depending on the application. The CS-NPs have been functionalized for various protein, drug, and gene delivery applications [127–130]. Modifications of chitosan nanocarriers with a variety of ligands have been shown to increase the recognition and uptake of these nanocarriers

into cells through receptor-mediated endocytosis. One of the common approaches to study the fate/interaction of nanoparticles with tissues/cells is by labeling the nanoparticles with fluorescein isothiocyanate (FITC) [131]. FITC could be covalently bound to the amine groups in the chitosan backbone using simple chemical reactions. The labeled CS-NP was predominantly internalized by endocytosis after a nonspecific interaction between the nanoparticles and cell membranes of A549 cells (human cell line derived from the respiratory epithelium). Another study also showed that labeling of CS-NPs with Texas Red complexes demonstrated the entry of nanoparticles into plasmids by endocytosis [132]. In antibacterial applications, a CS-NP has been mostly employed as a vehicle for the delivery of a range of antibiotics. The main advantage of using chitosan as a delivery agent is to circumvent the side effects of hepatic, nephrological, hematologic, or neurologic problems that are associated with systemic antibiotic administration [130, 133, 134]. Functionalization of chitosan micro-/nanoparticles with various tissue-specific molecules allows targeted activities [125, 135].

The versatility of chitosan to be modified into various sizes as well as chemical modifications could be advantageous for targeted antibacterial applications (killing bacteria specifically without damaging host cells). Furthermore, the nanoparticles with proper delivery strategies could be deposited into the anatomical complexities and dentinal tubules [84], where bacterial biofilms are known to persist subsequent to conventional disinfection strategies. Another salient property of chitosan is the biocompatibility and biodegradability that is of high priority for in vivo applications. Functionalization of CS with photosensitizers could be used to specifically target bacteria over human cells. In a photosensitizer-functionalized CS-NP, singlet oxygen released by the photoactivation of the photosensitizer potentiated the antibacterial activity of CS-NPs [135, 136]. Such bioactive functionalized nanoparticles could perform multiple functions, thereby overcoming issues related to antibiotics and bacterial resistance.

6.5.6 Nanoparticle Modification for Antimicrobial Photodynamic Therapy

Nanoparticle-based photosensitizers have been considered to potentiate the antimicrobial efficacy of photodynamic therapy (PDT) [137, 138]. Functionalized nanoparticles with photosensitizer molecules offer unique physicochemical properties such as ultrasmall sizes, large surface area/mass ratio, and increased physical/chemical reactivity. By using bioactive nanoparticles for functionalization, the favorable functions and physical/biological properties of the bioactive molecule can be utilized.

Certain metallic nanomaterials such as TiO_2, ZnO, and fullerenes as well as their derivatives showed the ability to generate singlet oxygen and induced bacterial elimination [139]. On the other hand, different strategies to combine nanoparticles and PSs have been suggested [67, 137]. As mentioned in the review by Kishen [67], combination of nanoparticles with photosensitizers could be achieved by (i) photosensitizers supplemented with nanoparticles, (ii) photosensitizers encapsulated within nanoparticles, (iii) photosensitizers bound or loaded to nanoparticles, and (iv) nanoparticles themselves serving as photosensitizers [67]. The combinations of nanoparticles with photosensitizers have been found to enhance antimicrobial PDT [137, 140]. Several factors could be attributed to this improved antimicrobial efficacy:

1. High concentration of photosensitizers per mass with resultant production of ROS
2. Reduced efflux of photosensitizers from the target cell, thereby decreasing the possibility of drug resistance
3. Possibility of targeting the bacteria due to greater interaction associated with the surface charge
4. Greater stability of photosensitizers after conjugation
5. Reduced physical quenching effect
6. Controlled release of ROS following photoactivation is possible [67, 137, 140]

The photosensitizer-supplemented nanoparticles such as methylene blue–loaded poly(lactic-co-glycolic) acid (PLGA) nanoparticles have been tested in vitro on *E. faecalis* biofilm and human dental plaque bacteria in combination with PDT [141]. The cationic methylene blue–loaded PLGA nanoparticles exhibited significantly higher bacterial phototoxicity in both planktonic and biofilm phases. These nanoparticles envelope the bacterial cells with higher concentration on the bacterial cell walls at all tested time points. It was concluded that cationic methylene blue–loaded PLGA nanoparticles have the potential to be used as carriers of photosensitizer PDT within root canals. Similarly, photosensitizer-bound polystyrene beads with Rose Bengal as a photosensitizer were used by Bezman et al. [142]. This modified photosensitizer was also postulated to bind to the cell membrane of bacteria with greater affinity as compared to the unmodified ones. The close association of photosensitizers following activation with light would favor improved bacterial elimination with ROS. The binding of Rose Bengal with silica nanoparticles, however, resulted in slower yield of ROS as compared to the free Rose Bengal [140]. The slower decay has been suggested to be advantageous in certain applications when light/dose fraction and/or deeper tissue or bacterial penetration is required [140].

Although photosensitizers have been conjugated with different readily available synthetic polymers and liposomes, these constructs possess limited biocompatibility when applied in vivo [81, 143]. The application of naturally occurring biopolymers such as chitosan may circumvent such issues of biocompatibility [143, 144]. This hydrophilic biopolymer with a large number of free hydroxyl and amino groups has been used for numerous chemical modifications and grafting [79, 81–83]. The biopolymer is wettable, which favors intimate contact between photosensitizers and the aqueous suspension of microorganisms. Shrestha et al. [135] used photosensitizers to functionalize CS-NPs that possessed the properties of both chitosan and Rose Bengal. This functionalized biopolymeric CSRBnp interacted greatly with bacteria and biofilms, leading to membrane damage and leakage of cellular constituents (Fig. 6.10). The increased uptake into biofilms of CSRBnp as compared to the photosensitizer alone

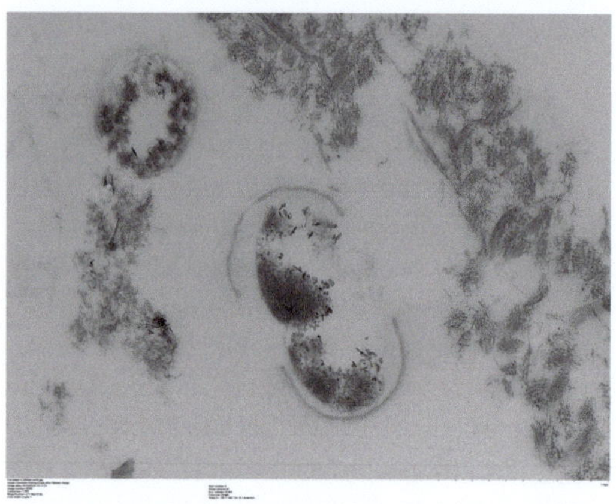

Fig. 6.10 Transmission electron microscopy image of *Enterococcus faecalis* bacteria and CSRBnp. The nanoparticles treatment resulted in disruption of bacterial membrane (★) and cell death. CSRBnp (ç) could be found inside the cells (Courtesy of Anil Kishen)

resulted in significant elimination of biofilm bacteria and disruption of structure following photoactivation. The functionalized CSRBnp demonstrated significantly lower cytotoxic properties toward fibroblast cells. One of the key issues of neutralization of antibacterial efficacy in presence of biological fluids and tissue was also overcome. The CSRBnp was found to maintain the antibacterial efficacy in the presence of bovine serum albumin, which is a strong inhibitor [145]. The higher affinity of cationic chitosan nanoparticles to bacterial cell surfaces and singlet oxygen release after photoactivation of RB provided a synergistic antibacterial mechanism for CSRBnp. Furthermore, the CSRBnp could be used for the treatment of infected dentin as it showed collagen cross-linking ability that could prevent degradation of dentin matrix. The photosensitizer-functionalized chitosan nanoparticles may serve as a single-step treatment of infected root dentin by combining the properties of chitosan and that of photosensitizers to eliminate bacterial biofilms.

6.6   Concluding Remarks

- For effective therapeutic efficacy with antibacterial nanoparticles, it is vital to tailor nanoparticles appropriately keeping in mind (1) the tissue-related factors at the location of infection and (2) the delivery method to be used to apply the nanoparticles in the tissue. Designing nanoparticles that can address all prerequisites is ideal for the management of infected tissue.

- Antibacterial nanoparticle-based treatment have the potential to improve antibacterial/antibiofilm efficacy. They have distinct advantages when applied in dentistry/endodontics.

- Tailoring of nanoparticles via functionalization could chemically modify nanoparticles, providing the opportunity to deliver therapeutic agents to the site of infection to selectively interact with (and penetrate) the biofilm and bacteria. Targeted antibacterial efficacy, provides an ideal antibacterial entity for treating infections in proximity to host tissues.

- Nanoparticle functionalization can result in higher drug efficacy as more active bioactive molecules could be loaded onto one nanoparticle.

- Newer multifunctional nanoparticles and their modifications should be developed based on the clinical requirements in collaboration with engineers, clinicians, and biologists.

- The whole concept of nanotechnology in health care should be accepted with positive zeal and caution for future development.

References

1. Taubes G. The bacteria fight back. Science. 2008;321:356–61.
2. del Pozo JL, Patel R. The challenge of treating biofilm-associated bacterial infections. Clin Pharmacol Ther. 2007;82:204–9.
3. Vakulenko SB, Mobashery S. Versatility of aminoglycosides and prospects for their future. Clin Microbiol Rev. 2003;16:430–50.
4. Cunha BA. Antibiotic resistance. Control strategies. Crit Care Clin. 1998;14:309–27.
5. Lewis K. Riddle of biofilm resistance. Antimicrob Agents Chemother. 2001;45:999–1007.
6. Finegold SM. Intestinal microbial changes and disease as a result of antimicrobial use. Pediatr Infect Dis. 1986;5:S88–90.

7. Prevention CfDCa. Antibiotic reistance threats in the United States. 2013.

8. Costerton JW, Lewandowski Z, DeBeer D, Caldwell D, Korber D, James G. Biofilms, the customized microniche. J Bacteriol. 1994;176:2137–42.

9. Boucher HW, Talbot GH, Benjamin Jr DK, Bradley J, Guidos RJ, Jones RN, et al. 10 x '20 Progress–development of new drugs active against gram-negative bacilli: an update from the Infectious Diseases Society of America. Clin Infect Dis. 2013;56:1685–94.

10. Cushing BL, Kolesnichenko VL, O'Connor CJ. Recent advances in the liquid-phase syntheses of inorganic nanoparticles. Chem Rev. 2004;104:3893–946.

11. Theodore LK, Robert G. Nanotechnology: environmental Implications and Solutions. Hoboken: Wiley; 2005.

12. Salata O. Applications of nanoparticles in biology and medicine. J Nanobiotechnol. 2004;2:3.

13. Stohs SJ, Bagchi D. Oxidative mechanisms in the toxicity of metal-ions. Free Radical Bio Med. 1995;18:321–36.

14. Yoon KY, Byeon JH, Park JH, Hwang J. Susceptibility constants of Escherichia coli and Bacillus subtilis to silver and copper nanoparticles. Sci Total Environ. 2007;373:572–5.

15. Sawai J. Quantitative evaluation of antibacterial activities of metallic oxide powders (ZnO, MgO and CaO) by conductimetric assay. J Microbiol Methods. 2003;54:177–82.

16. Yamamoto O. Influence of particle size on the antibacterial activity of zinc oxide. Int J Inorg Mater. 2001;3:643–6.

17. Reddy KM, Feris K, Bell J, Wingett DG, Hanley C, Punnoose A. Selective toxicity of zinc oxide nanoparticles to prokaryotic and eukaryotic systems. Appl Phys Lett. 2007;90.

18. Sawai J, Shoji S, Igarashi H, Hashimoto A, Kokugan T, Shimizu M, et al. Hydrogen peroxide as an antibacterial factor in zinc oxide powder slurry. J Ferment Bioeng. 1998;86:521–2.

19. Costerton JW, Stewart PS, Greenberg EP. Bacterial biofilms: a common cause of persistent infections. Science. 1999;284:1318–22.

20. Dufour D, Leung V, Levesque CM. Bacterial biofilm: structure, function, and antimicrobial resistance. Endod Topics. 2012;22:2–16.

21. Baumgartner C, Siqueira J, Sedgley C, Kishen A. Microbiology of endodontic disease. In: Ingle's endodontics. 6th ed. BC Decker Inc: Hamilton; 2008.

22. Costerton JW, Cheng KJ, Geesey GG, Ladd TI, Nickel JC, Dasgupta M, et al. Bacterial biofilms in nature and disease. Annu Rev Microbiol. 1987;41:435–64.

23. Kokare CR, Chakraborty S, Khopade AN, Mahadik KR. Biofilm: importance and applications. Indian J Biotechnol. 2009;8:159–68.

24. Flemming HC, Wingender J. The biofilm matrix. Nat Rev Microbiol. 2010;8:623–33.

25. Wells CL, Jechorek RP, Erlandsen SL. Evidence for the translocation of Enterococcus faecalis across the mouse intestinal tract. J Infect Dis. 1990;162:82–90.

26. Gilbert P, Das J, Foley I. Biofilm susceptibility to antimicrobials. Adv Dent Res. 1997;11:160–7.

27. Allison DG, Matthews MJ. Effect of polysaccharide interactions on antibiotic susceptibility of Pseudomonas aeruginosa. J Appl Bacteriol. 1992;73:484–8.

28. Nichols WW, Evans MJ, Slack MP, Walmsley HL. The penetration of antibiotics into aggregates of mucoid and non-mucoid Pseudomonas aeruginosa. J Gen Microbiol. 1989;135:1291–303.

29. Lewis K. Persister cells and the riddle of biofilm survival. Biochemistry (Mosc). 2005;70:267–74.

30. Stewart PS, Franklin MJ. Physiological heterogeneity in biofilms. Nat Rev Microbiol. 2008;6:199–210.

31. Davies DG, Parsek MR, Pearson JP, Iglewski BH, Costerton JW, Greenberg EP. The involvement of cell-to-cell signals in the development of a bacterial biofilm. Science. 1998;280:295–8.

32. Lee EW, Huda MN, Kuroda T, Mizushima T, Tsuchiya T. EfrAB, an ABC multidrug efflux pump in Enterococcus faecalis. Antimicrob Agents Chemother. 2003;47:3733–8.

33. Paulsen IT, Brown MH, Skurray RA. Proton-dependent multidrug efflux systems. Microbiol Rev. 1996;60:575–608.

34. Kakehashi S, Stanley HR, Fitzgerald RJ. The effects of surgical exposures of dental pulps in germ-free and conventional laboratory rats. Oral Surg Oral Med Oral Pathol. 1965;20:340–9.

35. Walton RE, Ardjmand K. Histological evaluation of the presence of bacteria in induced periapical lesions in monkeys. J Endod. 1992;18:216–27.

36. Marton IJ, Kiss C. Protective and destructive immune reactions in apical periodontitis. Oral Microbiol Immunol. 2000;15:139–50.

37. Siqueira Jr JF, Rocas IN. Community as the unit of pathogenicity: an emerging concept as to the microbial pathogenesis of apical periodontitis. Oral Surg Oral Med Oral Pathol Oral Radiol Endod. 2009;107:870–8.

38. Ramachandran Nair PN. Light and electron microscopic studies of root canal flora and periapical lesions. J Endod. 1987;13:29–39.

39. Ricucci D, Siqueira Jr JF. Biofilms and apical periodontitis: study of prevalence and association with clinical and histopathologic findings. J Endod. 2010;36:1277–88.

40. Carr GB, Schwartz RS, Schaudinn C, Gorur A, Costerton JW. Ultrastructural examination of failed molar retreatment with secondary apical periodontitis: an examination of endodontic biofilms in an endodontic retreatment failure. J Endod. 2009;35:1303–9.

41. Ricucci D, Siqueira Jr JF, Bate AL, Pitt Ford TR. Histologic investigation of root canal-treated teeth with apical periodontitis: a retrospective study

from twenty-four patients. J Endod. 2009;35: 493–502.

42. Chavez de Paz LE. Redefining the persistent infection in root canals: possible role of biofilm communities. J Endod. 2007;33:652–62.

43. Sundqvist G. Bacteriological studies of necrotic dental pulps [Dr Odont thesis]. Umea: University of Umea; 1976.

44. Fabricius L, Dahlen G, Ohman AE, Moller AJ. Predominant indigenous oral bacteria isolated from infected root canals after varied times of closure. Scand J Dent Res. 1982;90:134–44.

45. Sundqvist G. Taxonomy, ecology, and pathogenicity of the root canal flora. Oral Surg Oral Med Oral Pathol. 1994;78:522–30.

46. Sakamoto M, Rocas IN, Siqueira Jr JF, Benno Y. Molecular analysis of bacteria in asymptomatic and symptomatic endodontic infections. Oral Microbiol Immunol. 2006;21:112–22.

47. Fabricius L, Dahlen G, Sundqvist G, Happonen RP, Moller AJ. Influence of residual bacteria on periapical tissue healing after chemomechanical treatment and root filling of experimentally infected monkey teeth. Eur J Oral Sci. 2006;114:278–85.

48. Waltimo T, Trope M, Haapasalo M, Orstavik D. Clinical efficacy of treatment procedures in endodontic infection control and one year follow-up of periapical healing. J Endod. 2005;31:863–6.

49. Su L, Gao Y, Yu C, Wang H, Yu Q. Surgical endodontic treatment of refractory periapical periodontitis with extraradicular biofilm. Oral Surg Oral Med Oral Pathol Oral Radiol Endod. 2010;110:e40–4.

50. Siqueira Jr JF. Aetiology of root canal treatment failure: why well-treated teeth can fail. Int Endod J. 2001;34:1–10.

51. Sundqvist G, Figdor D, Persson S, Sjogren U. Microbiologic analysis of teeth with failed endodontic treatment and the outcome of conservative re-treatment. Oral Surg Oral Med Oral Pathol Oral Radiol Endod. 1998;85:86–93.

52. Nair PN, Henry S, Cano V, Vera J. Microbial status of apical root canal system of human mandibular first molars with primary apical periodontitis after "one-visit" endodontic treatment. Oral Surg Oral Med Oral Pathol Oral Radiol Endod. 2005;99:231–52.

53. Bystrom A, Claesson R, Sundqvist G. The antibacterial effect of camphorated paramonochlorophenol, camphorated phenol and calcium hydroxide in the treatment of infected root canals. Endod Dent Traumatol. 1985;1:170–5.

54. Bystrom A, Sundqvist G. Bacteriologic evaluation of the effect of 0.5 percent sodium hypochlorite in endodontic therapy. Oral Surg Oral Med Oral Pathol. 1983;55:307–12.

55. Gilbert GH, Tilashalski KR, Litaker MS, McNeal SF, Boykin MJ, Kessler AW. Outcomes of root canal treatment in Dental Practice-Based Research Network practices. Gen Dent. 2010;58:28–36.

56. Eriksen H, et al. Epidemiology of apical periodontitis. I. In: Ørstavik D, PittFord T, editors. Essential endodontology. Prevention and treatment of apical periodontitis. London: Blackwell Science; 1998. p. 179–91.

57. Lumley PJ, Lucarotti PS, Burke FJ. Ten-year outcome of root fillings in the General Dental Services in England and Wales. Int Endod J. 2008;41: 577–85.

58. Ng YL, Mann V, Gulabivala K. Outcome of secondary root canal treatment: a systematic review of the literature. Int Endod J. 2008;41:1026–46.

59. Ng YL, Mann V, Rahbaran S, Lewsey J, Gulabivala K. Outcome of primary root canal treatment: systematic review of the literature – part 1. Effects of study characteristics on probability of success. Int Endod J. 2007;40:921–39.

60. Abbott PV. Endodontics – current and future. J Conserv Dent. 2012;15:202–5.

61. Friedman S, Abitbol S, Lawrence HP. Treatment outcome in endodontics: the Toronto Study. Phase 1: initial treatment. J Endod. 2003;29:787–93.

62. Wang N, Knight K, Dao T, Friedman S. Treatment outcome in endodontics-the Toronto Study. Phases I and II: apical surgery. J Endod. 2004;30:751–61.

63. Farzaneh M, Abitbol S, Lawrence HP, Friedman S. Treatment outcome in endodontics-the Toronto Study. Phase II: initial treatment. J Endod. 2004;30: 302–9.

64. de Chevigny C, Dao TT, Basrani BR, Marquis V, Farzaneh M, Abitbol S, et al. Treatment outcome in endodontics: the Toronto study–phase 4: initial treatment. J Endod. 2008;34:258–63.

65. de Chevigny C, Dao TT, Basrani BR, Marquis V, Farzaneh M, Abitbol S, et al. Treatment outcome in endodontics: the Toronto study–phases 3 and 4: orthograde retreatment. J Endod. 2008;34:131–7.

66. Barone C, Dao TT, Basrani BB, Wang N, Friedman S. Treatment outcome in endodontics: the Toronto study–phases 3, 4, and 5: apical surgery. J Endod. 2010;36:28–35.

67. Kishen A. Advanced therapeutic options for endodontic biofilms. Endod Topics. 2010;22:99–123.

68. Habelitz S, Balooch M, Marshall SJ, Balooch G, Marshall Jr GW. In situ atomic force microscopy of partially demineralized human dentin collagen fibrils. J Struct Biol. 2002;138:227–36.

69. Prince AS. Biofilms, antimicrobial resistance, and airway infection. N Engl J Med. 2002;347:1110–1.

70. Zehnder M. Root canal irrigants. J Endod. 2006;32: 389–98.

71. Bystrom A, Sundqvist G. The antibacterial action of sodium hypochlorite and EDTA in 60 cases of endodontic therapy. Int Endod J. 1985;18:35–40.

72. Shuping GB, Orstavik D, Sigurdsson A, Trope M. Reduction of intracanal bacteria using nickel-titanium rotary instrumentation and various medications. J Endod. 2000;26:751–5.

73. Peters OA, Schonenberger K, Laib A. Effects of four Ni-Ti preparation techniques on root canal geometry assessed by micro computed tomography. Int Endod J. 2001;34:221–30.

74. Paque F, Sirtes G. Apical sealing ability of Resilon/Epiphany versus gutta-percha/AH Plus: immediate and 16-months leakage. Int Endod J. 2007;40:722–9.

75. Manzur A, Gonzalez AM, Pozos A, Silva-Herzog D, Friedman S. Bacterial quantification in teeth with apical periodontitis related to instrumentation and different intracanal medications: a randomized clinical trial. J Endod. 2007;33:114–8.

76. Muzzarelli RA, Isolati A, Ferrero A. Chitosan membranes. Ion Exch Membr. 1974;1:193–6.

77. Agnihotri SA, Mallikarjuna NN, Aminabhavi TM. Recent advances on chitosan-based micro- and nanoparticles in drug delivery. J Control Release. 2004;100:5–28.

78. Machida Y, Nagai T, Abe M, Sannan T. Use of chitosan and hydroxypropylchitosan in drug formulations to effect sustained release. Drug Des Deliv. 1986;1:119–30.

79. Kumar MN, Muzzarelli RA, Muzzarelli C, Sashiwa H, Domb AJ. Chitosan chemistry and pharmaceutical perspectives. Chem Rev. 2004;104:6017–84.

80. Tan W, Krishnaraj R, Desai TA. Evaluation of nanostructured composite collagen–chitosan matrices for tissue engineering. Tissue Eng. 2001;7:203–10.

81. Bonnett R, Krysteva MA, Lalov IG, Artarsky SV. Water disinfection using photosensitizers immobilized on chitosan. Water Res. 2006;40:1269–75.

82. Wang XH, Li DP, Wang WJ, Feng QL, Cui FZ, Xu YX, et al. Crosslinked collagen/chitosan matrix for artificial livers. Biomaterials. 2003;24:3213–20.

83. Everaerts F, Gillissen M, Torrianni M, Zilla P, Human P, Hendriks M, et al. Reduction of calcification of carbodiimide-processed heart valve tissue by prior blocking of amine groups with monoaldehydes. J Heart Valve Dis. 2006;15:269–77.

84. Shrestha A, Fong SW, Khoo BC, Kishen A. Delivery of antibacterial nanoparticles into dentinal tubules using high-intensity focused ultrasound. J Endod. 2009;35:1028–33.

85. Shrestha A, Shi Z, Neoh KG, Kishen A. Nanoparticulates for antibiofilm treatment and effect of aging on its antibacterial activity. J Endod. 2010;36:1030–5.

86. Muzzarelli R, Tarsi R, Filippini O, Giovanetti E, Biagini G, Varaldo PE. Antimicrobial properties of N-carboxybutyl chitosan. Antimicrob Agents Chemother. 1990;34:2019–23.

87. Kishen A, Shi Z, Shrestha A, Neoh KG. An investigation on the antibacterial and antibiofilm efficacy of cationic nanoparticulates for root canal disinfection. J Endod. 2008;34:1515–20.

88. Calvo P, Remunan Lopez C, Vila-Jato JL, Alonso MJ. Novel hydrophilic chitosan–polyethylene oxide nanoparticles as protein carriers. J Appl Polym Sci. 1997;63:125–32.

89. Rabea EI, Badawy ME, Stevens CV, Smagghe G, Steurbaut W. Chitosan as antimicrobial agent: applications and mode of action. Biomacromolecules. 2003;4:1457–65.

90. Qi L, Xu Z, Jiang X, Hu C, Zou X. Preparation and antibacterial activity of chitosan nanoparticles. Carbohydr Res. 2004;339:2693–700.

91. No HK, Park NY, Lee SH, Meyers SP. Antibacterial activity of chitosans and chitosan oligomers with different molecular weights. Int J Food Microbiol. 2002;74:65–72.

92. Liu XF, Guan YL, Yang DZ, Li Z, Yao KD. Antibacterial action of chitosan and carboxymethylated chitosan. J Appl Polym Sci. 2001;79:1324–35.

93. Zehnder M, Luder HU, Schatzle M, Kerosuo E, Waltimo T. A comparative study on the disinfection potentials of bioactive glass S53P4 and calcium hydroxide in contra-lateral human premolars ex vivo. Int Endod J. 2006;39:952–8.

94. Stoor P, Soderling E, Salonen JI. Antibacterial effects of a bioactive glass paste on oral microorganisms. Acta Odontol Scand. 1998;56:161–5.

95. Zehnder M, Baumgartner G, Marquardt K, Paque F. Prevention of bacterial leakage through instrumented root canals by bioactive glass S53P4 and calcium hydroxide suspensions in vitro. Oral Surg Oral Med Oral Pathol Oral Radiol Endod. 2007;103:423–8.

96. Zehnder M, Soderling E, Salonen J, Waltimo T. Preliminary evaluation of bioactive glass S53P4 as an endodontic medication in vitro. J Endodont. 2004;30:220–4.

97. Waltimo T, Mohn D, Paque F, Brunner TJ, Stark WJ, Imfeld T, et al. Fine-tuning of bioactive glass for root canal disinfection. J Dent Res. 2009;88:235–8.

98. Waltimo T, Brunner TJ, Vollenweider M, Stark WJ, Zehnder M. Antimicrobial effect of nanometric bioactive glass 45S5. J Dent Res. 2007;86:754–7.

99. Mortazavi V, Nahrkhalaji MM, Fathi MH, Mousavi SB, Esfahani BN. Antibacterial effects of sol-gel-derived bioactive glass nanoparticle on aerobic bacteria. J Biomed Mater Res A. 2010;94:160–8.

100. Fong J, Wood F. Nanocrystalline silver dressings in wound management: a review. Int J Nanomedicine. 2006;1:441–9.

101. Chernousova S, Epple M. Silver as antibacterial agent: ion, nanoparticle, and metal. Angewandte Chemie. 2013;52:1636–53.

102. Garcia-Contreras R, Argueta-Figueroa L, Mejia-Rubalcava C, Jimenez-Martinez R, Cuevas-Guajardo S, Sanchez-Reyna PA, et al. Perspectives for the use of silver nanoparticles in dental practice. Int Dent J. 2011;61:297–301.

103. Lansdown AB. Silver in health care: antimicrobial effects and safety in use. Curr Probl Dermatol. 2006;33:17–34.

104. Sotiriou GA, Pratsinis SE. Antibacterial activity of nanosilver ions and particles. Environ Sci Technol. 2010;44:5649–54.

105. Sondi I, Salopek-Sondi B. Silver nanoparticles as antimicrobial agent: a case study on E. coli as a model for Gram-negative bacteria. J Colloid Interface Sci. 2004;275:177–82.

106. Chaloupka K, Malam Y, Seifalian AM. Nanosilver as a new generation of nanoproduct in biomedical applications. Trends Biotechnol. 2010;28:580–8.

107. Melo MA, Guedes SF, Xu HH, Rodrigues LK. Nanotechnology-based restorative materials for dental caries management. Trends Biotechnol. 2013;31:459–67.

108. Hiraishi N, Yiu CK, King NM, Tagami J, Tay FR. Antimicrobial efficacy of 3.8 % silver diamine fluoride and its effect on root dentin. J Endod. 2010;36:1026–9.

109. Wu D, Fan W, Kishen A, Gutmann JL, Fan B. Evaluation of the antibacterial efficacy of silver nanoparticles against Enterococcus faecalis biofilm. J Endod. 2014;40:285–90.

110. Gomes-Filho JE, Silva FO, Watanabe S, Cintra LT, Tendoro KV, Dalto LG, et al. Tissue reaction to silver nanoparticles dispersion as an alternative irrigating solution. J Endod. 2010;36:1698–702.

111. Senges C, Wrbas KT, Altenburger M, Follo M, Spitzmuller B, Wittmer A, et al. Bacterial and Candida albicans adhesion on different root canal filling materials and sealers. J Endod. 2011;37:1247–52.

112. George S, Basrani B, Kishen A. Possibilities of gutta-percha-centered infection in endodontically treated teeth: an in vitro study. J Endod. 2010;36:1241–4.

113. Kayaoglu G, Erten H, Alacam T, Orstavik D. Short-term antibacterial activity of root canal sealers towards Enterococcus faecalis. Int Endod J. 2005;38:483–8.

114. Siqueira Jr JF, Favieri A, Gahyva SM, Moraes SR, Lima KC, Lopes HP. Antimicrobial activity and flow rate of newer and established root canal sealers. J Endod. 2000;26:274–7.

115. Orstavik D. Antibacterial properties of root canal sealers, cements and pastes. Int Endod J. 1981;14:125–33.

116. DaSilva L, Finer Y, Friedman S, Basrani B, Kishen A. Biofilm formation within the interface of bovine root dentin treated with conjugated chitosan and sealer containing chitosan nanoparticles. J Endod. 2013;39:249–53.

117. Barros J, Silva MG, Rodrigues MA, Alves FR, Lopes MA, Pina-Vaz I, et al. Antibacterial, physicochemical and mechanical properties of endodontic sealers containing quaternary ammonium polyethylenimine nanoparticles. Int Endod J. 2014;47:725–34.

118. Abramovitz I, Beyth N, Paz Y, Weiss EI, Matalon S. Antibacterial temporary restorative materials incorporating polyethyleneimine nanoparticles. Quintessence Int. 2013;44:209–16.

119. Kesler Shvero D, Abramovitz I, Zaltsman N, Perez Davidi M, Weiss EI, Beyth N. Towards antibacterial endodontic sealers using quaternary ammonium nanoparticles. Int Endod J. 2013;46:747–54.

120. Beyth N, Kesler Shvero D, Zaltsman N, Houri-Haddad Y, Abramovitz I, Davidi MP, et al. Rapid kill-novel endodontic sealer and Enterococcus faecalis. PLoS One. 2013;8:e78586.

121. Gao B, Zhang X, Zhu Y. Studies on the preparation and antibacterial properties of quaternized polyethyleneimine. J Biomater Sci Polym Ed. 2007;18:531–44.

122. Mohn D, Bruhin C, Luechinger NA, Stark WJ, Imfeld T, Zehnder M. Composites made of flame-sprayed bioactive glass 45S5 and polymers: bioactivity and immediate sealing properties. Int Endod J. 2010;43:1037–46.

123. Ravindran A, Chandran P, Khan SS. Biofunctionalized silver nanoparticles: advances and prospects. Colloids Surf B Biointerfaces. 2013;105:342–52.

124. Sanvicens N, Marco MP. Multifunctional nanoparticles–properties and prospects for their use in human medicine. Trends Biotechnol. 2008;26:425–33.

125. Veerapandian M, Yun K. Functionalization of biomolecules on nanoparticles: specialized for antibacterial applications. Appl Microbiol Biotechnol. 2011;90:1655–67.

126. Liu L, Xu K, Wang H, Tan PK, Fan W, Venkatraman SS, et al. Self-assembled cationic peptide nanoparticles as an efficient antimicrobial agent. Nat Nanotechnol. 2009;4:457–63.

127. Vila A, Sanchez A, Tobio M, Calvo P, Alonso MJ. Design of biodegradable particles for protein delivery. J Control Release. 2002;78:15–24.

128. Fonte P, Andrade JC, Seabra V, Sarmento B. Chitosan-based nanoparticles as delivery systems of therapeutic proteins. Methods Mol Biol. 2012;899:471–87.

129. Fonte P, Andrade F, Araujo F, Andrade C, Neves J, Sarmento B. Chitosan-coated solid lipid nanoparticles for insulin delivery. Methods Enzymol. 2012;508:295–314.

130. Duceppe N, Tabrizian M. Advances in using chitosan-based nanoparticles for in vitro and in vivo drug and gene delivery. Expert Opin Drug Deliv. 2010;7:1191–207.

131. Huang M, Ma Z, Khor E, Lim LY. Uptake of FITC-chitosan nanoparticles by A549 cells. Pharm Res. 2002;19:1488–94.

132. Ishii T, Okahata Y, Sato T. Mechanism of cell transfection with plasmid/chitosan complexes. Biochim Biophys Acta. 2001;1514:51–64.

133. Motwani SK, Chopra S, Talegaonkar S, Kohli K, Ahmad FJ, Khar RK. Chitosan-sodium alginate nanoparticles as submicroscopic reservoirs for ocular delivery: formulation, optimisation and in vitro characterisation. Eur J Pharm Biopharm. 2008;68:513–25.

134. Jain D, Banerjee R. Comparison of ciprofloxacin hydrochloride-loaded protein, lipid, and chitosan nanoparticles for drug delivery. J Biomed Mater Res B Appl Biomater. 2008;86:105–12.

135. Shrestha A, Hamblin MR, Kishen A. Photoactivated rose bengal functionalized chitosan nanoparticles produce antibacterial/biofilm activity and stabilize dentin-collagen. Nanomedicine. 2014;10:491–501.

136. Shrestha A, Kishen A. Antibiofilm efficacy of photosensitizer-functionalized bioactive nanoparticles on multispecies biofilm. J Endod. 2014;40(10):1604–10.
137. Perni S, Prokopovich P, Pratten J, Parkin IP, Wilson M. Nanoparticles: their potential use in antibacterial photodynamic therapy. Photochem Photobiol Sci. 2011;10:712–20.
138. Wilson RF. Nanotechnology: the challenge of regulating known unknowns. J Law Med Ethics. 2006;34:704–13.
139. Li Q, Mahendra S, Lyon DY, Brunet L, Liga MV, Li D, et al. Antimicrobial nanomaterials for water disinfection and microbial control: potential applications and implications. Water Res. 2008;42:4591–602.
140. Guo Y, Rogelj S, Zhang P. Rose Bengal-decorated silica nanoparticles as photosensitizers for inactivation of gram-positive bacteria. Nanotechnology. 2010;21:065102.
141. Pagonis TC, Chen J, Fontana CR, Devalapally H, Ruggiero K, Song X, et al. Nanoparticle-based endodontic antimicrobial photodynamic therapy. J Endod. 2010;36:322–8.
142. Bezman SA, Burtis PA, Izod TP, Thayer MA. Photodynamic inactivation of E. coli by rose bengal immobilized on polystyrene beads. Photochem Photobiol. 1978;28:325–9.
143. Moczek L, Nowakowska M. Novel water-soluble photosensitizers from chitosan. Biomacromolecules. 2007;8:433–8.
144. Nowakowska M, Moczek L, Szczubialka K. Photoactive modified chitosan. Biomacromolecules. 2008;9:1631–6.
145. Shrestha A, Kishen A. Photodynamic therapy for inactivating endodontic bacterial biofilms and effect of tissue inhibitors on antibacterial efficacy. Proc Spie. 2013;8566.

Nanoparticles for Dentin Tissue Stabilization

7

Anil Kishen and Annie Shrestha

Abstract

Nanotechnology has been applied to manage previously infected dentin. These treatment procedures are aimed for non-invasive elimination of residual bacterial biofilms, improve the resistance of dentin to enzymatic (host/bacterial-mediated) degradation and improve the mechanical integrity of dentin matrix. This chapter discusses the issues associated with previously infected dentin, strategies used to strengthen dentin tissue matrix and current progress/potential applications of various functional nanoparticles for the physical, chemical and mechanical stabilization of dentin. Nanoparticles of various materials (polymers, metals), size and shape as well as modifications are available depending on the requirement. Nanoparticles could be tailored to perform specific or multiple functions based on the tissue-specific requirements. Carefully tailored nanoparticles with sound scientific basis on the mechanism of action, safety and dose will find potential advantage in minimally invasive/non-invasive dentin tissue stabilization.

7.1 Introduction

The root-filled teeth are suggested to be more prone to vertical root fracture (VRF) as compared to their vital counterparts with intact crowns and no/minimal restoration [1–4]. Gathering evidence from various clinical surveys, the prevalence of VRF has not been well established till date. The prevalence of VRF in endodontically treated teeth depends on the method of assessment as well as the clinician's ability to detect VRF [5–11]. The studies that diagnosed VRF using radiographs [6, 9] reported a lower percentage of prevalence (2–5 %) in contrast to the studies that used extracted root-filled teeth to assess VRF (11–20 %) [10, 11]. VRF is attributed to the compromised dentin structure and other multitude of risk factors (pathologic, iatrogenic or physiologic) that initiated or propagated cracks in restored root-filled teeth [12].

A. Kishen, PhD, MDS, BDS
A. Shrestha, PhD, MSc, BDS (✉)
Department of Endodontics, Faculty of Dentistry,
University of Toronto, 124 Edward Street, Room 348,
Toronto, ON M5G1H4, Canada
e-mail: anil.kishen@utoronto.ca;
annie.shrestha@utoronto.ca

A. Kishen (ed.), *Nanotechnology in Endodontics: Current and Potential Clinical Applications*,
DOI 10.1007/978-3-319-13575-5_7, © Springer International Publishing Switzerland 2015

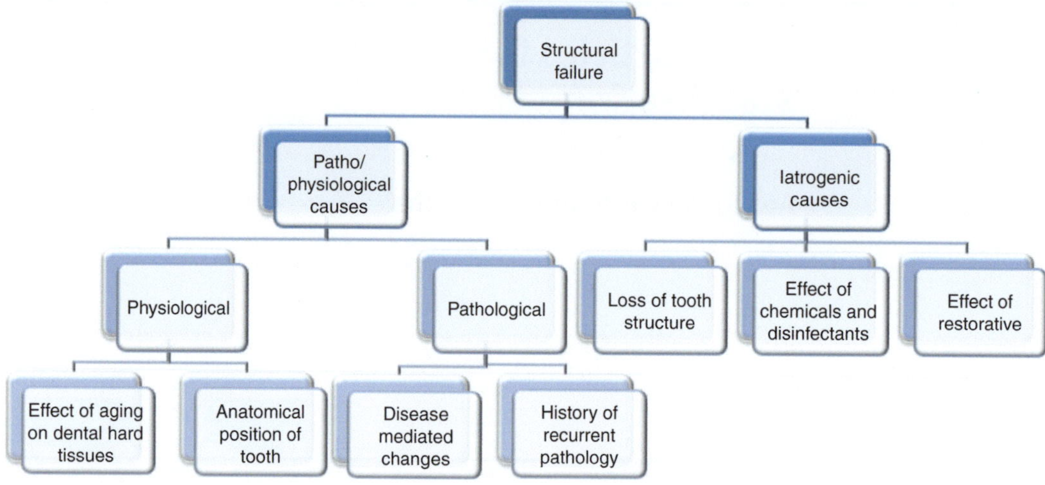

Fig. 7.1 Causes of structural failures in root-filled teeth [6]

The contributory factors for VRF can be broadly classified into (1) non-iatrogenic or patho-physiological changes and (2) iatrogenic or treatment-associated changes (Fig. 7.1). Dentin degradation/resorption during a disease process [13, 14] and post-treatment degradation of dentin surface collagen by bacteria and MMPs [15, 16] are the non-iatrogenic damages. Iatrogenic damage to the dentin occurs due to the use of strong and caustic root canal irrigants and medicaments inside the root canals [17, 18]. The collagens on the surface of root canal walls are exposed due to demineralization during a disease process as well as due to the application of caustic chemicals during the root canal treatment [12, 19]. Bacterial proteases and human matrix metalloproteinases degrade collagen and result in large resorptive defects inside the root canals [13] or surface collagen degradation of root canal dentin [15].

Dentin is a biological composite material consisting of inorganic phase (50 vol%), organic phase (30 vol%) and water (20 vol%) [20]. The loss of dentin and other age-, bacterial- and chemical-mediated changes to dentin matrix would compromise the mechanical integrity of root dentin and demands attention. In this line, optimization of dentin characteristics such as ultra-structural integrity, mainly from a mechanical/chemical perspective, in order to improve the mechanical properties and resistance to enzymatic degradation would facilitate the successful management of previously infected dentin tissue. Currently novel nanotechnology-based treatment approaches that produce significant antibiofilm efficacy and at the same time enhance the ultrastructural integrity of dentin tissue are being developed for the management of infected dentin during root canal treatment and minimally invasive dental caries management. This chapter will present the biomaterial aspects of dentin, issues related to dentin degradation and potential treatment methods available to stabilize the dentin matrix. Various stabilization methods and their effect on specific dentin properties will also be covered.

7.2 Dentin

Dentin at a structural level exists as a fiber-reinforced composite (Fig. 7.2). The less mineralized intertubular dentin forms the matrix, and the highly mineralized peritubular dentin forms the fibre reinforcements. It is traversed by dentinal tubules that run continuously through the bulk of the dentin. Thus, dentin microstructure is basically a scaffold of organic fraction on which inorganic fraction is well distributed [20]. The hierarchical order of dentin aids in understanding its structural and mechanical properties [21].

The inorganic component in dentin is mainly carbonated nanocrystalline apatite minerals. These minerals are closely associated with the collagen

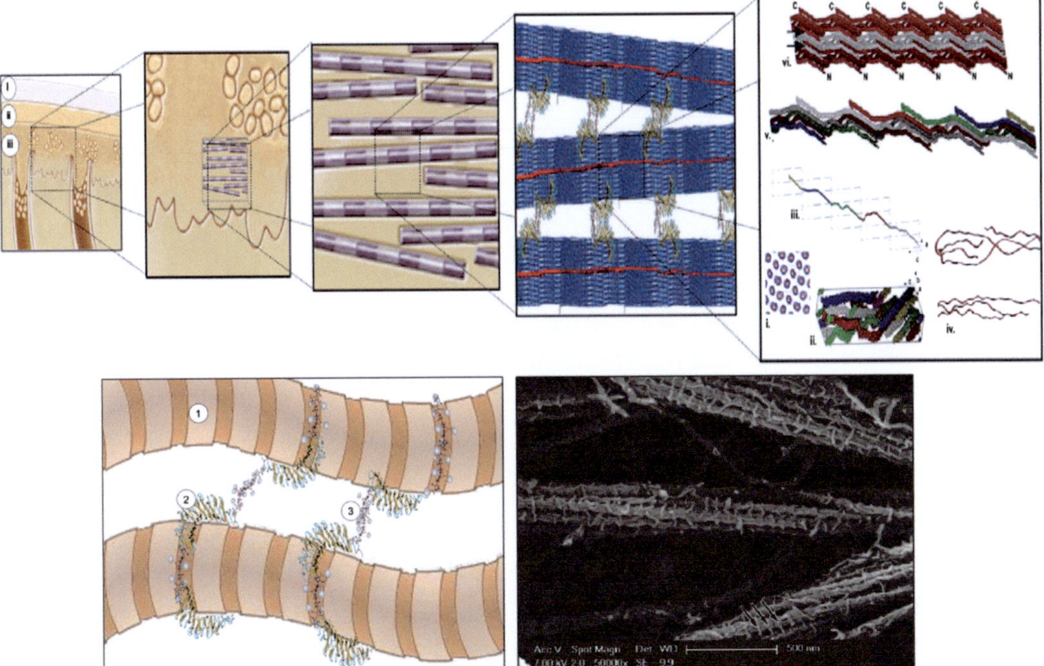

Fig. 7.2 Dentin presented in a hierarchical fashion. At structural level it exists as a fiber-reinforced composite. A not to scale schematic sketch of the interfibrillar supramolecular assemblies that interconnect collagen fibrils: (*1*) collagen fibril; (*2*) decorin protein core; (*3*) chondroitin 4-sulfate glycosaminoglycan. The known periodicity of these interfibrillar aggregates in register with the gap zones of collagen fibrils, present in most connective tissues, remains uncertain for mineralized tissues. (*C*) A high magnification image of a sample of acid-soluble collagen and decorin treated with cupromeronic blue, which reacts with glycosaminoglycans and demonstrates their assembly as interfibrillar co-aggregates (*arrows*) (Adapted from Bertassoni et al. [21]. With permission from Elsevier)

scaffold, either intrafibrillar or extrafibrillar [22]. The apatite crystals ($\approx$5 nm thick) are needle like near the pulp and plate like towards the dentino-enamel junction. Type I collagen is the major structural protein making up to 90 % of the organic fraction, and the rest are non-collagenous proteins [20]. Type I collagen, roughly 100 nm in diameter, exists as fibrils in dentin which are stabilized by endogenous covalent intermolecular cross-linking [23]. The collagen fibrils are randomly oriented in a plane perpendicular to the plane of dentin formation or axis of dentinal tubules [24]. In addition, the three-dimensional collagen network exists in an aqueous environment.

The 10 % of the organic phase consists of various phosphoproteins and other non-collagenous proteins. Proteoglycans and glycosaminoglycans (GAGs) are known to be closely associated with collagen fibrils via hydrogen bonds. The proteoglycans and the hydrophilic anionic GAG side chains interact with one another, forming interfibrillar bridges. These molecules absorb water and span the interfibrillar spaces, thereby regulating mechanical properties of the dentin collagen matrix [21]. The structural integrity of dentin provided by the inorganic and organic fraction is crucial to retain the function of a tooth (Fig. 7.3) [12]. In a root-filled tooth, other than the loss of dentin, changes in the physical properties of dentin matrix could also contribute to the increased propensity of fractures [12, 25].

The mechanical stability of biological composites depends on the optimum balance between toughness and stiffness [26]. The mechanical properties of dentin such as Young's modulus, strength, and fracture toughness are the result of the complex interactions of its constituents as well as the microstructural

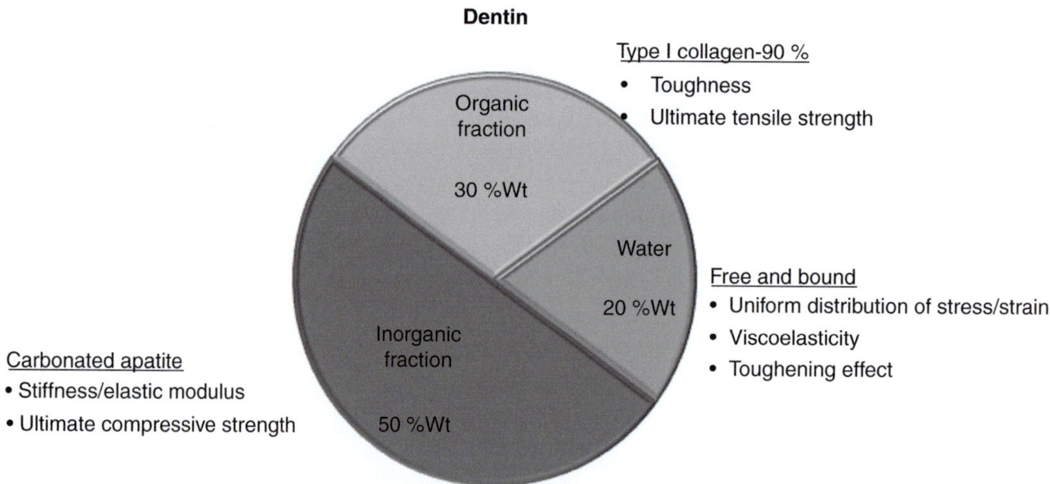

Fig. 7.3 The role of different constituents of dentin on its structural integrity

arrangement. Toughness is the total energy absorbed by a structure before it fractures (fracture resistance), whereas Young's modulus is the ratio of strength vs. strain within the elastic limit. The collagen fraction of dentin contributes to its toughness and ultimate tensile strength, while the mineral fraction contributes to its elastic modulus and compressive strength [12, 27]. The collagen fraction also acts as a cushion in between the inorganic elements of dentin [12, 28]. In bulk dentin, the alignment of the dentinal tubules also governs its mechanical properties [24].

The water content of dentin exists as a free or 'unbound' water and 'bound' water. The unbound water is found in dentinal tubules and other porosities in the dentin [29]. This water keeps the dentin matrix bathed in minerals such as calcium and phosphate. The bound water in dentin is associated with the inorganic apatite crystals and organic phase (collagen and non-collagenous proteins). This water forms a monolayer on the surface of hydroxyapatite via hydrogen bonds and van der Waals forces [30]. Bound water also forms an integral part to stabilize the triple helix of collagen molecules. Each tripeptide is known to be stabilized by two water molecules [31]. Loss of water in dentin reduces the elastic modulus of dentin from 23.9 to 20 GPa [28]. Partial dehydration at room temperature for 7 days would result in increase in stiffness and decrease in the toughness values significantly [29].

7.3 Physical and Mechanical Alterations in Dentin Matrix

7.3.1 Non-iatrogenic Causes

Degradation of dentin matrix is a common observation in endodontically treated teeth [15]. The disease processes/bacterial proteases degrade dentin collagen. It is evident with an increase in the years of clinical functioning [15]. Bacterial biofilms degraded the dentin surface collagen to various degrees depending on the growth environment [32, 33] (Fig. 7.4). Clinically, internal resorption of root dentin can occur in up to 75 % of teeth with post-treatment disease [34]. Bacterial collagenolytic enzymes and activation of host-derived matrix metalloproteinases MMPs are some of the possible reasons for resorption/degradation [35]. This dentin surface change has been linked to interfacial failure at the biomaterial–dentin interface of root-filled teeth [36].

The degradation of the exposed dentin collagen could be due to (a) extrinsic factors such as enzymatic hydrolysis by salivary leakage and/or (b) intrinsic factors such as bacterial and host-derived MMPs [15, 37–39]. These MMPs play

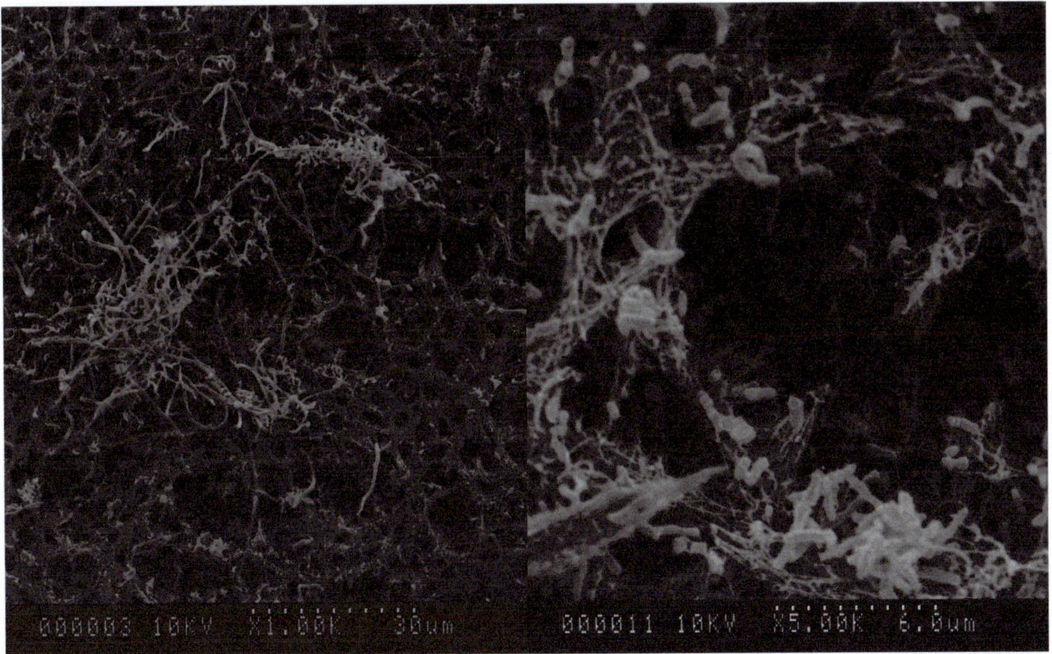

Fig. 7.4 Degradation of dentin by bacterial biofilms grown in vitro on dentin. Areas of demineralization as well as collagen degradation were evident on the surface

a vital role in degradation of exposed collagen fibrils created during dentin bonding procedures [40, 41]. Incompletely infiltrated collagen fibrils within the hybrid layers formed during dentin bonding were degraded by the host-derived proteases in the absence of bacteria [15, 42]. Commonly used acids such as EDTA and phosphoric acid are also capable of activating the host-derived protease activity [43]. Inhibitors of MMPs have been studied extensively to control the pathological degradation of ECM. Synthetic peptides, chemically modified tetracycline, bisphosphonates, natural compounds and chlorhexidine are considered as potential MMP inhibitors [43, 44].

7.3.2 Iatrogenic Causes

The iatrogenic factors usually complement the changes caused by the pathological process in dentin, which would further compromise the ultrastructural/mechanical integrity of teeth [10, 12, 13, 34, 45]. Commonly used root canal irrigants remove the inorganic phase and/or the

organic phase of the dentin [12, 17]. Removal of the organic and inorganic fractions of the dentin compromises the mechanical integrity of dentin. Overzealous application of chemicals/medicaments leads to erosion and microcracking of dentin [12, 15, 40]. EDTA and sodium hypochlorite (NaOCl) are commonly used for root canal debridement/disinfection. NaOCl is a deproteinating agent that results in the heterogeneous removal of organic substrate from dentin. Use of higher concentrations of NaOCl for longer duration could produce more deleterious changes to dentin matrix as compared to the application of lower concentrations of NaOCl for shorter duration due to their greater ability to interact with collagen matrix. This interaction would reduce the flexural strength of dentin [45]. Microhardness of dentin was significantly reduced when subjected to chemicals such as NaOCl, chlorhexidine, hydrogen peroxide and EDTA [46, 47].

The changes in dentin substrate following the application of common root canal irrigants resulted in almost 75 % decrease in the mechanical strength [17]. NaOCl resulted in the decrease of mechanical strength of root dentin up to 59 %

[48]. Similarly, the common intracanal medicament $Ca(OH)_2$ also reduced the mechanical strength by 32 % [48]. The caustic and strong alkalinity of these chemicals may denature the carboxylate and phosphate groups, which leads to the collapse of dentin structure. In addition to the chemical effects, routine instrumentation and root-filling procedures involves removal of bulk dentin, which may further jeopardize the mechanical integrity of root dentin [49]. Despite many attempts to improve the resistance to fracture of root-filled teeth with the help of a filling material, till date, there is no such treatment available that restores the mechanical integrity of the remaining dentin. Although the iatrogenic and non-iatrogenic changes occur in the inner aspect of root dentin, they could act as vulnerable areas of crack initiation/stress concentration, which can further increase the predilection of vertical root fractures in teeth [12, 50].

7.4 Dentin Tissue Stabilization

Tissue stabilization is the process of rendering the ultrastructure of a tissue more stable in order to provide or enhance its mechanical properties and resistance to chemical-mediated degradation. Dentinal collagen stabilization has been explored extensively to improve the resistance to degradation in hybrid layers, to improve the bond strength in case of adhesive restorations, to manage dentinal hypersensitivity and also to stabilize surface dentin of root canal walls. All these aspects have the same objective, which is to cross-link dentin collagen in order to improve its stability. Cross-linking the dentin collagens protects them from host-derived MMPs and bacterial proteases. Cross-linking of collagen could be achieved by several approaches, which can be broadly categorized as chemical or physical approaches (Table 7.1). The cross-linking process is meant to improve the mechanical and physical characteristics of collagenous tissues and scaffolds [51–53]. The following paragraphs will discuss various collagen cross-linking methods that have been used in biomaterials with relevance to dentistry.

7.4.1 Chemical Cross-Linking

Chemical cross-linking is one of the most widely used approaches. Aldehydes such as formaldehyde, glutaraldehyde, epoxy compounds, etc. have been used in fixing biological tissues including dentin collagen [52, 54]. Cross-linking of collagen using glutaraldehyde increases the tensile properties and stiffness of demineralized dentin [55, 56] and increases resistance to enzymatic degradation (collagenase) [57]. Dentin collagen cross-linking has also been reported as a treatment for hypersensitivity using various resins and glutaraldehyde [58]. Glutaraldehyde treatment decreases dentin sensitivity by several mechanisms: (1) Cross-linking of dentinal collagen. (2) The amino group–containing organic fraction in dentin reacts with GD, which results

Table 7.1 Various methods available to obtain crosslinking of collagen

Enzyme mediated	Chemical agents	Photoactivated	Others
Transglutaminase	1. Crosslinking of amino groups	1. Riboflavin + UVA [1–4]	1. Synthetic polymers
	Glutaraldehyde	2. Rose bengal + light (545 nm) [5–7]	Poly (vinyl alcohol)
	Glyceraldehyde	3. Tris(bipyridine) ruthenium(II) chloride + [8]	Poly (acrylic acid)
	Epoxy compounds		Polyethylene
	Genipin		Poly (vinyl pyrrolidone)
	2. Direct crosslinking of polypeptide chains		2. Hyaluronic acid
	Carbodiimides (EDC/NHS/MES)		3. Heparin
	3. Chlorhexidine		4. Chondroitin-6-sulfate

1-ethyl-3-(3-dimethylaminopropyl)carbodiimide (EDC), and N-hydroxysuccinimide (NHS)

in polymerization of HEMA. (3) Plasma proteins in dentinal fluid react with GD, which results in precipitation. However, one of the main disadvantages of glutaraldehyde is high cytotoxicity and calcification in the host tissue due to incomplete removal of toxic residues [59–61].

Use of various synthetic and biological chemical cross-linkers could address the issue of biocompatibility. Carbodiimides [57, 62, 63], genipin [56, 57], proanthocyanidin [56, 64, 65] and tannic acid [66] are few of the chemicals used for cross-linking collagen. Water-soluble carbodiimide such as l-ethyl-3-(3-dimethyl aminopropyl) carbodiimide hydrochloride (EDC) is commonly used to stabilize and strengthen collagen tissues and scaffolds [57, 63, 67, 68]. Cross-linking using EDC involves the activation of the carboxylic acid groups of glutamic or aspartic acid residues by EDC. EDC and N-hydroxysuccinimide (NHS) can link carboxylic acid and amino groups located within a very short distance of 1.0 nm from each other [69]. This poses a challenge when the functional groups located on adjacent collagen microfibrils are too far apart to be bridged by carbodiimides [67]. The increase in the use of such chemicals, however, results in a decrease in elasticity and toughness as well as biocompatibility to cells.

Proanthocyanidin, a grape seed extract (GSE), cross-linked collagen by covalent interaction, ionic interaction, hydrogen bonding interaction or hydrophobic interaction [70]. Due to these multiple interactions, GSE demonstrates better mechanical properties of cross-linked collagen as compared to the glutaraldehyde. Chemical cross-linking is a time-consuming process, which markedly limits its application in a clinical situation. Cross-linking agents, which are biocompatible or with low cytotoxicity that forms stable cross-linked collagen in a clinically feasible time period, are desired for dental applications.

7.4.2 Physical Cross-Linking

Physical cross-linking is achieved by use of heat, dehydrothermal treatment (DHT) and UV- and gamma- irradiation. The disadvantage of these methods is that they weaken the collagen due to thermal degradation [71] and collagen denaturation [72, 73]. These physical methods also consume hours to days for the cross-linking to take place. So far, these physical methods have not shown significant potential for dental applications [71].

7.4.3 Photodynamic Cross-Linking

Photodynamic therapy (PDT) involves the use of non-toxic dye or photosensitizer in combination with visible light, which in the presence of molecular oxygen leads to the production of cytotoxic oxygen radicals such as singlet oxygen. During PDT, the photosensitizers are capable of transferring the energy absorbed to other compounds, which in turn generate metastable species that are very reactive.

The light excited photosensitizer molecule can release its energy via (1) production of radical ions of oxygen due to electron transfer from photosensitizer triplet excited state to the substrate or (2) production of excited singlet oxygen due to energy transfer from photosensitizer triplet excited state to the ground-state molecular oxygen, which is responsible for the oxidation of various cellular constituents [74]. The latter one is a Type II pathway and is responsible for a majority of the PDT effects including photochemical cross-linking. The highly active singlet oxygen induces photo-oxidation of the amino acids such as cysteine, histidine, tyrosine and tryptophan and leads to the formation of covalent cross-links in a light-independent manner [75]. PDT results in photochemical cross-linking, which strengthened the mechanical properties of collagenous tissues and artificial scaffolds used in tissue engineering [62, 76–79].

Photo-cross-linking of collagen is a rapid process unlike chemical and physical cross-linking processes. In photo-cross-linking, the proteins and collagen molecules are covalently cross-linked (inter- and intramolecular) when illuminated in the presence of appropriate photosensitizers such as Rose Bengal (RB) [53, 80, 81] and riboflavin [77, 78, 82–84]. This close cross-linking results in increased ultimate tensile strength and elastic modulus of collagen. However, there are possibilities for the cross-linked collagen to be very stiff or brittle in nature. Incorporation of synthetic or natural polymers could serve as spacers/fillers in between the collagen fibrils and prevent undesired zero-length cross-linking [76, 85]. This cross-linking combined with the infiltration of polymeric fillers would reinforce the collagen structure, ensuing in the increased post-yield strain and fracture toughness [76, 86–88].

7.5 Nanoparticles and Their Role in Dentin Stabilization

Nanotechnology has been applied to dentistry as an innovative concept for the development of materials and/or treatment strategies. The scope of nanomaterials to eliminate biofilms, inhibit a demineralization process and promote remineralization of tooth structure to combat disease-causing bacteria and repair/stabilize previously diseased dentin matrix is being investigated (Fig. 7.5).

Nanoparticles of various materials (polymers, metals), sizes and shapes as well as modifications are available depending on the requirement or treatment needs. Nanoparticles could be tailored to perform specific or multiple functions based on the requirement of the infected tissue management such as root canal disinfection or minimally invasive management of dental caries. Careful selection of nanoparticles based on sound scientific grounds that encompasses the mechanism of action, safety and clinical requirements of dentin tissue stabilization will benefit endodontics and restorative dentistry.

Nanoparticles of bioactive polymers such as chitosan showed ability to enhance the mechanical properties of dentin collagen. Application of chitosan nanoparticles offers several attractive advantages such as the folowing: (1) They possess structural similarity to the extracellular matrix glycosaminoglycans [89]. Extracellular matrix proteins such as proteoglycans and glycosaminoglycans provide mechanical stability and compressive strength to the collagen by intertwining with the fibrous structure. (2) Chitosan composites with collagen could reinforce the collagen scaffolds as well as create a more suitable biomimetic environment for cells [90]. (3) Chitosan nanoparticles consist of reactive free amino and hydroxyl groups that can be utilized for chemical modifications and conjugation. Other reactive molecules/proteins could be attached to the chitosan nanoparticles to obtain a multifunctional nanoparticle. (4) Chitosan is known to be a non-toxic, biologically compatible polymer allowing its widespread use in biomedical applications. (5) The low solubility at a physiological pH of 7.4 can be taken care of by modification of its functional amino group. Conjugation of chitosan with Rose Bengal resulted in water-soluble particles even at higher pH [91]. (6) Chitosan nanoparticles and their derivatives interact with and neutralize MMPs or bacterial collagenase, thereby improving dentinal resistance to degradation [44, 92].

Nanoparticles of carbonated apatite, silica and bioactive glass have been used for the management of dentinal hypersensitivity [93–95]. Occlusion of exposed and patent dentinal tubules is an effective means to provide relief from dentinal hypersensitivity. The main advantage of using nanoparticles is their size effect, which allows these nanoparticles to be pushed into deeper aspects of the dentin tissue. Furthermore, these nanoparticles once modified or functionalized could promote remineralization of the demineralized dentin or occluded patent dentinal tubules [93, 96]. The topic of nanoparticles for dentinal sensitivity management has been covered in detail in the chapter 8 of this book.

Nanomaterials have been applied to improve the overall efficacy of PDT [97]. Towards this end, the key area of interest is using nanoparticles either in conjugation or in combination with photosensitizers depending on the treatment requirement. The nanoparticle-conjugated photosensitizers such as Rose Bengal–conjugated chitosan nanoparticles (CSRBnps) could

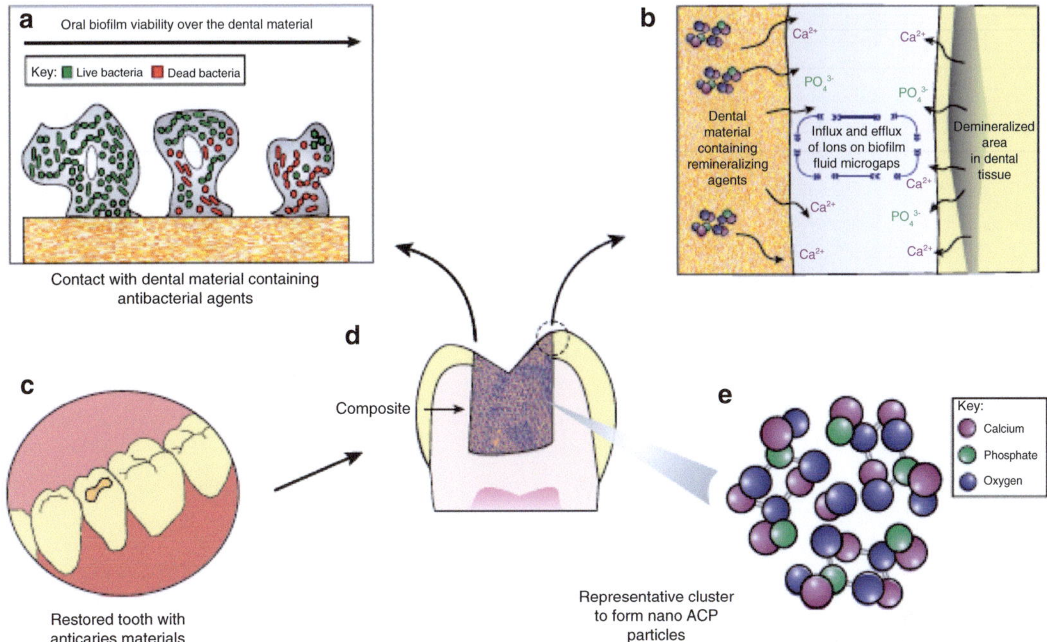

Fig. 7.5 Schematic illustration of the proposed anticaries procedures via restorative dental materials: antibacterial and remineralizing (ion-diffusion process) approaches. (**a**) The antibacterial approach involves release of nanostructured agents as described in the text. The strong antibacterial action is mainly attributed to the high surface:volume ratio that maximizes contact with the environment. These small particles easily penetrate through cell membranes and affect intracellular processes resulting in higher reactivity and antimicrobial activity. (**b**) In the remineralizing approach dental materials release calcium and phosphate to the dental plaque fluid present in microgaps between the tooth and restoration. The calcium and phosphate may be deposited into the tooth leading to gain of net mineral. (**c**) Clinical applicability of nanotechnology-based materials for dental caries management for restoring teeth with cavities. (**d**) Schematic drawing of a longitudinal section of a restored tooth showing the close contact of dental material with dental tissue. (**e**) Representative cluster to form amorphous calcium phosphate (*ACP*) nanofillers (*NACP*) with detail of molecular components (Adapted from Melo et al. [118]. With permission from Elsevier)

be used to stabilize dentin collagen [76]. Rose Bengal–conjugated chitosan in micro- and nanoparticles (CSRBnps) produce singlet oxygen upon photoactivation with a green light (540 nm wavelength). This singlet oxygen produces additional cross-linking between collagen molecules as well as collagen and chitosan nanoparticles. Chitosan incorporation into the photo-cross-linked dentin collagen improved the mechanical properties and resistance to collagenase degradation. In high-magnification images using transmission electron microscopy, CSnp incorporation into the collagen architecture could be evaluated (Fig. 7.6) [76]. CSnp that was incorporated on the dentin collagen could also act as an antibacterial coating, which prevents bacterial ingress along the tooth-filling interface and early biofilm formation. Similar concepts of coating surfaces with chitosan have been tested to reduce bacterial adherence and biofilm formation [98].

7.6 Effects of Stabilization on Dentin

Dentin stabilization is known to affect both mechanical properties and chemical characteristics of dentin matrix. Based on the various methods used to achieve dentin stabilization, the final properties of the matrix could vary. The following paragraphs will highlight the effects on specific properties of dentin after cross-linking procedures.

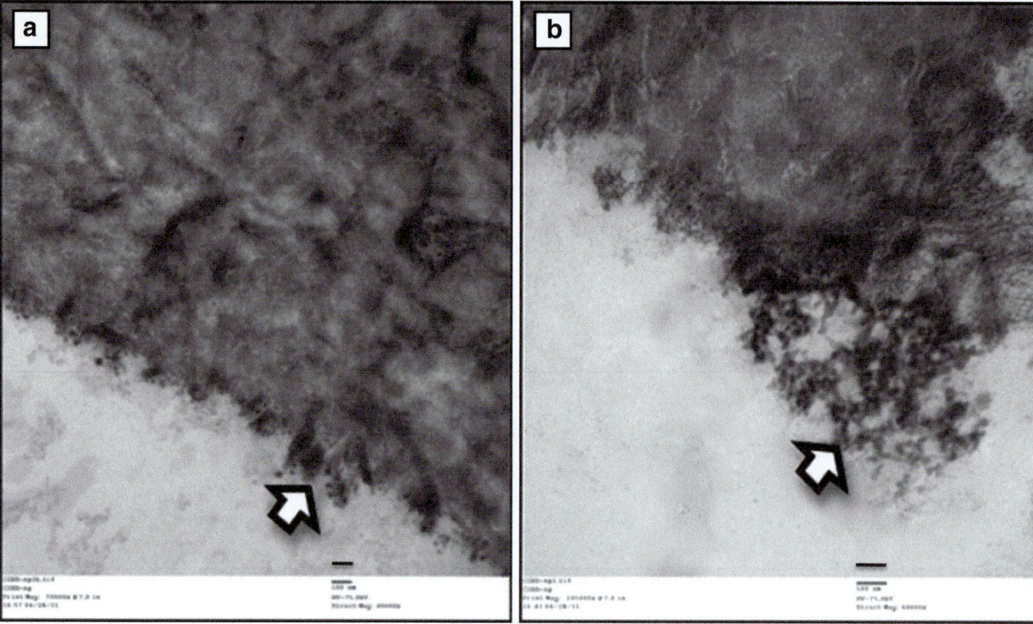

Fig. 7.6 (**a**, **b**) Transmission electron microscopy image of dentin collagen following photodynamic crosslinking using photosensitizer conjugated chitosan nanoparticles. *Arrow* indicate the nanoparticles lining the surface of the dentin collagen (Adapted from Shrestha [76]. With permission from Elsevier)

7.6.1 Effects on Mechanical Properties

As mentioned above, each constituent of dentin composite plays a distinct role in its mechanical properties. Collagen fibrils are the principal tensile stress–bearing component of dentin. The resistance to thermal and proteolytic cleavage and structural stability is provided by the high degree of intermolecular cross-linking and a tight mechanical weave [99, 100]. The procedures that involve cross-linking of collagen should typically show effects on the tensile properties and toughness of dentin. However, when the bulk of dentin is considered, even the compressive strength has been found to improve following cross-linking procedures. Dentin collagen cross-linked either with carbodiimides or photodynamic cross-linking using Rose Bengal showed significantly improved tensile strength and toughness. Generally, cross-linking of collagen using different agents increases the tensile properties, elastic modulus, hardness and toughness of demineralized dentin [55, 56, 62, 66, 76–79].

The incorporation of water-soluble chitosan during cross-linking further enhanced the mechanical properties of dentin [62]. The use of multifunctional chitosan-conjugated Rose Bengal nanoparticles helped achieve this photodynamic cross-linking in a single step [76]. CSRBnp-treated dentin collagen revealed marked improvements in the mechanical properties. Similarly, riboflavin is another potent photosensitizer that could produce collagen cross-linking [79, 82, 83] and showed improved mechanical properties of dentin. Proanthocyanidin (6.5 %) extracted from grape seeds showed increase of the elastic modulus from 8 to 40 times after 10 min and 4 h of treatment respectively [55]. The proanthocyanidin-treated dentin resisted the progression of root caries lesion depth to 80 μm as compared to 208 μm depth in untreated cases [101].

7.6.2 Effects on Chemical Stability

Cross-linking of collagen results in changes at the molecular level of collagen matrix. These changes could lead to the formation of new chemical bonds either between molecules of collagen or with other available reactive forms. The characteristic amide peaks of collagen reveal that cross-linking results in a change of ratio between amide I to amide II [62]. Addition of chitosan biopolymer could be confirmed by changes in the basic chemical spectra of collagen. Cross-linking renders the dentin collagen with increased resistance to degradation by collagenolytic and hydrolytic processes [57, 62, 76, 77]. The amounts of amino acids and specifically hydroxyproline (HYP) released following collagenase degradation of the cross-linked and non-cross-linked dentin collagen are significantly different [62, 77]. Transmission electron microscopy images of the cross-linked and non-cross-linked dentin collagen showed distinctly different patterns of degradation when subjected to high concentrations of bacterial collagenase. The collagenase treatment resulted in denaturation of collagen fibrils, the fibrillar arrangement was disrupted and the density of the fibrils had reduced, with open spaces resembling a moth-eaten appearance (Fig. 7.7). In contrast, the chemical and photodynamic cross-linked collagen resisted enzymatic degradation with no change in the appearance and density of collagen fibrils. Other than the need for proteases to break additional bonds in cross-linked collagen, the sites of collagenase attack may be hidden or modified after this treatment [54]. This contributes to the significant difference in the release of amino acid residues following enzymatic degradation.

7.6.3 Effects on MMPs

MMP-1, -2, -8 and -9 are found in dentin collagen matrix. These MMPs play a role in both physiological and pathological dentin matrix degradation. As mentioned in previous sections, activation of latent MMPs by various root canal irrigants and filling materials has been associated with dentin matrix degradation in a root canal–treated tooth. Treatment of infected dentin either in case of caries restoration or filling of root-treated teeth could benefit from neutraliza-

tion of these MMPs. Research for an effective MMP inhibitor is ongoing not only in dentistry but also in medicine. In medicine, these inhibitors are primarily used to manage cancer [102, 103]. MMP inhibitors that are being tested are synthetic peptides, chemically modified tetracyclines, bisphosphonates or compounds isolated from natural sources, chlorhexidine (CHX), benzalkonium chloride (BAC), chitosan, etc. [43, 44, 104]. Although these inhibitors are effective to a certain extent, they produce significant side effects to the host tissues [44].

CHX is widely used as an antimicrobial agent for disinfection due to its broad spectrum of activity against oral bacteria. This cationic bisbiguanide has also inhibited activities of both gelatinases (A and B), with MMP-2 inhibition more sensitive than MMP-9 [105]. CHX activity was affected in the presence of calcium chloride, suggesting that the mechanism of action is mainly via a cation-chelating mechanism. Thus, the final treatment of dentin matrices previously treated with EDTA/phosphoric acid with CHX showed increased resistance to degradation [40, 42, 105]. The dentin pretreatment with CHX inhibited all gelatinolytic activity (MMP-2 and -9) with both 0.2 and 2 % concentrations [106]. The CHX pretreatment is suggested to significantly lower the loss of bond strength and the nanoleakage observed following 2 years of ageing in acid-etched resin-bonded dentin. BAC is a nitrogenous cationic surface-acting agent containing a quaternary ammonium group, which is used mainly as an antimicrobial agent. Due to its similarity to CHX as a cationic antimicrobial agent, it was tested for its efficacy in MMP inhibition [104]. BAC at various concentrations produced complete inhibition of soluble recombinant MMP-2, -8 or -9 and partial inhibition of matrix-bound MMPs.

Chitosan and its modifications possess the ability to neutralize MMPs. Chitosan oligosaccharides (COS) effectively inhibited MMP-2 from human dermal fibroblasts mainly by inhibiting the activation of proMMP-2 [44]. The chelating property of chitosan resulted in effective binding capacity for Zn^{2+} that is vital for the activity of MMPs. The ability of water-soluble

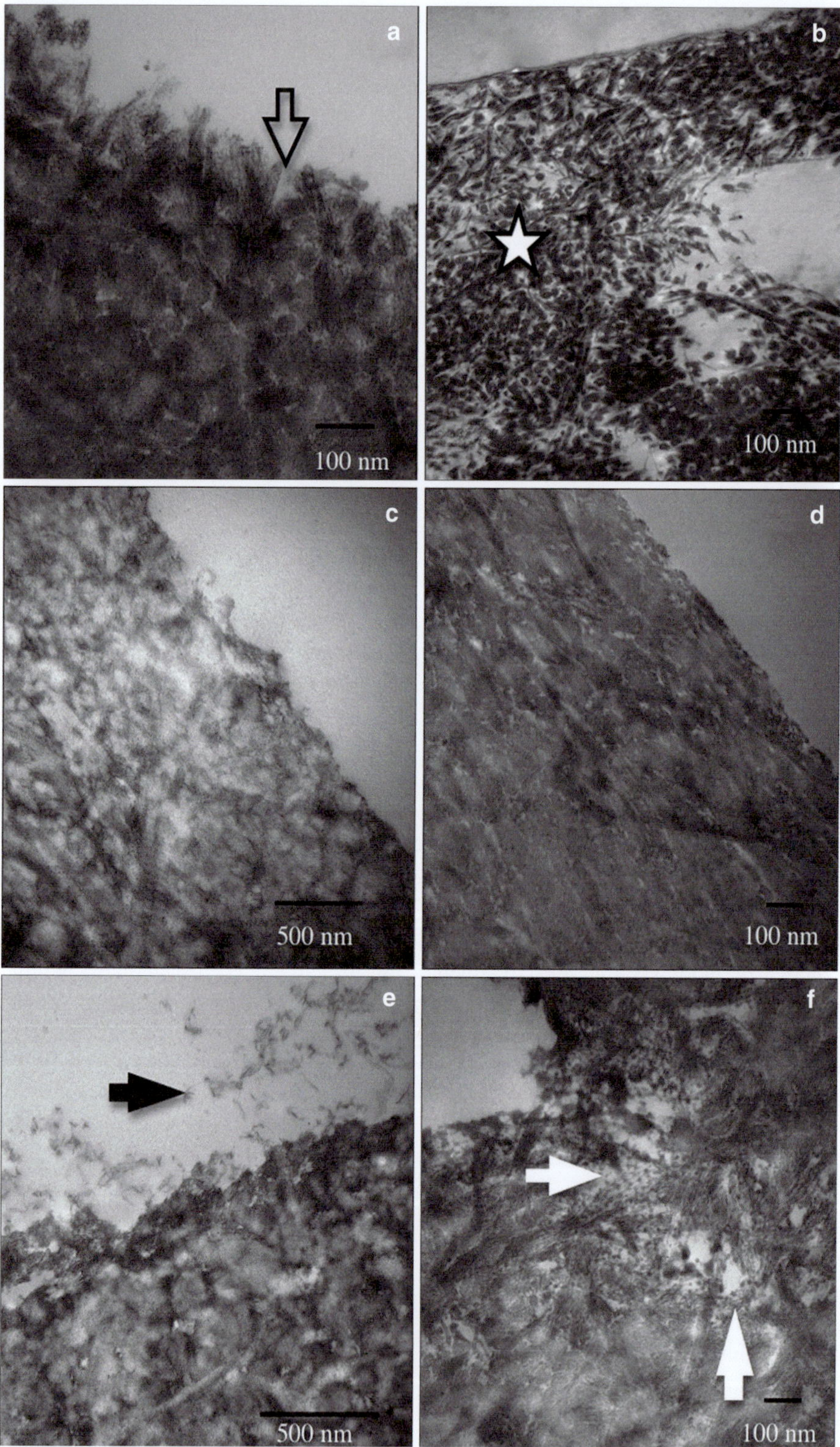
a
b
100 nm
100 nm
c
d
500 nm
100 nm
e
f
500 nm
100 nm

carboxymethylated chitosan (CMCS) to suppress MMP-1 and -3 gene expression in experimental osteoarthritis in turn would impart a protective effect on the degradation of cartilage [107]. Chitosan also has the ability to enhance the mechanical property of demineralized dentin matrix by cross-linking and serving as filler infiltrates within the collagen ultrastructure. MMP activity may be inhibited by cross-linking produced by UVA-activated 0.1 % riboflavin [108]. Cross-linking of dentin independently can inactivate MMPs by fixing their peptide chains and, after acid demineralization, hindered molecular mobility [109]. Chemical cross-linking using 1-ethyl-3-(3-dimethylaminopropyl) carbodiimide (EDC) at various concentrations and pretreatment times inactivated MMP-9 by 98–100 % compared with non-cross-linked controls [110]. EDC also showed positive mechanical effects in dentin beams by preserving elastic modulus, loss of dry mass and release of hydroxyproline when compared with the control without EDC cross-linking. The short treatment time with EDC makes it a favourable agent for clinical application in dentistry.

7.6.4 Effects on Bonding

Resin-based materials that bind to dentin matrix take care of the hydrophilic dentin and hydrophobic resin materials, facilitating a better interfacial seal at the core, sealer and dentin interfaces [111, 112]. The term *monoblock* has been used to describe the nature of bonding achieved by certain resin-based sealers. Monoblock literally means a single unit in which all the units function together. In case of teeth weakened by endodontic and restorative treatments, resin-based root filling material can strengthen the root dentin [113]. Immature root canals with thin circumferential dentin at the apices also need strengthening treatments. The dentin bonding agents present certain advantages; they increase the bond strength of resin-based restorative material to dentin substrate and aim to achieve a leakage-free restoration. These resin-based materials are known to gain retention through micromechanical retention with the collagen matrix of the intertubular dentin and dentinal tubules [114]. However, in the long term, issues such as geometry of the root canal, hydrolysis of the resin and degradation of the dentin collagen will lead to interfacial bond failure at hybrid layers or junctions between the restorative/filling material and dentin [113, 115, 116]. In a resin–dentin interface, a hybrid layer is the weakest and most susceptible component, and this was strengthened by cross-linking of collagen (Fig. 7.8). Cross-linking of collagen will provide improved resistance to degradation and bond strength at the resin–dentin interface even when challenged with hydrolysis [27, 56, 64, 77, 79]. Chemical modification of the dentin matrix with grape seed extract (GSE) and glutaraldehyde improved the bond strength significantly. Furthermore, the GD- and GSE-treated dentin showed stronger hybrid layers. The interfacial failure in these cases occurred on the adhesive or above the hybrid layer [117]. CHX treatment also showed improved bond strength over a period of

Fig. 7.7 TEM micrographs of collagen fibrils before and after enzymatic degradation for 5 days. All specimens were viewed along the cross-section. (**a**) Demineralized dentin matrix without any treatment showed the presence of well-arranged collagen fibrils. The edges of the specimens showed frayed open collagen fibrils due to sample preparation (*open arrow*). (**b**) After 5 days of degradation, the collagen fibrils were denatured, the fibrillar arrangement was disrupted, and the density of the fibrils had reduced, with open spaces resembling a moth-eaten appearance (*star*). The frayed edges of the specimens had smoothened. (**c**, **d**) The GD crosslinked specimens showed dense fibrillar arrangement and normal cross banding of the fibrils. There was no change in the appearance and density of the collagen fibrils after degradation. (**e**) Crosslinking in the presence of CMCS resulted in the incorporation of CMCS into the collagen matrix. The fibrillar arrangement and density were similar to those of the GD group. The CMCS formed a layer of polymeric film (*black arrow*) on the collagen surface. (**f**) Following degradation, the fibrillar arrangement and collagen architecture were well-maintained. Areas of CMCS polymer incorporated into the collagen fibrils were evident (*white arrows*) (Adapted from Shrestha et al. [62]. With permission from Elsevier)

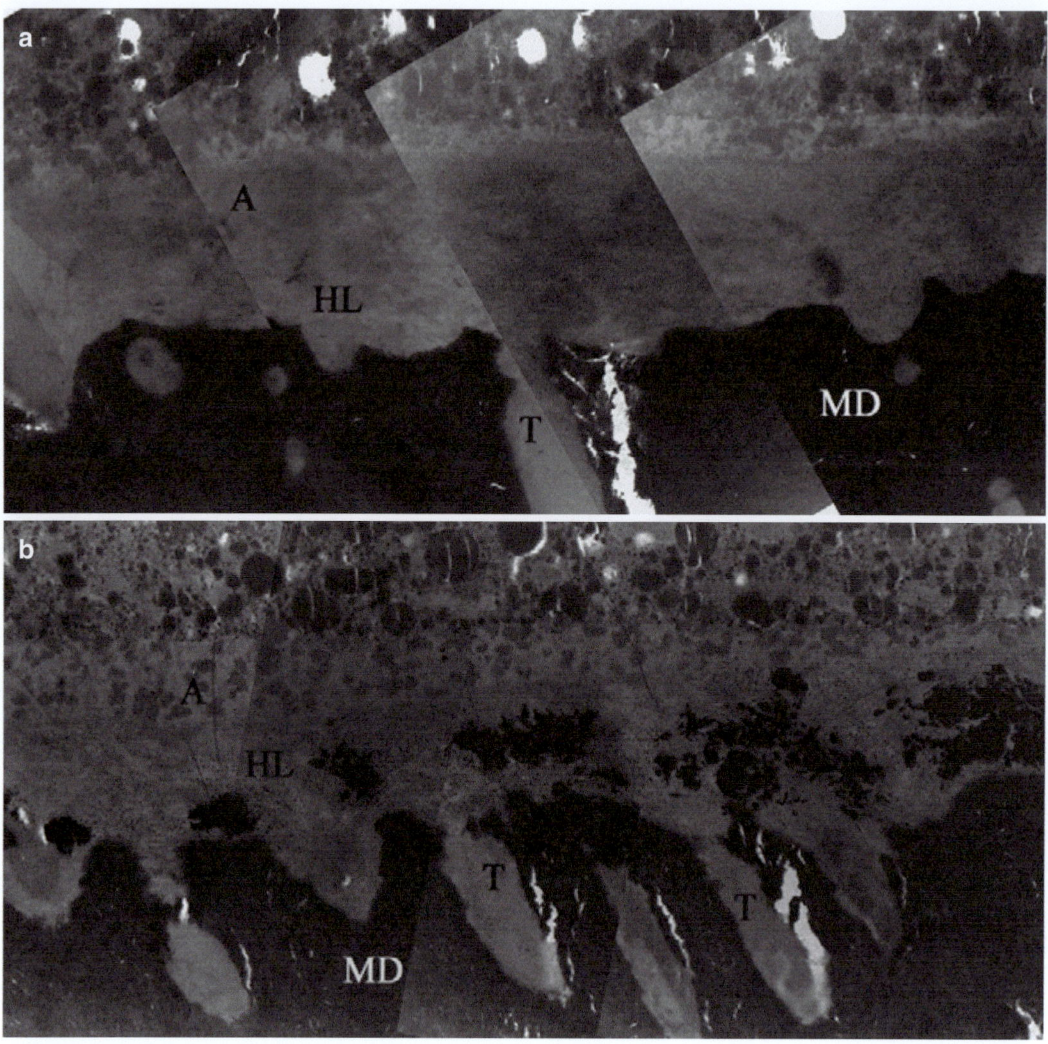

Fig. 7.8 TEM image obtained combining numerous micrographs of a representative specimen treated with 0.2 % CHX for 30 s, then bonded with SB1XT and stored for 2 years in artificial saliva at 37 °C. The adhesive (**a**) interface revealed only very few scattered particles of silver nanoleakage within the hybrid layers (*HL*). (**b**) Control specimen bonded with SB1XT and stored for 2 years in artificial saliva at 37 °C. This control adhesive interface reveals extensive interfacial silver nanoleakage due to individual silver grains and large clusters of silver deposits within the collagen fibrils of the hybrid layer (*HL*). *MD* mineralized dentin, *T* dentinal tubules, *A* filled adhesive. Bar=2 m (Adapted from Breschi et al. [106]. With permission from Elsevier)

time [36, 106]. On the same lines, photodynamic cross-linking using riboflavin improved the resin/dentin bond strength markedly [77]. At the microscopic level, riboflavin cross-linking with/without chitosan infiltration resulted in a uniform hybrid layer with well-formed resin tags. Nevertheless, at a chitosan:riboflavin ratio of 4:1, the hybrid layer was disrupted with short resin tags. This emphasizes that incorporation of a polymer during cross-linking of dentin is concentration dependent and increase in chitosan concentration in the cross-linking solution seriously compromised the bond strength.

7.7 Concluding Remarks

The adverse effect of disease processes and iatrogenic chemicals on dentin matrix reinforces the need for alternative treatment strategies that prevent further matrix deterioration and simultaneously enhance the ultrastructural stability and mechanical properties of dentin. Although many researches have focused on testing various treatment procedures ranging from cross-linking of collagen to neutralization of MMPs, there is a vast inconsistency when it comes to harnessing these techniques for clinical application.

Cross-linking of the dentin collagen presents as an attractive and clinically relevant option towards tissue stabilization. The wide variety of cross-linkers and cross-linking treatment options make it difficult to choose one method over the rest. Along these lines, nanoparticles, particularly biopolymeric/bioactivity chitosan nanoparticles due their multifunctional characteristics, present a potential treatment option to manage previously infected tissue. Cross-linking and incorporation of chitosan nanoparticles in dentin collagen could achieve stabilization of dentin with increased mechanical properties and resistance to bacterial enzymatic degradation as well as ability to neutralize host-derived MMPs. The use of bioactive/biopolymeric nanoparticles that are functionalized with photosensitizers presents a promising approach for disinfecting root dentin and simultaneously stabilizing the tissue. Based on the current findings, further optimization should be carried out to permit their clinical applications in the future.

References

1. Yang SF, Rivera EM, Walton RE. Vertical root fracture in nonendodontically treated teeth. J Endod. 1995;21:337–9.
2. Pitts DL, Matheny HE, Nicholls JI. An in vitro study of spreader loads required to cause vertical root fracture during lateral condensation. J Endod. 1983;9:544–50.
3. Chan CP, Tseng SC, Lin CP, Huang CC, Tsai TP, Chen CC. Vertical root fracture in nonendodontically treated teeth–a clinical report of 64 cases in Chinese patients. J Endod. 1998;24:678–81.
4. Yeh CJ. Fatigue root fracture: a spontaneous root fracture in non-endodontically treated teeth. Br Dent J. 1997;182:261–6.
5. Moule AJ, Kahler B. Diagnosis and management of teeth with vertical root fractures. Aust Dent J. 1999;44:75–87.
6. Testori T, Badino M, Castagnola M. Vertical root fractures in endodontically treated teeth: a clinical survey of 36 cases. J Endod. 1993;19:87–91.
7. Torbjorner A, Karlsson S, Odman PA. Survival rate and failure characteristics for two post designs. J Prosthet Dent. 1995;73:439–44.
8. Morfis AS. Vertical root fractures. Oral Surg Oral Med Oral Pathol. 1990;69:631–5.
9. Bergman B, Lundquist P, Sjogren U, Sundquist G. Restorative and endodontic results after treatment with cast posts and cores. J Prosthet Dent. 1989;61:10–5.
10. Fuss Z, Lustig J, Tamse A. Prevalence of vertical root fractures in extracted endodontically treated teeth. Int Endod J. 1999;32:283–6.
11. Coppens CRMD, DeMoor RJG. Prevalence of vertical root fractures in extracted endodontically treated teeth. Int Endod J. 2003;36:926.
12. Kishen A. Mechanisms and risk factors for fracture predilection in endodontically treated teeth. Endod Top. 2006;13.
13. Vier FV, Figueiredo JA. Internal apical resorption and its correlation with the type of apical lesion. Int Endod J. 2004;37:730–7.
14. Vier FV, Figueiredo JA. Prevalence of different periapical lesions associated with human teeth and their correlation with the presence and extension of apical external root resorption. Int Endod J. 2002;35:710–9.
15. Ferrari M, Mason PN, Goracci C, Pashley DH, Tay FR. Collagen degradation in endodontically treated teeth after clinical function. J Dent Res. 2004;83:414–9.
16. Hubble TS, Hatton JF, Nallapareddy SR, Murray BE, Gillespie MJ. Influence of Enterococcus faecalis proteases and the collagen-binding protein, Ace, on adhesion to dentin. Oral Microbiol Immunol. 2003;18:121–6.
17. Niu W, Yoshioka T, Kobayashi C, Suda H. A scanning electron microscopic study of dentinal erosion by final irrigation with EDTA and NaOCl solutions. Int Endod J. 2002;35:934–9.
18. Calt S, Serper A. Time-dependent effects of EDTA on dentin structures. J Endod. 2002;28:17–9.
19. Tay FR, Hosoya Y, Loushine RJ, Pashley DH, Weller RN, Low DC. Ultrastructure of intraradicular dentin after irrigation with BioPure MTAD. II. The consequence of obturation with an epoxy resin-based sealer. J Endod. 2006;32:473–7.
20. Habelitz S, Balooch M, Marshall SJ, Balooch G, Marshall Jr GW. In situ atomic force microscopy of partially demineralized human dentin collagen fibrils. J Struct Biol. 2002;138:227–36.
21. Bertassoni LE, Orgel JP, Antipova O, Swain MV. The dentin organic matrix – limitations of restorative dentistry hidden on the nanometer scale. Acta Biomater. 2012;8:2419–33.
22. Kinney JH, Pople JA, Marshall GW, Marshall SJ. Collagen orientation and crystallite size in human dentin: a small angle x-ray scattering study. Calcif Tissue Int. 2001;69:31–7.
23. Rivera EM, Yamauchi M. Site comparisons of dentine collagen cross-links from extracted human teeth. Arch Oral Biol. 1993;38:541–6.

24. Kinney JH, Marshall SJ, Marshall GW. The mechanical properties of human dentin: a critical review and re-evaluation of the dental literature. Crit Rev Oral Biol Med. 2003;14:13–29.

25. Gutmann JL. The dentin-root complex: anatomic and biologic considerations in restoring endodontically treated teeth. J Prosthet Dent. 1992;67:458–67.

26. Currey JD. Effects of difference in mineralization on the mechanical properties of bone. Philos Trans R Soc Lond Biol Sci. 1984;13:509–18.

27. Miguez PA, Pereira PN, Atsawasuwan P, Yamauchi M. Collagen cross-linking and ultimate tensile strength in dentin. J Dent Res. 2004;83:807–10.

28. Kinney JH, Habelitz S, Marshall SJ, Marshall GW. The importance of intrafibrillar mineralization of collagen on the mechanical properties of dentin. J Dent Res. 2003;82:957–61.

29. Jameson MW, Hood JA, Tidmarsh BG. The effects of dehydration and rehydration on some mechanical properties of human dentine. J Biomech. 1993;26:1055–65.

30. Pashley DH, Agee KA, Carvalho RM, Lee KW, Tay FR, Callison TE. Effects of water and water-free polar solvents on the tensile properties of demineralized dentin. Dent Mater. 2003;19:347–52.

31. Vincent J. Structural biomaterials. Princeton: Princeton University Press; 1990.

32. Kishen A, George S, Kumar R. Enterococcus faecalis-mediated biomineralized biofilm formation on root canal dentine in vitro. J Biomed Mater Res A. 2006;77:406–15.

33. George S, Kishen A, Song KP. The role of environmental changes on monospecies biofilm formation on root canal wall by Enterococcus faecalis. J Endod. 2005;31:867–72.

34. Vire DE. Failure of endodontically treated teeth: classification and evaluation. J Endod. 1991;17:338–42.

35. van Strijp AJ, Jansen DC, DeGroot J, ten Cate JM, Everts V. Host-derived proteinases and degradation of dentine collagen in situ. Caries Res. 2003;37:58–65.

36. Carrilho MR, Geraldeli S, Tay F, de Goes MF, Carvalho RM, Tjaderhane L, et al. In vivo preservation of the hybrid layer by chlorhexidine. J Dent Res. 2007;86:529–33.

37. Mazzoni A, Mannello F, Tay FR, Tonti GA, Papa S, Mazzotti G, et al. Zymographic analysis and characterization of MMP-2 and -9 forms in human sound dentin. J Dent Res. 2007;86:436–40.

38. Itoh T, Nakamura H, Kishi J, Hayakawa T. The activation of matrix metalloproteinases by a whole-cell extract from Prevotella nigrescens. J Endod. 2009;35:55–9.

39. Huang FM, Yang SF, Chang YC. Up-regulation of gelatinases and tissue type plasminogen activator by root canal sealers in human osteoblastic cells. J Endod. 2008;34:291–4.

40. Hebling J, Pashley DH, Tjaderhane L, Tay FR. Chlorhexidine arrests subclinical degradation of dentin hybrid layers in vivo. J Dent Res. 2005;84:741–6.

41. Hashimoto M, Tay FR, Ohno H, Sano H, Kaga M, Yiu C, et al. SEM and TEM analysis of water degradation of human dentinal collagen. J Biomed Mater Res B Appl Biomater. 2003;66:287–98.

42. Pashley DH, Tay FR, Yiu C, Hashimoto M, Breschi L, Carvalho RM, et al. Collagen degradation by host-derived enzymes during aging. J Dent Res. 2004;83:216–21.

43. Carrilho MR, Tay FR, Donnelly AM, Agee KA, Tjaderhane L, Mazzoni A, et al. Host-derived loss of dentin matrix stiffness associated with solubilization of collagen. J Biomed Mater Res B Appl Biomater. 2009;90:373–80.

44. Kim MM, Kim SK. Chitooligosaccharides inhibit activation and expression of matrix metalloproteinnase-2 in human dermal fibroblasts. FEBS Lett. 2006;580:2661–6.

45. Zhang K, Kim YK, Cadenaro M, Bryan TE, Sidow SJ, Loushine RJ, et al. Effects of different exposure times and concentrations of sodium hypochlorite/ethylenediaminetetraacetic acid on the structural integrity of mineralized dentin. J Endod. 2010;36:105–9.

46. Oliveira LD, Carvalho CA, Nunes W, Valera MC, Camargo CH, Jorge AO. Effects of chlorhexidine and sodium hypochlorite on the microhardness of root canal dentin. Oral Surg Oral Med Oral Pathol Oral Radiol Endod. 2007;104:e125–8.

47. Saleh AA, Ettman WM. Effect of endodontic irrigation solutions on microhardness of root canal dentine. J Dent. 1999;27:43–6.

48. White JD, Lacefield WR, Chavers LS, Eleazer PD. The effect of three commonly used endodontic materials on the strength and hardness of root dentin. J Endod. 2002;28:828–30.

49. Shemesh H, Bier CA, Wu MK, Tanomaru-Filho M, Wesselink PR. The effects of canal preparation and filling on the incidence of dentinal defects. Int Endod J. 2009;42:208–13.

50. Kishen A, Messer HH. Vertical root fractures: radiological diagnosis. In: Basrani B, editor. Endodontic radiology. 2nd ed. Ames: Wiley-Blackwell; 2012. p. 235–50.

51. Rao KP. Recent developments of collagen-based materials for medical applications and drug delivery systems. J Biomater Sci Polym Ed. 1995;7:623–45.

52. Sung HW, Chang Y, Liang IL, Chang WH, Chen YC. Fixation of biological tissues with a naturally occurring crosslinking agent: fixation rate and effects of pH, temperature, and initial fixative concentration. J Biomed Mater Res. 2000;52:77–87.

53. Spikes JD, Shen HR, Kopeckova P, Kopecek J. Photodynamic crosslinking of proteins. III. Kinetics of the FMN- and rose bengal-sensitized photooxidation and intermolecular crosslinking of model tyrosine-containing N-(2-hydroxypropyl)methacrylamide copolymers. Photochem Photobiol. 1999;70:130–7.

54. Jayakrishnan A, Jameela SR. Glutaraldehyde as a fixative in bioprostheses and drug delivery matrices. Biomaterials. 1996;17:471–84.

55. Bedran-Russo AK, Pashley DH, Agee K, Drummond JL, Miescke KJ. Changes in stiffness of demineralized dentin following application of collagen crosslinkers. J Biomed Mater Res B Appl Biomater. 2008;86B:330–4.

56. Bedran-Russo AK, Pereira PN, Duarte WR, Drummond JL, Yamauchi M. Application of crosslinkers to dentin collagen enhances the ultimate tensile strength. J Biomed Mater Res B Appl Biomater. 2007;80:268–72.

57. Sung HW, Chang WH, Ma CY, Lee MH. Crosslinking of biological tissues using genipin and/or carbodiimide. J Biomed Mater Res A. 2003;64:427–38.

58. Qin C, Xu J, Zhang Y. Spectroscopic investigation of the function of aqueous 2-hydroxyethylmethacrylate/glutaraldehyde solution as a dentin desensitizer. Eur J Oral Sci. 2006;114:354–9.

59. Loke WK, Khor E. Validation of the shrinkage temperature of animal tissue for bioprosthetic heart valve application by differential scanning calorimetry. Biomaterials. 1995;16:251–8.

60. Simmons DM, Kearney JN. Evaluation of collagen cross-linking techniques for the stabilization of tissue matrices. Biotechnol Appl Biochem. 1993;17(Pt 1):23–9.

61. Beauchamp Jr RO, St Clair MB, Fennell TR, Clarke DO, Morgan KT, Kari FW. A critical review of the toxicology of glutaraldehyde. Crit Rev Toxicol. 1992;22:143–74.

62. Shrestha A, Friedman S, Kishen A. Photodynamically crosslinked and chitosan-incorporated dentin collagen. J Dent Res. 2011;90:1346–51.

63. Madhavan K, Belchenko D, Motta A, Tan W. Evaluation of composition and crosslinking effects on collagen-based composite constructs. Acta Biomater. 2010;6:1413–22.

64. Liu Y, Chen M, Yao X, Xu C, Zhang Y, Wang Y. Enhancement in dentin collagen's biological stability after proanthocyanidins treatment in clinically relevant time periods. Dent Mater. 2013;29:485–92.

65. Bedran-Russo AK, Castellan CS, Shinohara MS, Hassan L, Antunes A. Characterization of biomodified dentin matrices for potential preventive and reparative therapies. Acta Biomater. 2011;7:1735–41.

66. Bedran-Russo AK, Yoo KJ, Ema KC, Pashley DH. Mechanical properties of tannic-acid-treated dentin matrix. J Dent Res. 2009;88:807–11.

67. Rafat M, Li F, Fagerholm P, Lagali NS, Watsky MA, Munger R, et al. PEG-stabilized carbodiimide crosslinked collagen-chitosan hydrogels for corneal tissue engineering. Biomaterials. 2008;29:3960–72.

68. Olde Damink LH, Dijkstra PJ, van Luyn MJ, van Wachem PB, Nieuwenhuis P, Feijen J. Cross-linking of dermal sheep collagen using a water-soluble carbodiimide. Biomaterials. 1996;17:765–73.

69. Staros JV, Wright RW, Swingle DM. Enhancement by N-hydroxysulfosuccinimide of water-soluble carbodiimide-mediated coupling reactions. Anal Biochem. 1986;156:220–2.

70. Han B, Jaurequi J, Tang BW, Nimni ME. Proanthocyanidin: a natural crosslinking reagent for stabilizing collagen matrices. J Biomed Mater Res A. 2003;65:118–24.

71. Itoh S, Takakuda K, Kawabata S, Aso Y, Kasai K, Itoh H, et al. Evaluation of cross-linking procedures of collagen tubes used in peripheral nerve repair. Biomaterials. 2002;23:4475–81.

72. Weadock KS, Miller EJ, Bellincampi LD, Zawadsky JP, Dunn MG. Physical crosslinking of collagen fibers: comparison of ultraviolet irradiation and dehydrothermal treatment. J Biomed Mater Res. 1995;29:1373–9.

73. Billiar K, Murray J, Laude D, Abraham G, Bachrach N. Effects of carbodiimide crosslinking conditions on the physical properties of laminated intestinal submucosa. J Biomed Mater Res. 2001;56:101–8.

74. Redmond RW, Gamlin JN. A compilation of singlet oxygen yields from biologically relevant molecules. Photochem Photobiol. 1999;70:391–475.

75. Dubbelman TM, Haasnoot C, van Steveninck J. Temperature dependence of photodynamic red cell membrane damage. Biochim Biophys Acta. 1980;601:220–7.

76. Shrestha A, Hamblin MR, Kishen A. Photoactivated rose bengal functionalized chitosan nanoparticles produce antibacterial/biofilm activity and stabilize dentin-collagen. Nanomedicine. 2014;10(3):491–501.

77. Fawzy AS, Nitisusanta LI, Iqbal K, Daood U, Beng LT, Neo J. Chitosan/Riboflavin-modified demineralized dentin as a potential substrate for bonding. J Mech Behav Biomed Mater. 2013;17:278–89.

78. Daood U, Iqbal K, Nitisusanta LI, Fawzy AS. Effect of chitosan/riboflavin modification on resin/dentin interface: spectroscopic and microscopic investigations. J Biomed Mater Res A. 2013;10:1846–56.

79. Fawzy AS, Nitisusanta LI, Iqbal K, Daood U, Neo J. Riboflavin as a dentin crosslinking agent: ultraviolet A versus blue light. Dent Mater. 2012;28:1284–91.

80. Chan BP, Chan OC, So KF. Effects of photochemical crosslinking on the microstructure of collagen and a feasibility study on controlled protein release. Acta Biomater. 2008;4:1627–36.

81. Chan BP, Amann C, Yaroslavsky AN, Title C, Smink D, Zarins B, et al. Photochemical repair of Achilles tendon rupture in a rat model. J Surg Res. 2005;124:274–9.

82. Wollensak G, Iomdina E. Long-term biomechanical properties of rabbit cornea after photodynamic collagen crosslinking. Acta Ophthalmol. 2009;87:48–51.

83. Ibusuki S, Halbesma GJ, Randolph MA, Redmond RW, Kochevar IE, Gill TJ. Photochemically crosslinked collagen gels as three-dimensional scaffolds for tissue engineering. Tissue Eng. 2007;13:1995–2001.

84. Wollensak G, Iomdina E, Dittert DD, Salamatina O, Stoltenburg G. Cross-linking of scleral collagen in the rabbit using riboflavin and UVA. Acta Ophthalmol Scand. 2005;83:477–82.

85. Everaerts F, Torrianni M, van Luyn M, van Wachem P, Feijen J, Hendriks M. Reduced calcification of bioprostheses, cross-linked via an improved carbodiimide based method. Biomaterials. 2004;25:5523–30.

86. Taravel MN, Domard A. Collagen and its interactions with chitosan, III some biological and mechanical properties. Biomaterials. 1996;17:451–5.
87. Chen J, Li Q, Xu J, Huang Y, Ding Y, Deng H, et al. Study on biocompatibility of complexes of collagen-chitosan-sodium hyaluronate and cornea. Artif Organs. 2005;29:104–13.
88. Sionkowska A, Wisniewski M, Skopinska J, Kennedy CJ, Wess TJ. Molecular interactions in collagen and chitosan blends. Biomaterials. 2004;25:795–801.
89. Chandy T, Sharma CP. Chitosan–as a biomaterial. Biomater Artif Cells Artif Organs. 1990;18:1–24.
90. Tan W, Krishnaraj R, Desai TA. Evaluation of nanostructured composite collagen–chitosan matrices for tissue engineering. Tissue Eng. 2001;7:203–10.
91. Moczek L, Nowakowska M. Novel water-soluble photosensitizers from chitosan. Biomacromolecules. 2007;8:433–8.
92. Persadmehr A, Torneck CD, Cvitkovitch DG, Pinto V, Talior I, Kazembe M, et al. Bioactive chitosan nanoparticles and photodynamic therapy inhibit collagen degradation in vitro. J Endod. 2014;40:703–9.
93. Tian L, Peng C, Shi Y, Guo X, Zhong B, Qi J, et al. Effect of mesoporous silica nanoparticles on dentinal tubule occlusion: an in vitro study using SEM and image analysis. Dent Mater J. 2014;33:125–32.
94. Lee SY, Kwon HK, Kim BI. Effect of dentinal tubule occlusion by dentifrice containing nano-carbonate apatite. J Oral Rehabil. 2008;35:847–53.
95. Mitchell JC, Musanje L, Ferracane JL. Biomimetic dentin desensitizer based on nano-structured bioactive glass. Dent Mater. 2011;27:386–93.
96. Besinis A, van Noort R, Martin N. Infiltration of demineralized dentin with silica and hydroxyapatite nanoparticles. Dent Mater. 2012;28:1012–23.
97. Allison RR, Mota HC, Bagnato VS, Sibata CH. Bio-nanotechnology and photodynamic therapy–state of the art review. Photodiagnosis Photodyn Ther. 2008;5:19–28.
98. Decraene V, Pratten J, Wilson M. An assessment of the activity of a novel light-activated antimicrobial coating in a clinical environment. Infect Control Hosp Epidemiol. 2008;29:1181–4.
99. Nakabayashi N. Bonding mechanism of resins and the tooth. Kokubyo Gakkai Zasshi. 1982;49:410.
100. Nakabayashi N, Kojima K, Masuhara E. The promotion of adhesion by the infiltration of monomers into tooth substrates. J Biomed Mater Res. 1982;16:265–73.
101. Walter R, Miguez PA, Arnold RR, Pereira PN, Duarte WR, Yamauchi M. Effects of natural cross-linkers on the stability of dentin collagen and the inhibition of root caries in vitro. Caries Res. 2008;42:263–8.
102. Nam KS, Shon YH. Suppression of metastasis of human breast cancer cells by chitosan oligosaccharides. J Microbiol Biotechnol. 2009;19:629–33.
103. Shen KT, Chen MH, Chan HY, Jeng JH, Wang YJ. Inhibitory effects of chitooligosaccharides on tumor growth and metastasis. Food Chem Toxicol. 2009;47:1864–71.
104. Tezvergil-Mutluay A, Mutluay MM, Gu LS, Zhang K, Agee KA, Carvalho RM, et al. The anti-MMP activity of benzalkonium chloride. J Dent. 2011;39:57–64.
105. Gendron R, Grenier D, Sorsa T, Mayrand D. Inhibition of the activities of matrix metalloproteinases 2, 8, and 9 by chlorhexidine. Clin Diagn Lab Immunol. 1999;6:437–9.
106. Breschi L, Mazzoni A, Nato F, Carrilho M, Visintini E, Tjaderhane L, et al. Chlorhexidine stabilizes the adhesive interface: a 2-year in vitro study. Dent Mater. 2010;26:320–5.
107. Liu SQ, Qiu B, Chen LY, Peng H, Du YM. The effects of carboxymethylated chitosan on metalloproteinase-1, -3 and tissue inhibitor of metalloproteinase-1 gene expression in cartilage of experimental osteoarthritis. Rheumatol Int. 2005;26:52–7.
108. Cova A, Breschi L, Nato F, Ruggeri Jr A, Carrilho M, Tjaderhane L, et al. Effect of UVA-activated riboflavin on dentin bonding. J Dent Res. 2011; 90:1439–45.
109. Pashley DH, Tay FR, Breschi L, Tjaderhane L, Carvalho RM, Carrilho M, et al. State of the art etch-and-rinse adhesives. Dent Mater. 2011; 27:1–16.
110. Tezvergil-Mutluay A, Mutluay MM, Agee KA, Seseogullari-Dirihan R, Hoshika T, Cadenaro M, et al. Carbodiimide cross-linking inactivates soluble and matrix-bound MMPs, in vitro. J Dent Res. 2012;91:192–6.
111. Schwartz RS. Adhesive dentistry and endodontics. Part 2: bonding in the root canal system-the promise and the problems: a review. J Endod. 2006;32:1125–34.
112. Schwartz RS, Fransman R. Adhesive dentistry and endodontics: materials, clinical strategies and procedures for restoration of access cavities: a review. J Endod. 2005;31:151–65.
113. Tay FR, Pashley DH. Monoblocks in root canals: a hypothetical or a tangible goal. J Endod. 2007;33:391–8.
114. Gwinnett AJ. Quantitative contribution of resin infiltration/hybridization to dentin bonding. Am J Dent. 1993;6:7–9.
115. Garcia-Godoy F, Tay FR, Pashley DH, Feilzer A, Tjaderhane L, Pashley EL. Degradation of resin-bonded human dentin after 3 years of storage. Am J Dent. 2007;20:109–13.
116. Tay FR, Hashimoto M, Pashley DH, Peters MC, Lai SC, Yiu CK, et al. Aging affects two modes of nanoleakage expression in bonded dentin. J Dent Res. 2003;82:537–41.
117. Al-Ammar A, Drummond JL, Bedran-Russo AK. The use of collagen cross-linking agents to enhance dentin bond strength. J Biomed Mater Res B Appl Biomater. 2009;91:419–24.
118. Melo MAS, Guedes SFF, Xu HKK, et al. Nanotechnology-based restorative materials for dental caries management. Trends Biotechnol. 2013;31(8):459–67.

Nanoparticles in Restorative Materials

8

Grace M. De Souza

Abstract

Nanotechnology has made significant progress in the past 20 years. Particles as small as 3 nm are being employed in restorative materials in attempts to improve their functional performance. There are currently many commercial brands with different particle size distribution; some of them are termed *nanohybrids*, where nanoparticles (minimum size ~3 nm) are associated with particles larger than 100 nm. Materials called *nanofill* contain nanoparticles with a more even distribution (smaller than 100 nm). Amongst the particles used, some of them are applied to enhance the material's bioactivity, which may control or reduce viable bacterial count on the tooth surface or on the tooth–restoration interface. Some examples of those particles are titanium dioxide (TiO_2), chlorhexedine-hexametaphosphate (CHX-HMP) and silver (Ag). Nanofillers are also used to improve the material's clinical performance, by either strengthening the restoration or enhancing its aesthetic characteristics, such as translucency and polishability. Zirconium dioxide (ZrO_2), colloidal platinum and zirconia–silica nanoparticles are examples in this category of nanofillers. Amongst the desirable characteristics of nano-based restorative materials are higher mechanical properties; enhanced ion release of glass ionomer cements; development of bioactive adhesives, to provide antibacterial effect within the restoration or at the tooth–restoration interface; polishability and stable optical properties of resin composites; phase stability of high–crystalline content ceramics and lesser chipping of dental porcelains. The main goal of this chapter is to provide an overview of the advancements in the field of restorative materials with the application of nanoparticles. Nonetheless, it

G.M. De Souza, DDS, MSc, PhD
Clinical Sciences Department, Faculty of Dentistry,
University of Toronto,
124 Edward St, Toronto, ON M5G1G6, Canada
e-mail: grace.desouza@dentistry.utoronto.ca

A. Kishen (ed.), *Nanotechnology in Endodontics: Current and Potential Clinical Applications*,
DOI 10.1007/978-3-319-13575-5_8, © Springer International Publishing Switzerland 2015

is worth mentioning that any progress reported here is very novel and has not been fully investigated, and more investigations are required before new restorative materials can be widely disseminated as a permanent solution to a given clinical problem.

Abbreviations

4-META/MMA-TBB	4-methacryloxyethyl trimellitate anhydride in methyl methacrylate initiated by tri-n-butyl borane
ACP	Amorphous calcium phosphate
Ag	Silver
Al_2O_3	Aluminum trioxide *or* alumina
AlF_3	Aluminum fluoride
$BaSO_4$	Barium sulfate
Bis-EMA	Ethoxylated bisphenol A glycol dimethacrylate
Bis-GMA	Bisphenol-glycidil methacrylate
Bz	Benzoate
Ca/P	Calcium phosphate
CAD-CAM	Computer-aided design/Computer-aided manufacturing
CDHA	Calcium-deficient hydroxyapatite
CeO_2	Cerium dioxide
Ce-TZP	Cerium-stabilized tetragonal zirconia polycrystal
CFU	Colony forming units
CHX	Chlorhexedine
CHX-HMP	Chlorhexedine-hexametaphosphate
CPN	Colloidal platinum nanoparticles
CQ	Camphorquinone
DEB	Dentin enamel body
DNA	Deoxyribonucleic acid
F	Fluoride
FA	Fluorapatite
$FeCl_3$	Ferric chloride
FSS	Filtek Supreme Standard
FST	Filtek Supreme Translucent
GA	Glyoxylic acid
GIC	Glass ionomer cement
HA	Hydroxyapatite
HEMA	2-hydroxyethyl methacrylate
HNT	Halloysite nanotubes
MgO	Magnesium oxide
MMP	Matrix-metalloproteinase
MOD	Mesio occlusal distal
MPTMS	3-methacryloxypropyl trimethoxy silane
N	Zinc
N	Newton
nACP	Nano-amorphous calcium phosphate
nAg	Nano-silver
$nCAF_2$	Calcium fluoride nanoparticles
n-CDHA	Nano calcium-deficient hydroxyapatite
nDCPA	Nano-dicalcium phosphate anhydrous
nHA	Nano-hydroxyapatite
nm	Nanometer
NVP	N-vinylpyrrolidone
PMAA	Polymetracrylic acid
PO_4	Phosphate
ppm	Parts per million
QADM	Quaternary ammonium dimethacrylate
QA-PEI	Quaternary ammonium polyethylenimine
SEM	Scanning electron microscopy
TEGDMA	Trithylene glycol dimethacrylate
TiO_2	Titanium dioxide
UDMA	Urethane-dimethacrylate
wt%	By weight percent
YbF_3	Ytterbium fluoride
YSZ	Ytrium-stabilized zirconia
ZnO	Zinc oxide
ZrO_2	Zirconium dioxide *or* zirconia
ZTA	Zirconia toughened alumina
μm	Micrometer
μm/m.K	Micrometer per meter Kelvin

8.1   Introduction

Nano is a Greek word synonymous with *dwarf*, which means extremely small [1]. Nanotechnology consists mainly of the processing, separating, consolidating and deforming of materials by one atom or molecule [2]. It is also known as *molecular technology* or *molecular engineering*, by which it is possible to produce materials and structures in the size range of 0.1–100 nm [3]. This recent technology has generated a wave of developments in the field of dental materials. Nanoparticles may be effective to control the oral biofilm formation due to their biocidal, anti-adhesive and drug delivery capabilities [4]. In an attempt to reduce bacterial and fungal adhesion to oral devices, such as removable partial dentures and prostheses, silver nanoparticles have been recently incorporated into denture materials [5] and orthodontic adhesives [6].

The inorganic fillers employed in dental materials can be classified according to their size range, varying from macrofillers to nanofillers. The particle size can be controlled by either a *top-down* or a *bottom-up* manufacturing approach. In the *top-down* technique, conventional milling procedures are employed to reduce the particle, generally at a level not below 100 nm (100 nm=0.1 µm). To develop particles below 100 nm, direct molecular assembly is used in the *bottom-up* approach. In this approach, fillers are built from atoms and molecular precursors into progressively larger structures and transformed into nanosized fillers for application in dentistry [7].

The use of nanoparticles in restorative materials has a wide range of applications. They have been employed to develop a new generation of restorative composites, which ideally are highly translucent, retain polishability [2] and have mechanical properties similar to those of micro-hybrid composites [3]. However, cracking of the resin matrix and clumping of the nanoparticles have been reported [3]. Bioactive nanoparticles incorporated into dental materials are an area of increasing interest since the presence of pathogenic germs at the marginal gap of restorations may damage the hard dental tissue by development of secondary caries. Specially formulated metal oxide nanoparticles have good antibacterial activity [8] because the size of metallic nanoparticles ensures that a significantly large surface area of the particles is in contact with the microorganism effluent, and laboratory tests show that bacteria, viruses and fungi may be killed within minutes of contact [9]. However, the physical and mechanical properties of restorative materials as well as the setting or polymerization reaction must not be affected by the addition of these active ingredients. The antibacterial efficacy of nanoparticles is further discussed in Chap. 7.

As previously mentioned, nanoparticles have a significantly larger surface area as opposed to their actual size. The high concentration of surface free energy causes the nanoparticles to bond strongly to other materials and to each other [1]. One of the greatest challenges for the application of nanotechnology in restorative materials is therefore to prevent the association of those small particles, which tend to form agglomerates [3, 10]. When particles are weakly bonded together by van der Waals forces, they are considered *'agglomerates'*, while *'aggregates'* are particles bonded together by solid bridges like calcination [11]. The tendency of nanoparticles to agglomeration is highly dependent on their zeta potential [12, 13]. Zeta potential is the electrostatic potential at the electric double layer surrounding a nanoparticle in solution [14]. In simple words, it is the potential difference between the dispersion medium and the stationery layer of fluid attached to the dispersed particle. The manufacturing process employed for obtaining nanoparticles has a significant effect on their zeta potential [12], and the higher the zeta potential, the larger the size of the agglomerates [13]. The manufacturing process also affects crystallinity, shape and size of the nanoparticles [15], with higher degree of crystallinity being related to improved hardness and strength [16].

In spite of the revolutionary approaches in *bionanotechnology* – an area that has emerged since *biotechnology* and *nanotechnology* became integrated – there is a general concern with regard to the biosafety of nanoparticles. The occurrence of future diseases caused by the incorporation of nanoparticles in materials used in the human body and the consequent adsorption of nanoparticles by live tissue requires investigation. Much of the current work is using *in vitro* methodologies because concerns regarding the nanoparticles' biocompatibility have not been fully addressed as yet [4].

8.2 Nanoparticles in Glass Ionomer Materials

Glass ionomer cements are water-based cements, also known as polyalkenoate cements [17]. Their generic name is based on the interaction between silicate glass and polyacrylic acid, whereby an acid–base reaction occurs [18]. Mechanical and optical properties of glass ionomer cements have been improved by the development of resin-modified glass ionomers, in which some water content is replaced by water-soluble methacrylate monomers, such as 2-hydroxyethyl methacrylate (HEMA) [19]. Although commercial materials vary widely in composition, resin-modified glass ionomer materials undergo an initial set by light-activated polymerization followed by a slower acid-based setting reaction [20]. The photo-polymerizable version of glass ionomer cement (GIC) has been extensively employed because it is less water sensitive than the conventional one and presents translucency and chemical bonding to the tooth structure [18, 21], but clinical problems are still observed [22]. GIC is also biocompatible and presents potential for release and uptake of fluoride ions [23, 24]. However, the clinical implication of local fluoride release in caries prevention and tooth structure remineralization has not been fully clarified [25, 26].

Even though GIC has very unique properties, its application as a restorative material is limited due to some major drawbacks. The material presents low fracture toughness and a high rate of occlusal wear when compared to amalgam and modern composite resin systems [21], being not currently recommended for permanent restorations in posterior teeth. The material is also brittle – a property that causes early bulk fracture when it is subjected to tensile stress [27].

The addition of nanoparticles into glass ionomer composition has different rationales. Salt- or oxide-based nanoparticles may react with the polyacrylic acid in the matrix and increase the overall resistance of the material. Additionally, the fact that the particles are significantly smaller than GIC fillers makes them capable of filling up interstitial spaces, strengthening the cement even more. According to Xie et al. [27], there is an inverse size–strength relationship for GIC fillers, and the matrix interaction with the filler is critical for the mechanical properties of the cement [27].

8.2.1 Experimental Materials

To investigate the performance of nano-modified glass ionomer cements, the most common approach is to add nanoparticles into the powder of a commercial GIC, and then the powder:liquid ratio is recalculated [10]. In 2006, Prentice et al. [10] evaluated the effect of ytterbium fluoride (YbF_3 – ~25 nm) and barium sulfate ($BaSO_4$ – <10 nm) on mechanical properties and reactivity of a commercial GIC (Riva SC – SDI). The authors expected an improvement in the compressive strength considering that YbF_3 would act similarly to aluminum fluoride (AlF_3) in glass ionomer cements due to their similar chemical structure. The greater availability of the aluminum ion caused by the addition of AlF_3 improved compressive strength of an experimental GIC [28]. However, nano-scale particles perform differently from their micro-scale equivalents [29, 30]. Therefore, contrary to what was expected, 1–2 % YbF_3 addition had no effect on compressive strength, and 5 % had a deleterious effect. $BaSO_4$ reduced compressive strength even in a concentration as low as 1 % [10]. It is known that bivalent cations (such as Ca^{+2}, Sr^{+2}, Ba^{+2}) contribute less to strength than trivalent cations (Al^{+3}, La^{+3}, Yb^{+3}) [31]. An additional drawback of the experimental material was the insufficient dispersion of the nanoparticles in the glass ionomer powder, which led to "cotton-wool" formation within the cement, visible under light microscope.

Another structure that has been investigated to improve mechanical properties of GIC is hydroxyapatite (HA). HA is a type of calcium phosphate, and it is the main mineral of the human dental enamel and dentine, with very similar composition and crystal structure [32]. It is also present in the inorganic matrix of human bone and contains both phosphate and hydroxyl ions (HA structure: $Ca_{10}(PO_4)_6(OH)_2$). Nanohydroxyapatite and nanofluorapatite (FA – average particle size 100–200 nm) were mixed to a commercial GIC (5 wt%, Fuji II, GC Corp.), and mechanical properties were investigated [32]. Both experimental GICs (HA and NA) presented better mechanical behaviour than the control group (commercial GIC) at different storage periods. The addition of HA and FA had an effect on both the setting reaction mechanism and the degree of polysalt bridge formation, improving the mechanical properties of the cement after its final set. Bond strength to dentine was also improved in the experimental GIC, probably due to the formation of ionic and hydrogen bonds between GIC and the tooth structure. In another research, HA and FA nanoparticles (50–100 nm) were added to the powder and N-vinylpyrrolidone (NVP) to the liquid of a commercial GIC (Fuji II, GC International, Japan) [33]. Overall, the addition of nanoparticles (HA and FA) was more effective to improve properties such as biaxial flexural strength, diametral tensile and compressive strength than the modification of the polymeric chain. Nano-FA/ionomer had higher mechanical properties than Nano-HA/ionomer, probably due to the higher stability and lower dissolution rate of FA in distilled water.

The HA particle size effect on GIC has also been investigated by comparing micro- and nano-HA (nHA) addition to RelyX luting cement (3M ESPE). Nano-HA-modified GIC presents higher bond strength to tooth structure as well as increased resistance to demineralization when observed under scanning electron microscopy (SEM). However, nHA exceeds the maximum setting time recommended by the *International Organization for Standardization* (ISO) [34]. The association of nHA with Yttria-stabilized zirconia (YSZ) (20 wt% nHA/80 wt% YSZ) in the powder composition (5 wt%) of a commercial GIC shows that the presence of zirconia also reduced dissolution of the cement during soaking periods [24].

An alternative to HA is the calcium-deficient hydroxyapatite (CDHA), which is a variant of HA with a Ca/P ratio between 1.67 and 1.33. Its composition and structure are similar to those of HA but with low thermal stability and higher solubility [35]. Nano-CDHA (n-CDHA – average particle size 24 nm) added in different concentrations (5, 10 and 15 wt%) to the powder of Fuji II LC (GC International) results in higher solubility of the GIC and higher weight loss when samples are stored in both water or acidic solution. Surface hardness decreases with the increase of n-CDHA. Although the compressive strength increases with n-CDHA addition, it decreases after immersion in liquid media, probably due to the weight loss and ionic release [23]. Even though n-CDHA is a promising alternative to improve clinical performance of GIC, stability of the material in the mouth may be a limiting factor.

Titanium dioxide (TiO$_2$) nanoparticles are inorganic additives with promising properties such as chemical stability and biocompatibility [36] and have been suggested as reinforcing fillers for dental materials. TiO$_2$ nanoparticles (average size ~21 nm) added to the powder of a commercial GIC (Kavitan, SpofaDental, Czech Republic) in 3 wt% results in overall better mechanical performance of the material. However, 7 wt% addition reduces flexural strength, compressive strength and surface hardness due to the insufficient amount of polyacrylic acid to react and bond to the nanoparticles. The association of 7 wt% TiO$_2$ and poly(acrylamide-co-sodium acrylate) copolymer also compromises the compressive strength of GIC [37]. Microtensile bond strength, antibacterial activity and fluoride release are not affected by the proportion of TiO$_2$ [38]. The smaller size of the TiO$_2$ particles allows for a wider range of particle size distribution, occupying empty spaces between the larger GIC particles, which results in better mechanical properties due to the additional bonding sites between the polyacrylic acid and the glass particles [38]. Inhibition of the bacterial growth when TiO$_2$ is employed is related to the bactericidal effect of TiO$_2$ nanoparticles, which cause quick intracellular damage to the infiltrated bacteria [39].

An antimicrobial agent that has been recently added to the composition of GIC is chlorhexidine (CHX). Nanoparticles of CHX-hexametaphosphate (CHX-HMP) added to the powder of a commercial GIC (Diamond Carve, Kemdent) in ratios of 0, 1, 2, 5, 10, 20 and 30 % by mass indicates that 30 % modified samples are difficult to handle and have to be discarded.

Diametral tensile strength was similar between experimental materials and the control. CHX and fluoride release were observed for up to 33 days. Interestingly, experimental GIC releases the most fluoride, especially at greater proportions of CHX-HMP. It is hypothesized that CHX-HMP alters the setting reaction and this renders fluoride more mobile in the cement lattice [40]. The results of CHX-HMP addition to GIC are very promising, and further investigations to fully characterize the performance of experimental materials should be performed in a near future.

8.2.2 Commercial Alternatives

Ketac Nano (3M ESPE) is a two-paste commercial resin-modified glass ionomer advertised as having aggregated 'nanoclusters' in the 1 μm size range composed of 5–20 nm spherical particles as well as non-agglomerated silica fillers and acid-reactive glass fillers in its powder (3M internal data) [41]. However, naming clusters of nanoparticles as 'nanoclusters' is not accurate. In spite of consisting of loosely bound aggregates of engineered nanofiller particles, the final dimensions of those clusters are in the 'micro-size' range, with clusters as big as 5 μm [42]. Therefore, such clusters in nanotechnology should be reported as *nano-filled clusters* or *clusters of nanoparticles*. The comparison between Ketac Nano, a conventional resin-modified GIC (Vitremer), a nano-filled resin composite (Filtek Z350, 3M) and a microhybrid resin composite (TPH Spectrum, Dentsply) showed that nano-based GIC presents significantly higher roughness after 7 days biofilm degradation than the other groups, which is also significantly higher than that of its own control group (stored in humidity). Severe biodegradation is also observed under SEM micrographs (Fig. 8.1). According to the authors, the monomeric system, containing HEMA, bisphenol glycidyl methacylate (Bis-GMA) and trithylene glycol dimethacrylate (TEGDMA), makes the material less resistant to biodegradation due to the microphase separation of Bis-GMA/HEMA

that happens in the presence of water [44, 45]. However, the presence of the nano-filled clusters increases the resistance of the material to bio-mechanical degradation (biofilm + three-body abrasion) since nano-sized particles are taken out of the matrix instead of the removal of the entire cluster, making nano-based GIC (Ketac Nano) about three times more resistant to that specific challenge than other GICs (Vitremer, 3M ESPE; Ketac Molar Easymix, 3M ESPE; Fuji IX, GC).

The fluoride release of Ketac Nano (3M) has also been compared to other commercially available GICs for a period of 360 days. The amount and pattern of fluoride release rate of nano-ionomer is similar to that of other resin-modified (Fuji II LC, GC and Vitremer, 3M) [20] and conventional (Fuji II, GC) GICs [46], with a burst effect of fluoride release at shorter times. The application of a primer as surface protection does not affect the fluoride release rate [20]. Although smaller glass particles have a larger surface area, which would theoretically increase the acid-based reactivity and release fluoride from the powder more effectively, that effect has not been clearly demonstrated. In spite of the smaller particle size, the filler content on Ketac Nano (69 %) is lower than that on Fuji II LC (76 %) and Vitremer (71 %). Therefore, mechanical properties as well as fluoride uptake/release characteristics are somewhat lower than those presented by other resin-modified GICs [47].

The clinical performance of Ketac Nano (3M) has been assessed and compared to a resin-modified GIC (Fuji II LC, GC America) and a nanofilled restorative composite (Filtek Supreme, 3M). Non-carious cervical lesions were restored and evaluated after 6 and 12 months. For Ketac Nano, marginal staining after 1 year of clinical service was significantly worse than baseline and also worse than Fuji II LC within the same period of time. Marginal adaptation and colour match of nano-based GIC were also significantly inferior than its baseline and other materials [48]. Indeed, the micromechanical interaction between Ketac Nano and dentine/enamel is very superficial, without structural evidence of demineralization or hybridization since the monomer and

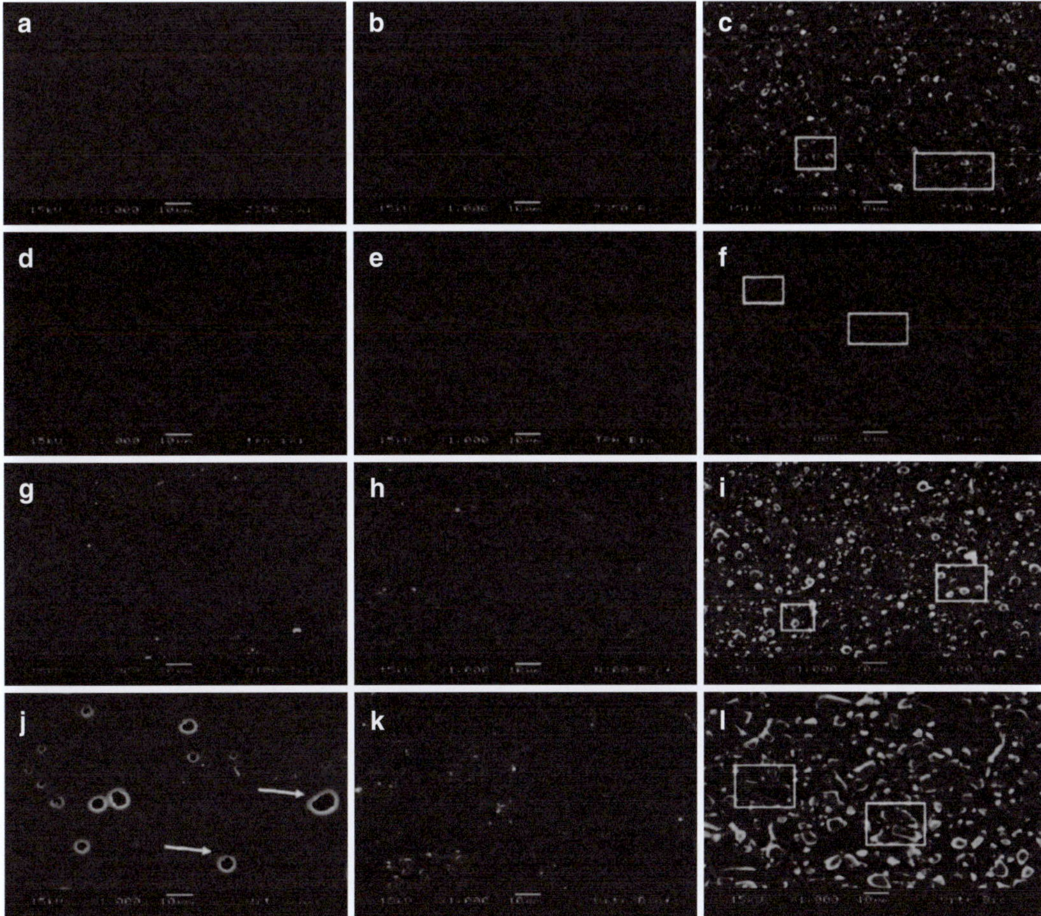

Fig. 8.1 Scanning electron micrographs of microhybrid composites Filtek Z350 (**a–c**) and TPH Spectrum (**d–f**); nano-based Ketac N100 (**g–i**), and resin-modified GIC Vitremer (**j–l**). The first column shows the relative humidity storage groups (**a, d, g, j**), with porosities (small spherical and irregular shapes) indicated by *arrows*. The second column represents the *S mutans* biofilm storage groups (**b, e, h, k**), with a severe degraded aspect of the matrix. The third column corresponds to biofilm storage plus abrasion groups (**c, f, i, l**), with many exposed particles at the surface of materials (*squares*) (Reprinted from De Paula et al. [43]. With permission from Operative Dentistry, Inc.)

photoinitiator of the primer form a resin coating that avoids deeper infiltration of GIC [49].

Another application of nanotechnology in glass ionomer cements is the development of a nano-filled light curing varnish (G-Coat Plus, GC Europe), which is applied onto the surface of a highly viscous GIC (Fuji IX GP Extra, GC Europe). This combination has been commercially branded as EQUIA ('Easy-Quick-Unique-Intelligent-Aesthetic'). The main purpose is to provide surface protection in the early maturation phase of the cement to avoid both water uptake and dehydration [50]. The longer the protective varnish is in contact with the restoration, the smaller the chance that GIC will have its mechanical properties reduced [51]. A retrospective cohort study evaluated anatomical form, surface texture, marginal integrity and marginal discoloration of EQUIA restorations in class I and class II preparations in either molars or premolars. A significant correlation between cavity size and volume loss of material was observed. Chipping and volume loss were mainly observed in the proximal surface, where varnish could not be applied. Wear of the occlusal surface was not significant during the 2 year observation period. Although the authors concluded that the application of EQUIA restorations was considered acceptable for class I and small class II preparations [50], 2 years is a relatively short period of evaluation, and results should be examined with care.

The influence of the nano-filled surface varnish (G-Coat Plus, GC) on wear resistance and flexural strength has also been investigated in coated and uncoated conventional GICs (Fuji IX GP Extra, GC; Ketac Molar, 3M) [52]. Flexural strength of Fuji IX is significantly improved by the application of the nano-filled varnish, but it is still lower than the flexural strength of Ketac Molar, either coated or uncoated. There is no chemical interaction between Ketac Molar and the varnish layer, forming a deposited layer that is easily removed by the wear test. Although the nano-filled varnish coating may be considered very promising to extend longevity of GIC restorations, decrease in wear rate as a function of time was observed for all groups, with or without coat [53]. In spite of being called 'nano varnish', there is no information available about particle size distribution neither in any of the research papers investigated nor in the manufacturer's brochure, which makes it difficult to correlate the performance of the material with the presence of nanofillers.

Some authors observe that GIC is still a very interesting option in public health systems, especially considering the economic aspects. Besides being an aesthetic and safe alternative for those patients who refuse the placement of an amalgam restoration, the bulk placement of the material reduces clinical time and, consequently, the cost of the procedure. Additionally, fluoride release is a very unique characteristic of this class of material. Therefore, there is still demand for the improvement of glass ionomer cements in an attempt to develop a 'permanent' restorative material, instead of the current status of the GICs as being semi-permanent restorations.

8.3 Application of Nanoparticles in Resin-Based Materials

By definition, *composite materials* (also called resin composites or shortened to *composites*) are materials made from two or more constituent materials with significantly different physical or chemical properties that, when combined, produce a material with characteristics different from the individual components. The current chemistry of restorative composites involves a Bis-GMA organic matrix combined with some other monomers (TEGDMA, HEMA, UDMA – urethane dimethacrylate and others) [54] and inorganic–organic fillers of different sizes [55]. The treatment of the fillers' surface with 3-methacryloxypropyl trimethoxy silane (MPTMS) allows the formation of covalent links between the inorganic filler particles and the organic resin matrix [55]. Typically, nano-based restorative composites contain 13–30 wt% of polymerizable organic matrix and 70–87 wt% mixture of different inorganic fillers [54] in addition to a photoinitiator system or other curing systems [56].

The types of nanofillers in dental composites include silica [57, 58], tantalum thoxide [59], zirconia-silica [3], alumina [60], nano-fibrillar silicate [61] and titanium oxide [62], among others. Those nanoparticles may be used as the sole filler of the adhesive [63, 64] or in combination with other types of fillers for composites [65]. However, the wide spectrum of fillers and dimensions has made it difficult to classify different materials within the same category. Resin composites with particles smaller than 100 nm have been conveniently named *nanocomposites* [66, 67]. Nonetheless, this classification is controversial, and some authors name the same material *nanonybrids* [68] based on the association of nanoparticles, nano-filled clusters and microparticles. Neither information provided by the manufacturers nor research articles have been successful in reporting the precise amount of nanofillers in the final composition of resin composites, informing only the overall amount of fillers (Table 8.1). Considering the composition of the materials containing nanoparticles that are currently available, if the filler loading is composed only of particles smaller than 100 nm, the resin composite shall be considered *nanofill* or *nanocomposite*. Restorative composites containing filler particles smaller than 100 nm associated with microfillers (particles larger than 0.1 µm) are hybrid materials and may be precisely considered *nanohybrids*.

The key challenges in the development of dental nanocomposites include (1) effective dispersion of nanoparticles in the resin to avoid agglomeration and (2) achieving high loading of nanofiller level to

Table 8.1 Materials currently available making use of nanotechnology

Category	Brand name	Classification	Particles	Characteristics (reference #)
Glass ionomer materials	*Ketac Nano* *3M Espe* St Paul, MN, US	Nano-GIC	1 μm clusters containing 5–25 nm spherical particles [41]	Good resistance to biomechanical degradation [44] Fluoride release similar to other GIC's [46] Marginal staining under clinical conditions [48]
	G-Coat Plus *GC America* Alsip, IL, USA	Nano-filled light curing varnish	No information available	May reduce occlusal wear [50] and improve flexural strength of some GIC's [53]
Enamel/ dentin bonding	*Prime&Bond NT* Dentsply International York, PA, USA	Nano-filled adhesive	7–12 nm SiO_2 particles	Higher microtensile bond strength to dentin than one-bottle self-etch adhesive systems [69] Low microtensile bond strength to feldspathic porcelain [70]
Composite resin	*Filtek DEB* 3M Espe St Paul, MN, USA	Nanofill	Zirconia-silica nano-filled clusters (0.6–1.4 μm; 90 %) and nanoparticles (5–20 nm; 10 %) dispersed in the matrix *79 wt%*	Fracture strength lower than for ormocer-based composite [66] Higher polymerization shrinkage, water sorption and solubility [65] Decrease in flexural strength and flexural modulus after 30 days water storage [65] Higher chance of occlusal staining [71], Relatively high monomer elution [72] Low clinical roughness values [67], Reduced polishability after cyclic loading [73] Decreased Knoop hardness after 6 months water storage [74] Color instability after additional heat post-curing [75] Good resistance to biomechanical degradation [43]
	Filtek translucent 3M Espe	Nanofill	Nanoparticles (~75 nm) and minor amount of silica nano-filled clusters (0.6–1.4 μm; 50 %) *70 wt%*	High polymerization shrinkage, water sorption and solubility [65] Decreases flexural strength and flexural modulus after 30 days water storage [65]
	Tetric EvoCeram Dentsply	Nanohybrid	Barium glass (1 μm). Ba-aluminum-silicate glass (0.4–0.7 μm) and Ytterbium trifluoride (550 nm) *82–83 wt%*	Fracture strength lower than for ormocer-based composite [67] Low roughness values [67] Reduced polishability after cyclic loading [73]
	Grandio/Grandio Nano Voco Briarcliff Manor, NY, USA	Nanohybrid	Silica dioxide (20–60 nm) and barium-aluminaborosilicate (0.1–2.5 μm) *87 wt%*	Minimum polymerization shrinkage when compared to nanofill and Tetric EvoCeram, lower water sorption and solubility, high and stable flexural strength and flexural modulus after 30 days water storage [65] Higher degree of conversion and lower color stability than microhybrid materials [76]
	Clearfil Majesty Dentsply	Nanohybrid	Silanated Barium glass filler and pre-polymerized organic filler including nanoparticles (0.2–100 μm) with average particle size of 0.7 μm *78 wt%*	Color instability after additional heat post-curing [75]
Ceramic	*NanoZr* Matsushita Electric Works, Tokyo, Japan	Nanocomposite	ZrO_2, 67.9 mass%; Al_2O_3, 21.5 mass%; CeO_2, 10.6 mass%; MgO, 0.06 mass%; TiO_2, 0.03 mass%	High flexural strength after airborne-particle abrasion and acid etch [77] Good shear bond strength to veneer material, without the need of a liner [78]

reduce polymerization shrinkage while maintaining good handling characteristics and manufacturing costs [79]. It is also paramount to understand the effect of nanofiller size, morphology, composition and filler hybridization on composite properties as well as the long-term durability of those restorations *in vivo* [79].

The adhesion between a resin-based restorative material and tooth structure is required to avoid microleakage, secondary caries, postoperative sensitivity and discoloration [80]. The quality of the bonding between tooth structure and restoration relies on the impregnation of the fluid resin into a superficially decalcified zone of enamel rods and dentine collagen fibrils [81] to form the hybrid layer [82]. Due to the humidity and dynamic characteristics of dentine substrate, adhesion to dentine is more complex than to enamel [83]. A high-quality and durable hybrid layer can be only achieved if the demineralized dentine collagen matrix is fully resin infiltrated [84]. When collagen is not fully enveloped by the adhesive or porosity is present in the hybrid layer, micro- and nano-pathways can expedite the process of interfacial degradation [85, 86], reducing the resin–dentine interface durability [87, 88].

There are currently two methods to create dentine bonding and promote the hybrid layer formation: etch-and-rinse and self-etch techniques. Regardless of the selected strategy, the mechanism available for adhesive resin infiltration is diffusion of the resin into whatever fluid is in the spaces of the substrate and along the collagen fibrils [89]. The elastic modulus of the hybrid layer formed at the resin–dentine interface is the lowest compared to dentine and resin composite [80]. This low elastic modulus can cause the failure of the restoration when the occlusal loading is higher than the strength of the layer [90]. The mechanical properties of the hybrid layer may be improved by the incorporation of fillers, which also increase viscosity and radiopacity [91]. However, filler aggregates resultant of nanoparticle clusters block the penetration of the nanoparticles into the collagen network, thus compromising the mechanical properties at the hybrid layer [92].

Different rationales justify the application of nanotechnology in adhesive systems: (1) to stimulate the remineralization of the demineralized/etched collagen below the hybrid layer [85]; (2) to improve mechanical properties, avoiding the early failure of the adhesive interface subjected to cyclic loading, and (3) to develop a bioactive resin-based adhesive capable of eliminating residual bacteria in the prepared tooth cavity as well as bacteria invading the tooth–restoration interface *via* microleakage [63].

Are the restorative resin composites resistant enough to endure a long-lasting clinical application? This is an important clinical question posed while optimizing the bonding technique. Till date, none of the composite resin materials has been shown to have the ability to meet both the functional requirements of a posterior Class I or II restoration and the superior aesthetics required for anterior restorations [3, 93]. The clinical requirements for a successful composite resin restoration are so many [55] that manufacturers can barely meet half of them and the average longevity of a composite resin restoration is about 4 years [94].

The inorganic fillers incorporated into resin composites directly affect the material's radiopacity, wear resistance, flexural modulus and thermal coefficient of expansion [55, 95]. Considering that the polymerization shrinkage of the organic matrix is largely correlated with the volume fraction of fillers in the composite [56], the synergistic use of releasing nanofillers and reinforcing fillers may result in nanocomposites with the potential of having both stress-bearing and caries-inhibiting capabilities, a combination not available in current dental materials [79, 96].

8.3.1 Dentin Bonding Systems

As previously mentioned, are added to the composition of dentine bonding systems in an attempt to stabilize the hybrid layer by improving its mechanical properties [90, 97]. Although nano-filler particles are supposed to present a colloidal behaviour, by which each particle dispersed is surrounded by oppositely charged ions called the 'fixer layer', the filler content of the bonding agents frequently agglomerate into clusters [92]. Then, the demineralized dentine serves as a screener that prevents from resin penetration in the underneath dentine by collecting fillers on the top surface of the hybrid layer [98]. This outcome is somewhat controversial, as shown by Wagner et al. [15], who observed that nano-filled agglomerates were dispersed throughout the adhesive layer instead of being deposited on the top of the demineralized dentine layer (Fig. 8.2). The presence of additional solvent (water or ethanol) underneath the hybrid layer resultant from a self-etch technique may also generate filler aggregation. The effect of solvent on cluster formation in several commercial self-etch adhesive systems (three two-step and three all-in-one self-etching systems) modified with nanoparticles has been investigated. When adhesive resin is placed on wet dentine, the residual solvent existing within the dentine (either water or ethanol) will produce filler aggregation, with cluster size ranging from 19.6 to 103 nm [92], whilst interfibrilar collagen space on etched dentine surface is in the range of 12.1–19.8 nm when total etch technique is applied [99]. Even when smaller nanofillers are employed (average size 7–12 nm) in commercial (Prime&Bond NT, Dentsply) or

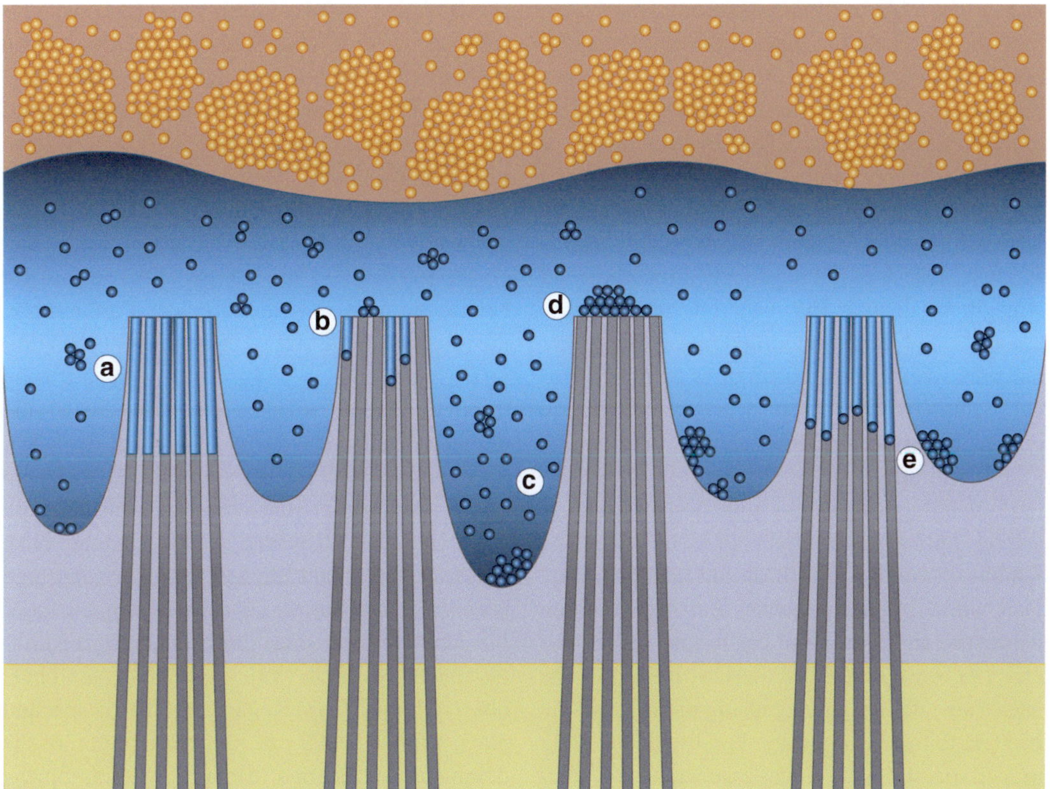

Fig. 8.2 Schematic representation of the dentin/adhesive hybrid layer (*orange* composite resin, *blue* adhesive resin, *gray* demineralized dentin, *yellow* intact dentin) with the effect of different nanoparticles application: (*A*). Wagner et al. 2013: Hydrothermally-grown HA with smaller surface area are embedded within the adhesive and not deposited on top of the hybrid layer. (*B*) Besinis et al. 2012: Hydrophobic diluent combined with smaller nHA improves infiltration ability of demineralized dentin. (*C*) Hoshika et al. 2010: Longer and denser resin tags obtained when CPN nanoparticles are added to the primer composition, by either improving monomer infiltration or enhancing monomer conversion. (*D*) Lohbauer et al. 2010: Spherical zirconia nanoparticles are unable to penetrate the interfibrilar spaces, compromising hybrid layer formation. (*E*) Osorio et al. 2011: Zinc oxide nanoparticles infiltrate the interfibrilar spaces and remain at the bottom of the hybrid layer, in contact with the demineralized dentin (Courtesy of Grace M De Souza and published with permission from University of Toronto)

experimental adhesives, there is some controversy as to whether the particles can penetrate into the restricted space of the hybrid layer or not [100]. Fluoralumino-silicate glass filler nanoparticles also present larger cluster size and more cluster frequency than silanated colloidal silica [92]. This is probably due to the thermal instability and other unknown factors of the nanoparticle–organosilane bond [101].

A quaternary ammonium dimethacrylate (QADM) has been recently synthesized and presents strong antibacterial characteristics without compromising the mechanical properties of the restorative resin [102, 103]. In an attempt to reduce the residual bacteria counting in the surface of the tooth cavity preparation as well as to limit the invasion of bacteria at the tooth–restoration interface, quaternary ammonium salt monomers were copolymerized in resins to yield antibacterial activities. The incorporation of QADM and nano-silver particles (nAg – ~2.7 nm) to the primer and/or the adhesive of a commercial adhesive system (Adper Scotchbond Multi-Purpose, 3M) does not negatively affect the bond strength to dentine [63, 64]. Furthermore, the association of the two modified materials (primer and adhesive) has the potential to combat residual bacteria in both the cavity preparation, due to primer diffusion to deeper layers, and at the tooth–restoration interface, where invading bacteria are reduced due to the more potent and long-lasting antibacterial effect [63, 64]. QADM inhibits *S. mutans* growth on the material's surface, while nAg-resin inhibits *S. mutans* both on its surface and away from the surface [104]. The effect of Ag is related to the inactivation of the bacteria's vital enzymes, causing the DNA in the bacteria to lose its replication ability and leading to cell death [105]. The association of both QADM and nAg in primer suspension also reduces the bacteria count by three orders of magnitude (Fig. 8.3) [106, 107]. This novel antibacterial product has two advantages: it acts as a cavity disinfectant without compromising bond strength and has long-term antibacterial effect in the cured state [106, 107]. Considering that there is no noticeable difference in colour when Ag concentrations are up to 0.2 % mass fraction and

that the mass fraction currently being used is in the order of 0.1 %, there is limited concern with regard to colour alteration in the final restoration [63, 64].

Besides QADM and nAg incorporation into dentine adhesives, nanoparticles of amorphous calcium phosphate (nACP) have the potential to release Ca and P ions to remineralize tooth structure [108]. The association of nACP with QADM and nAg in dentine adhesives is an attempt to obtain the best of different approaches [109]. It remineralizes demineralized collagen and reduces bacterial counting at the adhesive interface [110]. A new experimental bonding agent containing 0.1 % nAg, 10 % QADM and 40 % nACP (% by mass) reduces biofilm viability and CFU (colony-forming units) to about one-third of the control without affecting bond strength [110]. The incorporation of nACP and nAg to a commercial bonding system (Scotchbond Multi-Purpose, 3M) also does not affect bond strength and presents reduced lactic acid produced by biofilm [109]. NACP may efficiently release Ca and P ions at lower filler levels for composites [96], but this effect has not yet been investigated in dental adhesives. Furthermore, the relatively large size of the particles (~100 nm) [109] represents a problem to their diffusion through dentinal tubules until they reach the demineralized collagen layer. Indeed, even HA nanoparticles with size varying from 20 to 70 nm do not penetrate into the hybrid layer even though they are well dispersed in the matrix [15]. However, high bond strength values are observed because nHA promotes cohesive reinforcement of the adhesive layer since they are homogeneously distributed over it [15]. Besinis et al. [111] observed that besides particle size, dispersion plays a fundamental role in the infiltration of the demineralized dentine matrix as well as on the enamel matrix [112]. The authors demonstrated that the use of a hydrophobic diluent such as acetone for the dilution of nHA instead of distilled water increases the infiltrative ability of nHA in solution by displacing water molecules. Hydrophobic diluent and smaller particles obtained by sol-gel methods appear to be the key to successful infiltration of the demineralized dentine (Fig. 8.2) [111].

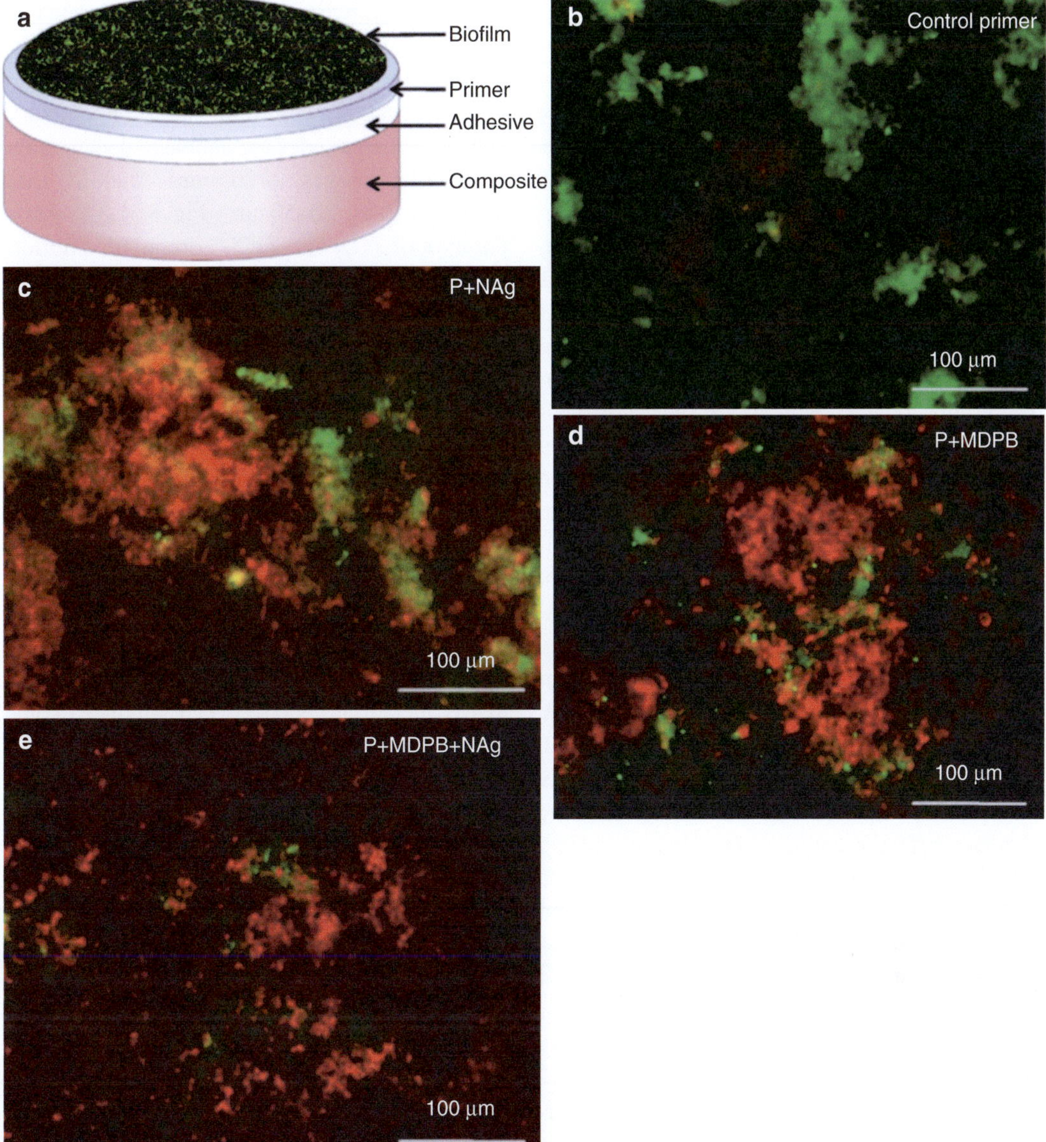

Fig. 8.3 Live/dead bacterial staining assay of biofilms adherent on layered discs: (**a**) schematic experimental setup; (**b–e**) representative images of biofilms for control primer, P+nAg, P+MDPB, and P+MDPB+nAg. Live bacteria were stained green, and dead/compromised bac-teria were stained red. Live and dead bacteria in close proximity of each other yielded *yellow/orange* colours (Reprinted from Zhang et al. [106]. With permission from Elsevier)

Synthesized HA nanorods have also the potential of reinforcing the bonding in one-bottle dentine adhesive systems [113], but there is an inverse correlation between filler concentration and bond strength [15, 113]. With particle length between 30 and 145 nm, 50 % of dentine tubules are fully occluded and, 40 % are partially occluded by the nanorods [114]. The low bond strength may also be related to the incomplete cure of the adhesive [113]. Investigators observed that HA has the potential to remineralize the collagen demineralized by the acid etch. However, the size of the nanorods and their consequent inability to penetrate through the dentinal tubules is a serious limitation for this class of particles.

Colloidal platinum nanoparticles (CPN) are reported as the next generation of antioxidants [115] as they efficiently quench reactive oxygen species [116]. They have been used to manufacture cosmetics and to treat pulmonary inflammation [116]. It is also a low-allergy metal and is widely used as a catalyst in diverse applications [117]. The possibility of increasing the bond strength twice as much as that presented by control groups bonded with 4-META/MMA-TBB (4-acryloyloxyethyl trimellitate anhydride/ methyl methacrylate-tri-n-butylborane)–based resin, when the dentine surface was treated with CPN (average particle size ~2 nm), was first evidenced by Nagano et al. [118]. The effect of CPN application on bond strength between 4-META/ MMA-TBB–based adhesive (Super-Bond C&B, Sun Medical, Japan) and tooth structure, either prior to or after acid etch (10–3 solution: 10 % citric acid and 3 % $FeCl_3$), has been investigated. The highest bond strength is observed when CPN is applied before etching, evidencing that the platinum effect remained after etching [119]. Denser and longer resin tags are associated with CPN-treated groups [117], and high bond strength may be a consequence of one or more factors: increased monomer infiltration, enhanced adhesive resin conversion in the hybrid layer or better micromechanical interlocking caused by denser and longer resin tags (Fig. 8.2) [117].

The improvement of the mechanical properties in the adhesive/hybrid layer is an alternative to stabilize the bonding, which would then withstand the cyclic loading over the restorative interface [120]. Nevertheless, the incorporation of nanofillers must not compromise initial bond strength values. Spherical zirconia (ZrO_2) nanoparticles (size range ~20–50 nm) have been added to either primer or adhesive of a commercial system (Adper Scotchbond Multi-Purpose, 3M Espe) in different concentrations and result in significantly higher bond strength probably due to the reinforcing effect of the adhesive system promoted by filler incorporation, which acts as a barrier to crack propagation initiating from the

bottom of the hybrid layer during a microtensile strength test [91]. The incorporation of ZrO_2 into the primer aimed to reinforce the hybrid layer by filler infiltration within the collagen network. However, the particles are unable to penetrate the interfibrilar spaces (~20 nm) due to their increased size (~20–50 nm – Fig. 8.2). The incorporation of ZrO_2 and $ZrCl_2$ (zirconium chloride) nanoparticles into a self-priming/dentine adhesive (Adper Single Bond Plus, 3M) stabilizes the hybrid layer by maintaining steady values of nanohardness and nanoelasticity after 3 months of storage [120].

Zinc (Zn) is another metal of interest in the dental field. It is widely used as a component of toothpastes, denture retention adhesives, oral rinsing solutions, dental amalgams and others [121]. It has been speculated that Zn and some other divalent metals [122] may inhibit some proteinases. Zn-doped adhesives indeed exert a therapeutic/protective process by which the matrix metalloproteinase (MMP)–induced collagen degradation in the hybrid layer is prevented due to the local release of Zn^{+2} [120, 123], resulting in stable bond strength by reducing degradation of exposed collagen over time [123], and the effect is more pronounced on self-etch systems, where lesser collagen is left unprotected due to the mildness of the self-etchant [85]. The effect of zinc chloride ($ZnCl_2$) on human dentine evidenced that the collagen MMP–mediated degradation is inhibited not only by zinc's role in enzyme catalysis but also in protein folding/stability [124]. Interestingly, zinc oxide (ZnO) nanoparticles preferentially remain at the bottom of the hybrid layer in direct contact with the demineralized dentine (Fig. 8.2) [121], but no information about the size of the nanoparticles employed was provided.

The effect of poly(methacrylic acid)-grafted-nanoclay (PMAA-g-nanoclay) as a reinforcing filler and the potential chemical interaction with Ca^{+2} ions from hydroxyapatite has also been investigated [125–127]. Although significantly high diametral tensile strength and flexural strength [126] as well as superior bond strength to dentine [125] are observed when 0.5 % PMAA-g-nanoclay is added to a commercial dentine bonding, higher concentration of particles should be avoided as this causes agglomeration of PMAA-g-nanoclay, which acts as weak points prone to crack initiation, compromising mechanical properties of the material as well as its bond strength to dentin [125–127]. The agglomeration also results in an inverse correlation between microshear bond strength and concentration of PMAA-g-nanoclay in the adhesive [127].

Apart from filler addition, the polymeric chain of the adhesive may also be reinforced by adding functionalized prepolymers into the adhesive formulation. Nanogels are 10- to 100-nm cross-linked globular particles that can be swollen by and dispersed in monomers such as Bis-GMA and HEMA [128] and are anticipated to carry nanoparticles into the demineralized dentin when added to adhesives [129]. Experimental dentin adhesives containing either UDMA or ethoxylated bisphenol-A dimethacrylate (BisEMA)–based nanogel (particle size range ~10–80 nm) evidenced that the size of the nanogel particles allows their dispersion into dentinal tubules and sometimes into the interfibrilar spaces [128]. The nanogel addition restricts oxygen diffusion rates at the material's surface, which may reduce oxygen inhibition during adhesive layer polymerization [129]. It also provides significantly higher bond strength to dentin and stable mechanical properties (flexural modulus and flexural strength) after 7 days water storage [129].

The literature search evidences that the results of incorporating bioactive and reinforcing nanofillers to dental adhesives are very promising. Based on the information presented above, the ideal bonding agent should provide high and stable mechanical properties to the hybrid layer and have some level of bioactivity, which would somewhat decrease the incidence of pulpal damage and the occurrence of secondary caries. The high surface area of nanoparticles as opposed to microparticles can contribute to a higher level of ion release and a highly bioactive restorative material. Just for illustration, considering a particle size of 53 nm, the specific surface area is 35.5 m^2/g, whereas in a hypothetical 1 µm-sized particle, the specific surface area would be 1.9 m^2/g [130]. However, the high surface area of those particles results in two major outcomes: agglomeration of particles that lay on the top of the hybrid layer obliterating the interfibrilar spaces and therefore, avoiding monomer infiltration and hybridization; and the higher the ion release, the less stable is the material. Much effort is still required to develop a material based on ideal ratios between nanoparticles with different sizes and characteristics to ensure high bond strength values and mechanical properties at the hybrid layer that will remain stable over time while reducing bacterial count on the restorative interface.

8.3.2 Restorative Composite Resins, Sealants and Luting Systems

In general, dental composites contain a mixture of at least two different fillers so that reasonable mechanical properties may be combined with radiopacity and some degree of fluoride release [56, 103]. Different manufacturing techniques are employed to obtain the inorganic fillers. Radiopaque glass, quartz or ceramics are generally ground from larger particles until a size between 0.2 and 5 µm is obtained. Other fillers are prepared by the sol-gel route, starting from tetraalkyl orthosilicates or metal alkoxides such as titanium and zirconium ethoxide [131] or different mixtures [56]. The shape of the filler is dependent on the manufacturing process employed [56, 131] and influences the characteristics of the resin-based material [132]. The sol-gel method, for example, results in spherical particles of approximately 5–100 nm (referred to as *nanoparticles*) with a large specific surface area, which can show a pronounced thickening effect of the resin matrix [72, 132]. This effect can be reduced by using unassociated nanoparticles well dispersed in the matrix [3]. Furthermore, the large surface area still hinders the incorporation of great amounts of nanoparticles, compromising the mechanical properties of the restorative composite [132, 133]. So, it seems necessary to associate nanofillers with other particles capable of reinforcing the matrix and guaranteeing good handling properties [96].

The current composites available for posterior restorations do not release ions capable of remineralizing the tooth structure such as Ca, PO_4 and F. In contrast, restorative materials that release those ions are relatively weak and cannot be employed in stress-bearing areas [79, 133]. The incorporation of octyl alkylated quaternary ammonium poly (ethylene imine) (QA-PEI) particles in dental restorative composites has been investigated as a possible strategy to develop antibacterial composites [134, 135]. QA-PEI (1 % w/w) powder (~320 nm) added to a commercial resin composite (Filtek Flow, 3M) demonstrated at least 3 months antibacterial effect against *S. mutans*, probably due to damage to the bacterial cellular membrane [135], which happens at the material's surface and not by releasing cytotoxic components [134]. *S. aureus* inoculated on 1 % QA-PEI nanohybrid composite was eliminated after 24 h, probably due to the loss of their defined cell boundaries, indicating that QA-PEI-based composites are bactericidal and not bacteriostatic [136]. The *in situ* analysis of QA-PEI-based commercial composites showed biofilm formation on both surfaces. Nonetheless, the biofilm on QA-PEI-based materials had fewer bacteria and indistinct membranes, indicating that cell death at the surface may have triggered a mechanism named 'programmed cell death' in the surrounding bacteria [137]. Biocompatibility of QA-PEI nanoparticles has already been evidenced by cell viability and agar diffusion tests [135, 136].

As previously mentioned for dental adhesive systems, amorphous calcium phosphate (ACP) has the potential to remineralize the surrounding tooth structure by transferring mineral ions (Ca and PO_4) into the body of the lesion and therefore restoring the mineral lost to acid attack [138]. However, CaP-based composites have low mechanical properties and are not adequate to be employed as bulk restorative materials [108]. The larger surface area of nano-ACP particles (nACP: 17.76 m^2/g) [96] as opposed to the surface area of micro-ACP particles (0.5 m^2/g) [139] allows for the incorporation of significantly smaller amounts of nACP, leaving room for the placement of reinforcing non-releasing glass fillers [96]. Evaluation of different ratios of nACP indicates that flexural strength and modulus of experimental nanohybrid composites containing either 10 or 20 % nNACP fillers are initially similar to those of hybrid commercial materials [96, 140] and superior after 2 years water storage [141]. Furthermore, the higher the concentration of nACP, the higher the ion release [96], acid neutralization capacity and the bactericidal effect [140]. Interestingly, those materials present a 'smart' performance, whereby greater ion leaching is observed in cariogenic pH of 4.0 [96]. Results indicate that experimental nACP-based composites would withstand mechanical challenges of clinical applications where hybrid composites have been recommended [96], but this requires specific clinical investigations. Moreau et al. [140] observed that nACP-based composites may be indicated to areas where complete removal of caries tissue is contraindicated, in early carious lesions and in patients at high caries risk as in the case of patients undergoing radiation therapy and consequently presenting dry mouth [141].

When the resin matrix is modified with quaternary ammonium dimethacrylate (QADM), a strong anti-bacterial monomer, in addition to nACP particle incorporation, pronounced antibiofilm capacity is observed up to 180 days [102]. The degree of conversion of the experimental material is also higher than that of nACP-based composite without QADM, probably due to the low viscosity of QADM, which improves the mobility of the reactive species. Elastic modulus and flexural strength of nACP-QADM composites are similar to those of commercial composites (Renamel, Cosmedent – microhybrid; Heliomolar, Ivoclar – microfill) after 180 days water immersion [102]. To maximize the caries-inhibiting capabilities of a restorative composite resin, silver nanoparticles (nAg) have been added to nACP-QADM-based composites. The nACP+QADM+nAg composite greatly reduces *S. mutans* biofilm growth, metabolic activity, colony-forming unit (CFU) viable bacterial counts and lactic acid production when compared to commercial materials (Renamel and Heliomolar) and to QADM and nAg separately.

Mechanical properties are comparable to those of the commercial materials [142], but further studies are needed to investigate the long-term effect of that association (nACP-QADM- nAg).

Silver is a safe antibacterial metal that is not toxic for eukaryotic cells, while it is severely toxic and lethal to prokaryotic cells [143]. Addition of nAg to commercial microhybrid composite resins (P90 and Z250 – 3M) increases the contact angle at the materials' surface, thereby limiting the adhesion and proliferation of bacterial pathogens [144]. However, the synthesis and dispersion of silver nanoparticles are challenging because of the strong propensity for particle agglomeration, reducing the surface energy [145]. The synthesis of cross-linking dimethacrylate/silver nanohybrid composites by coupling photopolymerization particles evidenced that ~3 nm particles were well dispersed throughout the matrix [130]. However, only concentrations as low as 0.02 % do not affect mechanical properties due to the aggregation of nanoparticles in higher nAg concentrations. Higher concentrations are not beneficial to the antibactericidal effect of the resin composites as well [130] and tend to compromise the degree of conversion of the light-cured restorative material [146]. Chemically cured composites with nAg incorporated *via* benzoate (Bz) may present higher degrees of conversion and higher release of Ag^+ due to the slower curing process, which allows for a greater number of nAg nucleation sites to form, thereby generating more particles, a better dispersion of the particles and smaller particle sizes [146]. Light-cured nAg-based materials seem not to be biologically safe since a small concentration of nAg triggers the release of (co) monomers TEGDMA, BisEMA and camphorquinone (CQ) [147]. It is hypothesized that the reflection and scattering of the light by the silver particles would be responsible for the less amount of light reaching out the surface, which reduces the degree of conversion [146, 147].

The effect of nano-zinc oxide (nZnO) incorporation into dental composites may also have a clinically relevant antibacterial effect. Although an experimental material presents mechanical properties similar to those of hybrid materials and *S. mutans* strains are inhibited at the surface, the penetration of visible light is compromised by the opacity of ZnO, adversely affecting the curing process [148].

Another alternative to control the occurrence of recurrent caries is to increase the dissolution resistance of the tooth structure by the incorporation of fluoride (F) released from restorative materials [149], which is capable of promoting remineralization and inhibiting microbial growth and metabolism [26]. Resin-based materials containing calcium fluoride nanoparticles ($nCaF_2$) present high F release rates within a period of 10 weeks [150]. The stability of CaF_2-based composite resins has been recently investigated [151]. The average particle size distribution is about 10 nm, but particles are likely to aggregate, forming clusters as large as 100–300 nm. The mechanical properties of the experimental material are highly dependent on the glass filler particle levels; the higher the glass filler level, the higher the strength during thermal cycling and 2-year water ageing and the better the wear resistance [151]. The $nCaF_2$ paste has a greyish colour due to the metal ion incorporated during the spray-drying process, indicating that the technique needs to be optimized before those particles can be incorporated into clinical restorations [150]. Chlorhexidine (CHX) particles have been combined with $nCaF_2$ and nACP in experimental composites [142] to release CHX in an attempt to achieve a restorative material that promotes remineralization, is antibacterial and withstands load-bearing applications [152, 153]. However, all experimental materials presented decreased flexural strength from day 1 to day 28 in water storage [142]. CHX release rate after 30 days (about 2 %) is similar between nACP- and $nCaF_2$-based composites. This concentration is optimal to slow down or eliminate bacterial growth and to reduce acid production.

Whiskers are single crystals possessing high degrees of structural perfection and high mechanical properties [154]. Silicon carbide whiskers have the potential to reinforce the resin matrix by pinning and bridging cracks during flexural strength tests [29, 155]. The impregnation of whiskers with nano-dicalcium phosphate anhydrous (nDCPA) results in fast release of Ca and PO_4 ions due to the large surface area of the nDCPA particles [29], which has been shown to reprecipitate and form hydroxyapatite inside tooth lesions outside the tooth–restoration interface [156]. The smaller the particle size, the larger the surface area and the higher the amount of ion released [30, 157]. The employment of unsilanized nDCPA also releases significantly more ions than silanized nDCPA because the silane coupling agent hinders the diffusion of water and ions through the matrix/filler interface [158]. However, the effect of incorporating unsilanized particles to the restorative materials is not known yet, and it may vary from compromising the mechanical properties to the degradation of the resin matrix due to water/saliva seepage. There appear to be three main factors influencing the Ca and PO_4 ion release from the composite: a higher volume fraction of nDCPA in the composite increases (1) the source of ions and (2) the filler–matrix interfacial area, which serves as a path for water and ion diffusion; (3) the resin matrix may have a slightly lower polymerization conversion [29, 79]. Additionally, the opacity of an experimental composite with silica-impregnated whiskers caused by the refractive index mismatch between the whiskers and the resin matrix limits the application of the material [79].

Hannig and Hannig [159] have recently hypothesized that prior to the less abrasive diet of modern days, nano-sized crystallites were found in the oral cavity as a result of physiological wear of the enamel surface due to abrasion and attrition with the food bolus. Those crystallites would be responsible for promoting tooth remineralization and 'bio-film management'. In a biomimetic approach, hydroxyapatite nanocrystals which resemble the structure of the nano-scale abraded enamel have been employed to reduce bacterial adherence and have an impact on biofilm formation [160]. The comparison among HA whiskers, nanoparticles (~50–100 nm) and sphere particles (~1–3 um diameter) evidenced that the type of HA filler employed has a significant effect on mechanical properties of experimental dental composites and the higher the concentration of HA, the lower the concentration of glass particles and the lower the hardness [161]. Nano-HAs are not homogeneously distributed throughout the matrix, and it is suggested that different sizes of HA fillers should be combined in future studies so that ideal handling characteristics and high mechanical properties can be achieved [161].

Glyoxylic acid (GA), a natural product present in fruits, leaves and sweet beet, has been used to modify the surface of nHA and reduce formation of agglomerates in the resin matrix [133]. Although the nanoparticle dispersion is slightly better in comparison to untreated nHA, agglomerates are noticed when concentration of treated nHA increases to 5 %. The treatment with GA resulted in higher water solubility of the resin matrix, probably due to the weak hydrogen bond established between carboxyl and hydroxyl groups on the GA-modified nano-HA and the resin matrix [133].

Nanotechnology has opened a wide variety of possibilities to improve the clinical performance of resin-based materials. For instance, the association of antibacterial monomers and bioactive glass nanoparticles (~45 nm) may improve fluoride release and reduce microleakage scores of experimental dental sealants [146]. Halloysite nanotubes (HNTs), which are naturally occurring minerals, are easy to purify and available in nature in a large quantity. They are safe and biocompatible and have a diameter of tens of nanometers and length ranging from ~200 nm to 1–2 µm [132]. HNT mixed to an experimental resin matrix indicates that increase in HNT concentration results in higher polymerization shrinkage and low mechanical properties while associated with cluster formation [132]. The effective dispersion of HNT and a consequent improvement of the mechanical properties require further investigation.

The application of nanotechnology into resin-based luting agents has also been investigated. The incorporation of nanoparticles in dental cements is somewhat desirable since it may improve the mechanical properties of the luting systems without increasing the film thickness of the material. However, the incorporation of 2.5 % silica nanoparticles (~7 nm) compromises the flexural strength and film thickness [162]. This effect is probably due to the particle entanglement and agglomeration. Considering that spherical particles would have only one point of contact between one another, researchers expect the tendency to agglomerate to be reduced since less energy would be needed to break those interparticle interactions [162]. Furthermore, poor connective forces within the clusters allow them to act as spots of stress concentration. If strong particle-to-particle connective forces could be provided, those agglomerates might have a protective effect on the material, which is not yet observed [162].

Commercial Alternatives

One of the most difficult challenges in restorative materials is to combine strength, handling characteristics and optical properties similar to the tooth structure. Two types of resin composites have been recently developed employing nanotechnology: Filtek Supreme Standard (FSS – 3M Espe), with zirconia–silica clusters (0.6–1.4 μm) and nanoparticles (average particle size ~20 nm), and Filtek Supreme Translucent (FST – 3M Espe), with nanoparticles (~75 nm) and a minor amount of silica clusters (0.6–1.4 μm) [3]. Due to the spheroidal shape of the particles and clusters, low stress concentration is observed in the resin matrix in comparison to composites containing irregular-shaped fillers with sharp edges, and the material exhibits a distinct fracture mechanism with enhanced damage tolerance, whereby cracks are dissipated into the cluster porosities [42].

Different brand names from the same manufacturer often have been assigned to materials with the same composition. An example is the 3M Espe nanocomposites (Filtek™ Supreme XTE, Filtek™ Z350 XT and Filtek™ Supreme Ultra). Within a brand, different opacities may be available, which will have an impact on the material's composition. For example, the aforementioned Filtek™ brands are available in four opacities: Dentin (D), Enamel (E), Body (B) and Translucent (T). The filler composition for the D, E and B shades are all identical, representing 78.5 % of the total weight, with clusters containing nanofillers comprising 90 % of the filler loading. They differ, primarily, in opacity.

The T-Shades utilize a different filler composition and are formulated to 72.5 wt% filler, with 50 % of its filler loading being comprised of clusters containing nanofillers. The first and second generation of this composite (Filtek™ Supreme and Filtek™ Supreme Plus/Filtek™ Supreme XT and Filtek™ Z350, respectively) used the same aforementioned shading strategy. Therefore, to facilitate correlation of different research data, Filtek composites (3M Espe) will be referred to as 'Filtek DEB' (dentin-enamel-body) and 'Filtek Translucent'.

In 1984, organic–inorganic polymeric hybrids were introduced by Schmidt [163, 164]. They were first called 'ormosils' (organically modified silicates) and later named 'ormocers' (organically modified ceramics). Ormocers based on urethane or carboxy-functionalized methacrylate alkoxysilanes have been used in dentistry [165] in an attempt to reduce the necessity for diluents and, therefore, increase the filler particle loading [166]. The association of ormocers and nano-SiO_2-ZrO_2 particles (~250 nm) resulted in experimental materials with increased viscosity and therefore reduced double-bond conversion. In another study, an ormocer-based commercial material (CeramX, Dentsply) presented significantly lower monomer elution and thus lower toxicity in comparison to a nano-filled composite (Filtek DEB) and to a chemically cured resin for core build-up (Clearfil Core, Kuraray) [72]. Considering that the filler loading was similar among all the materials employed, this may be an indication of a different chemical interaction between the surface of the fillers and the resin matrix [68]. The higher the degree of conversion of a composite resin, the lower the monomer elution from the restorative system [147]. Therefore, it may be speculated that the degree of conversion of an ormocer is higher than the one of a nanofill composite. The fracture strength of premolars with MOD preparations restored with either ormocer-based composites (Admira, Voco; CeramX, Dentsply), a nanofilled composite (Filtek DEB) or a microhybrid composite (Tetric Ceram – Ivoclar) was significantly better when teeth were restored with an ormocer-based material (CeramX) [67].

The evaluation of water sorption and its effect on selected properties of resin-based materials (Tetric EvoCeram; Grandio, Voco – classified as nanohybrids; Filtek DEB and Filtek translucent, nanofill) indicates that polymerization shrinkage, water sorption and solubility are critical for nanocomposites, probably due to the lower filler content (Table 8.1). Flexural strength and flexural modulus also present a significant decrease after 30 days storage in water or saliva for nanocomposites [65]. Knoop hardness of a nanocomposite is also compromised by 6 months water storage [74]. Surprisingly, biomechanical degradation (three-body wear test + *S. mutans* biofilm degradation) does not affect the surface roughness of a nanocomposite (Filtek DEB) in comparison to microhybrid and glass ionomer materials [43], probably due to the composition of the material where clusters and nanofillers are associated, filling in the interstitial spaces and protecting the softer matrix [43]. When the nanocomposite undergoes toothbrush abrasion, only nano-sized particles are plucked away, leaving the surface with defects smaller than the wavelength of light [3].

The polishability of Tetric EvoCeram, referred to as *nanohybrid* [66] due to the presence of nanofillers and microparticles, has been compared to a nanocomposite (Filtek translucent) (Table 8.1). Both materials present low values of roughness, with no signs of dislodged nano-filled clusters or nanofilled prepolymer particles [66]. However, when surface smoothness is evaluated after cyclic loading, the nanocomposite (Filtek Translucent) does not retain the polishability as much as the microhybrid material does (Opallis, FGM) [73].

With regard to the aesthetic potential of nano-based materials, the colour stability as a function of polishing systems and additional curing source has been evaluated for a nanocomposite (Filtek DEB) and a nanohybrid (Clearfil Majesty, Dentsply) material. Additional heat post curing causes colour change for both materials, which is significant after 7 months water storage [75], but better results are presented by Filtek DEB. This phenomenon is probably related to the degradation of the matrix due to temperature rise, increasing the chance of discoloration due to absorption of pigments during the storage time. The immersion of samples of microhybrid (Arabek, Voco) and nanohybrid (Grandio Nano, Voco) composites in coffee, tea or Coke and the investigation of the colour change indicates that nanohybrid composite presents less colour stability in spite of the higher degree of conversion. The staining susceptibility of the nanohybrid composite may be related to the hydrophilic nature of the matrix, which is capable of absorbing fluids and pigments [76]. Considering that low degree of conversion results in the elution of cytotoxic monomers [147], it is necessary to find a midpoint between biosafety and aesthetic stability to guarantee the success and longevity of resin-based restorations.

To predict the clinical performance of dental restorations, it is crucial to make an early assessment of new materials by using *in vitro* investigations. However, the most conclusive and validated data are obtained from clinical studies, whereby all potential variables influencing the restorative's material performance can be taken into account. The evaluation of occlusal restorations made either with ormocer-based composite (Admira), nanohybrid composite (Tetric EvoCeram, Ivoclar), nanofill resin composite (Filtek DEB) or microhybrid resin composite (Tetric Ceram, Ivoclar) indicates no difference among all restorations after 2 years. Only one ormocer and one microhybrid composite restorations failed after 2 years, but the clinical performance of all the materials was similar [167]. Similar findings are observed when occlusal maxillary cavities are performed either with Admira, Filtek DEB or a hybrid composite (Renew, Bisco) with regard to secondary caries and post-operative sensitivity [168]. The assessment of ormocer-based (Admira) or nanofill composites (Filtek DEB) lined or not with flowable composites has also been performed in occlusal restorations [169]. The clinical performance of both materials was similar after 2 years [169]. Another 2-year clinical evaluation of nanofill (Filtek DEB) and microhybrid (Tetric Ceram) materials using the split-mouth study design, where similar cavities are prepared on both sides of the arch either in molars or premolars (from 2 to 4 surface fillings) indicated similar performance between them, considering variables such as marginal adaptation, anatomical form, secondary caries, colour match and possible complaints [71]. However, the authors observed a higher chance of occlusal staining when nanofill composite is employed. Clinical studies are fundamental for characterization of the performance of a given material, but the results presented up to date must be cautiously considered since the longest observation period of nano-based composites published up to the preparation of this chapter was 2 years.

A promising application of nanotechnology would be to develop a new class of sealants. Sealants are fluid resins with or without fillers and have been used for nearly 40 years to prevent caries [170]. The antibacterial property of some monomers such as methacryloyloxydodecyl pyrimidinium bromide (MDPB) [171] combined with bioactive glass nanoparticles significantly increases fluoride release and recharge of sealants and reduces microleakage scores when compared to commercial sealants (Fluroshield, Dentsply; Clinpro, 3M Espe; and SeLECT Defense, Element 34 Technology). However, as previously mentioned for other classes of materials, the surface sectioning evidenced nanoparticle aggregation [172], and the impact of those on the mechanical properties of sealants has not been reported as yet.

Expanding the application of resin-based nanocomposites, there has been some development on resin-based luting systems employing nanoparticles. Although the filler loading is lower for resin cements (~60 wt%) [173] in comparison to restorative composites (~75 wt%) [75], the incorporation of nanoparticles may be beneficial to absorb stress generated under cemented crowns/prostheses, and the reduced size of the particles may facilitate the control of the film thickness, which is critical when luting indirect restorations. Cement type is one of the factors that may influence the longevity of an indirect restoration [174]. The water ageing of the resin cement reduces the load-bearing capacity to interfacial debonding of a lithium disilicate crown by 45 % [11]. The fracture mode of the restoration is also influenced by cement ageing, and it switches from surface chipping to radial cracking, evidencing how critical is the cement layer for the mode of failure and longevity of an indirect restoration [175]. The development of an experimental resin-based cement with silica particles (~7 nm) indicated that the incorporation of up to 2.5 % is beneficial to the mechanical properties. Beyond that, mechanical properties are compromised by particle entanglement and agglomeration, acting as spots of localized stress concentration [162].

8.4 Indirect Restorative Systems

The failure process in clinically unsuccessful restorations is known to follow a sequence of events: (1) cracks develop underneath or near the occlusal contact area due to a repetitively applied occlusal force in the presence of saliva; (2) the developed cracks cause subcritical crack growth in the presence of saliva, leading to fracture of the veneering porcelain; (3) when the cracks extend to the coping interface, the veneering porcelain chips, exposing the veneer-coping interface [176–178]. There have been many attempts to inhibit the fracture and chipping of the veneering porcelain to improve the mechanical reliability of the restoration. The particle size of the powders employed in the indirect restorative material plays an important role in determining the microstructure and mechanical properties of the ceramics [179]. Ideal powders should have small particle size and narrow particle size distribution. For ceramic materials, *nanocomposites* are ceramics in which nanometer-sized secondary particles, with a composition distinct from the primary particles, are dispersed within the ceramic matrix grains and/or at the grain boundaries. Nanocomposites have been developed and investigated as an alternative to improve the reliability of the indirect restoration [180]. There is a significant improvement in strength due to a decrease in flaw size associated with the intragranular nano-dispersion. However, the technology to employ nanoparticles in ceramic-based restorations is not fully developed. The high temperatures required to fully densify ceramic powders result in large grain sizes when traditional sintering techniques are used due to a phenomenon named 'Ostwald ripening', by which crystals or particles dissolve and are deposited onto larger crystals [181]. This makes it extremely difficult to obtain dense materials with nanometric and submicronic grain sizes [182].

Yttria–tetragonal zirconia polycrystal (Y-TZP) is a structural ceramic with high crystalline content and outstanding mechanical properties. Nonetheless, the material is unstable at room temperature due to crystalline changes from tetragonal to monoclinic when internal compressive stresses are generated in the surroundings of the grains [183]. To develop a new machinable ceramic with improved mechanical properties and more stable tetragonal phase, Y-TZP particles (average particle size 0.3 μm) were coated with alumina (Al_2O_3) nanopowder (average particle size 20 nm). Alumina nanoparticles were added as a way to inhibit grain growth of tetragonal zirconia (Zr) in an attempt to combine high strength and toughness, and phase transformation was indeed controlled. This effect may be better optimized by homogeneous distribution of alumina grains in the Zr matrix. According to the authors, the new ceramic powder developed is well suited to fabricate all-ceramic inner-core frameworks using computer-aided design/computer-aided manufacturing (CAD/CAM) technique [184].

The association of Y-TZP (40 nm) and Al_2O_3 (~3.6 μm) grains has also been investigated in the development of a zirconia-toughened alumina (ZTA) skeleton infiltrated with a thin glass film [181]. The higher the zirconia content in the skeleton, the higher the bending strength and fracture toughness, and those are probably related to the transformation-toughening mechanism of zirconia's tetragonal phase, by which local tetragonal–monoclinic crystalline changes occur, sealing the tip of a crack [185]. Although the glass film infiltrated the whole depth of the experimental samples, the penetration depth may be a critical and limiting factor due to reduced size of the pores [181].

Another alternative to the Yttrium-stabilized zirconia is the use of Cerium dioxide as stabilizer (CeO_2). CeO_2-stabilized tetragonal zirconia polycrystal (Ce-TZP) presents high toughness and higher resistance to low-temperature degradation, but strength and hardness still require improvement [186, 187]. Many attempts have been made to increase the material's strength while maintaining high toughness values [188, 189]. The association of Ce-TZP, titanium dioxide (TiO_2) and alumina powders in different proportions and sintered under different conditions resulted in nanoparticles of alumina (average size 10–100 nm) trapped within the zirconia grains, which were expected to increase the tetragonal grains' stability by reducing flaw size due to a homogeneous intragranular nano-dispersion. An optimized nanocomposite (0.05 mol% TiO_2-doped 10Ce-TZP/30 vol% Al_2O_3) resulted in both high strength and fracture toughness, and agglomerates are not observed under SEM [189].

Diatomite is a biocompatible porous silicate structure with high adsorption capacity, low density and low cost [190]. An experimental technique using diatomite particles impregnated with ZrO_2 nanoparticles (~80 nm) has also been recently investigated, and the experimental ceramic materials present reduced porosity with homogeneous distribution of small-sized particles, resulting in improved fracture toughness in comparison to ZrO_2-free materials [11]. Other techniques have been employed to develop a ceramic material with nanocrystalline structures [191, 192], but they are still in a developmental phase, and the mechanical characterization of the materials and their clinical stability need to be investigated.

The chipping of porcelain veneer layer is the most frequent failure of all-ceramic prostheses [193]. Leucite glass ceramics are used in restorative dentistry as a veneer material for all-ceramic restorations. However, properties such as brittleness, which is related to early fracture, and wear of the opposing tooth limit its clinical application [194]. Milling of an aluminosilicate using the top-down approach and heat treatment allows for the nucleation and growth of nano-scale leucite crystals inside the glass matrix, which results in higher biaxial flexural strength than that of commercial materials (e.g. IPS Empress and Ceramco-3) [195]. However, the mechanical characterization of the material under cyclic loading and the optical properties have not been evaluated yet and require further investigation before the material's clinical application.

Another approach to reduce the chipping by increasing the toughness of the veneer layer is the incorporation of silver nanoparticles (Ag ~10 nm) into a dental porcelain (Noritake Super, Noritake Dental Supply) in different concentrations (100, 200, 500 and 1,000 ppm). The higher the Ag concentration, the higher the hardness and the toughness of the experimental material; 500 ppm of Ag resulted in the highest Young's modulus. However, the colour difference is significantly greater than the perceptibility threshold [196]. The coefficient of thermal expansion of Ag is higher than that of the matrix glass. So, tensile stresses are developed in the radial direction and compressive stresses in the tangential direction. When the crack extends towards the silver particles, they are inhibited by the compressive stress in the tangential direction, causing crack deflection and possibly reducing porcelain chipping [197].

8.4.1 Commercial Alternatives

NanoZr is the brand name of a Ce-TZP/Al$_2$O$_3$ (ZrO$_2$, 67.9 mass%; Al$_2$O$_3$, 21.5 mass%; CeO$_2$, 10.6 mass%; MgO, 0.06 mass%; TiO$_2$, 0.03 mass% – Matsushita Electric Works) nanocomposite. Interestingly, static biaxial flexural strength and cyclic fatigue resistance of NanoZr are significantly higher than those of Y-TZP [77]. The application of 6 mechanical cycles to NanoZr and Y-TZP after different surface treatments (airborne particle abrasion and acid etching) indicates that although a slight increase in monoclinic content is noticeable, it does not influence fatigue strength [77]. According to the authors, the crystalline change may have had a 'transformation toughening effect', by which cracks were sealed due to local grain growth [77]. However, it is valid to consider that the monoclinic content presented a steady increase and, in the clinical scenario, other challenges may be present, such as pH and temperature oscillations, variation in mastication loads and

realistic corresponding time (1–2 years) whereby all the factors combined might maximize the risk of uncontrolled crystalline change. Considering that the mechanical properties of an experimental Ce-TZP/Al$_2$O$_3$ are somewhat comparable to those of Y-TZP [198], the reliability of Ce-TZP/Al$_2$O$_3$ (NanoZr) nanocomposite and Y-TZP veneered framework has been investigated. Results indicated higher fracture strength of Y-TZP. However, when Ce-TZP/Al$_2$O$_3$ is used as infrastructure, higher incidence of porcelain chipping occurs instead of bulk fracture, indicating that Ce-TZP/Al$_2$O$_3$ presents higher resistance to crack propagation and, therefore, higher fracture toughness [199].

Fischer et al. (2009) observed that superior mechanical properties are not the only prerequisite for the clinical success of indirect restorations and that the compatibility between core material and veneering ceramic is also critical [200]. The investigation of the compatibility between different veneering ceramic materials and NanoZr coping in single crowns and the fracture strength results evidenced a strong correlation between the thermal properties of the dentin material of the veneering ceramics and fracture strength. The coefficient of thermal expansion of the veneering material should ideally be slightly below (1 μm/m.K) that of the coping material [78, 200, 201] so that the crack propagation towards the interface is impeded by the compressive stresses, therefore reducing incidence of porcelain chipping [78]. However, low coefficient of thermal expansion results in excessive compressive stress and decreases fracture strength. In the same way, high coefficient of thermal expansion of the veneer material leads to tensile stress at the interface, leading to early or even spontaneous cracking [202]. Furthermore, the results of fracture strength of NanoZr crowns (574–1,009 N) are higher than the maximum *in vivo* biting forces, which are in the range of 400 N in the molar region [201, 203]. Additional investigation on the compatibility between Ce-TZP/Al$_2$O$_3$ coping (NanoZr) and porcelain evidences that airborne particle abrasion [78, 201] and the application of a liner [78] do not improve bond strength between coping and veneer material, the properties of a specific zirconia/veneer combination may not apply to other combinations and every core/veneer combination (e.g. association of different brands) should be tested before clinical application [78].

Lava Ultimate (3M) is an indirect restorative material advertised as a resin nanoceramic, which contains nanoparticles (silica: 20 nm; ZrO_2: 4–11 nm) [204] in a highly cured urethane dimethacrylate (UDMA)–based matrix [205]. It is expected to have intermediate mechanical properties, slightly higher than direct composites and lower than crystalline ceramics. However, a comparison of the fatigue resistance of ceramic-based systems and indirect restorative composites evidenced that the latter may be securely indicated only for low mastication load areas, where a similar mechanical behaviour is expected between indirect and direct composites [204]. Nonetheless, even for those areas, high–crystalline content zirconia and lithium disilicate–based ceramics would be more suitable based on the fact that longer restoration lifetimes may be expected with those materials [204]. Although Lava Ultimate has not been fully characterized yet, it seems to present good fluorescence and light transmission for characterization of anterior teeth [205]. A clear advantage of indirect restorative composites is that they allow further characterization of the restoration either in the model or in the mouth by the application of dyes and translucent composites, resulting in more individualized aesthetics, as may be seen in Fig. 8.4.

8.5 Future Perspectives of Nanoparticles in Restorative Materials

8.5.1 Resin Composites

Although the application of nanotechnology into dentistry has increased consistently in the past two decades, the application of nanoparticles to restorative materials requires further investigation. Theoretically, the smaller the particle size, the smoother the surface will be and more polish will be retained with less plaque accumulation on the surface of the restoration as well as at the tooth–restoration interface. However, the technology available up to this date indicates that nano-based composites present water sorption,

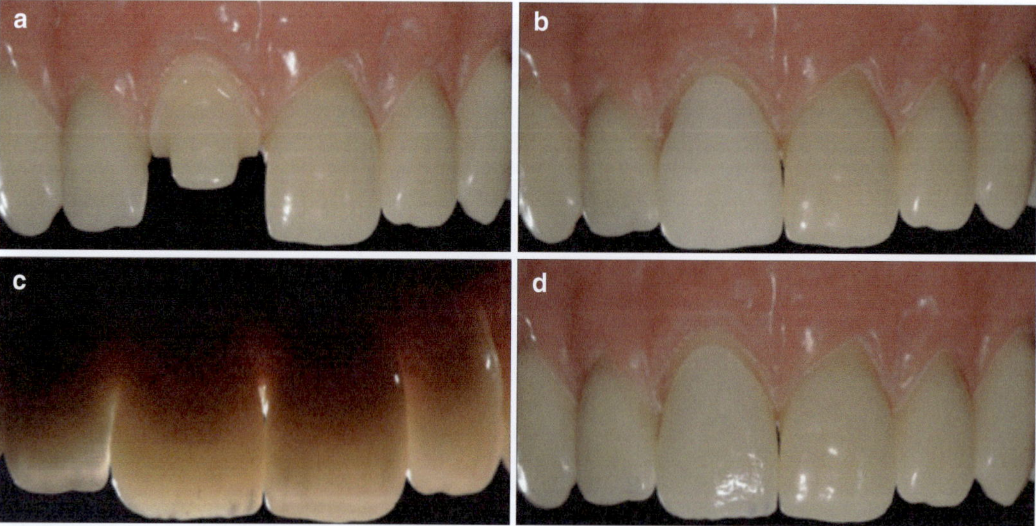

Fig. 8.4 Clinical case of a crown cementation on tooth 11 using indirect restorative composite, also known as resin nanoceramic: (**a**) Crown preparation on tooth 11; (**b**) Initial appearance after crown was milled; (**c**) Final result evidencing translucence of the indirect restoration; (**d**) Labial view of crown (Reprinted from Koller et al. [206]. With permission from Quintessence Publishing Co Inc.)

solubility and therefore low colour stability. Restorations also do not retain polishability as much as the microhybrid composites. Those results raise two concerns: the first one is that the benefit of nanoparticles in restorative composites has not been evidenced as yet; the second one is that the inferior properties of nano-based composites are mostly related to the hydrophilic nature of the matrix, where unreacted C=C and amines may be linked to the material's discoloration. The monomeric systems used also have an impact on water sorption and discoloration. TEGDMA, for example, is more prone to discolour than UDMA. Hence, the clinical limitations of the nano-based materials are not only linked to the size of the particles employed and their tendency to agglomerate but also to the overall degradation of the material, indicating that there is a current need for the investigation of new monomeric systems. Nanoparticle dispersion is still difficult to control, and this may be overcome with the application of other polymers in dentistry. The final purpose would be to associate high filler loading well dispersed within the matrix, optimized handling characteristics and less affinity to humidity.

8.5.2 Further Applications to Hydroxyapatite

Further to the improvement of mechanical properties and bioactivity of resin-based materials, research in nanotechnology has showed that HA nanoparticles are capable of working as building blocks to develop dental enamel [207]. Nanosized HA presents a biomimetic capacity more applicable than the conventional approach of biomineralization proposed with micro-sized HA. Nano-HA (~20 nm) can be self-assembled to form enamel-like prisms *in vitro*. They can also adsorb onto human enamel surface strongly and be fully integrated with enamel structure, increasing the enamel's capacity to resist acidic challenges. This concept has many applications in preventive/restorative dentistry, from surface sealants with nano-HA to the facilitation of macroscopic enamel remineralization. A few studies

have also demonstrated that HA has the capacity to reinforce collagen demineralized by acid etching. However, considering the interfibrilar space within the demineralized dentin (~20 nm), only non-aggregated particles smaller than that (~5–15 nm) would be of interest to enhance the stability of the hybrid layer. Furthermore, the application of those particles in desensitizing agents may offer a longer-lasting solution to cervical non-carious lesions either by sealing the dentinal tubules or by reinforcing the exposed collagen, making it less susceptible to future acidic challenges.

8.5.3 Concluding Remarks

The technology of incorporating nanoparticles into ceramic materials is complex. As previously mentioned, the high temperatures required to fully densify ceramic powders results in dissolution of the crystals, which are then deposited onto larger crystals. This is the reason why nanotechnology has not evolved in ceramic-based materials as much as it has in direct restorative materials. Initial results evidence that the association of nanoparticles around micro-sized zirconia crystals allows for control of tetragonal-to-monoclinic phase transformation. Additionally, the understanding of the crack propagation mechanism through porcelain veneer materials indicates that high content of nanoparticles within the veneer material would increase the complexity of the crack propagation by modifying the direction of the subcritical crack growth consequently either stopping it or significantly postponing it, increasing the longevity of the indirect restoration.

References

1. Mohamed Hamouda I. Current perspectives of nanoparticles in medical and dental biomaterials. J Biomed Res. 2012;26(3):143–51.
2. Shetty NJ, Swati P, David K. Nanorobots: future in dentistry. Saudi Dent J. 2013;25(2):49–52.
3. Mitra SB, Wu D, Holmes BN. An application of nanotechnology in advanced dental materials. J Am Dent Assoc. 2003;134(10):1382–90.

4. Allaker RP. The use of nanoparticles to control oral biofilm formation. J Dent Res. 2010;89(11): 1175–86.

5. Monteiro DR, Gorup LF, Takamiya AS, Ruvollo-Filho AC, de Camargo ER, Barbosa DB. The growing importance of materials that prevent microbial adhesion: antimicrobial effect of medical devices containing silver. Int J Antimicrob Agents. 2009; 34(2):103–10.

6. Ahn SJ, Lee SJ, Kook JK, Lim BS. Experimental antimicrobial orthodontic adhesives using nanofillers and silver nanoparticles. Dent Mater Off Publ AcadDent Mater. 2009;25(2):206–13.

7. Saunders SA. Current practicality of nanotechnology in dentistry. Part 1: focus on nanocomposite restoratives and biomimetics. Clin Cosmet Investig Dent. 2009;1:47–61.

8. Stoimenov PK, Rosalyn L, Marchin GL, Klabunde KJ. Metal oxide nanoparticles as bactericidal agents. Langmuir. 2002;18:6679–86.

9. Rezaei-Zarchi S, Javed A, Ghani MJ, Soufian S, Firouzabadi FB, et al. A novel extracellular synthesis of monodisperse gold nanoparticles using marine alga, Sargassum wightii Greville. Colloids Surf B Biointerfaces. 2010;57:97–101.

10. Prentice LH, Tyas MJ, Burrow MF. The effect of ytterbium fluoride and barium sulphate nanoparticles on the reactivity and strength of a glass-ionomer cement. Dent Mater Off Publ AcadDent Mater. 2006;22(8):746–51.

11. Lu X, Xia Y, Liu M, Qian Y, Zhou X, Gu N, et al. Improved performance of diatomite-based dental nanocomposite ceramics using layer-by-layer assembly. Int J Nanomedicine. 2012;7:2153–64.

12. Patil S, Sandberg A, Heckert E, Self W, Seal S. Protein adsorption and cellular uptake of cerium oxide nanoparticles as a function of zeta potential. Biomaterials. 2007;28(31):4600–7.

13. Berg JM, Romoser A, Banerjee N, Zebda R, Sayes CM. The relationship between pH and zeta potential of ~30 nm metal oxide nanoparticle suspensions relevant to *in vitro* toxicological evaluations. Nanotoxicology. 2009;3(4):276–83.

14. Clogston JD, Patri AK. Zeta potential measurement. In: McNeil SE, editor. Characterization of nanoparticles intended for drug delivery methods in molecular biology, vol. 697. New York: Humana Press; 2011. p. 63–70.

15. Wagner A, Belli R, Stotzel C, Hilpert A, Muller FA, Lohbauer U. Biomimetically- and hydrothermally-grown HAp nanoparticles as reinforcing fillers for dental adhesives. J Adhes Dent. 2013;15(5): 413–22.

16. Sadat-Shojai M, Khorasani MT, Dinpanah-Khoshdargi E, Jamshidi A. Synthesis methods for nanosized hydroxyapatite with diverse structures. Acta Biomater. 2013;9(8):7591–621.

17. Crisp S, Ferner AJ, Lewis BG, Wilson AD. Properties of improved glass-ionomer cement formulations. J Dent. 1975;3(3):125–30.

18. McLean JW, Nicholson JW, Wilson AD. Proposed nomenclature for glass-ionomer dental cements and related materials. Quintessence Int. 1994;25(9):587–9.

19. McKinney JE, Antonucci JM, Rupp NW. Wear and microhardness of glass-ionomer cements. J Dent Res. 1987;66(6):1134–9.

20. Mitra SB, Oxman JD, Falsafi A, Ton TT. Fluoride release and recharge behavior of a nano-filled resin-modified glass ionomer compared with that of other fluoride releasing materials. Am J Dent. 2011;24(6): 372–8.

21. Smith DC. Development of glass-ionomer cement systems. Biomaterials. 1998;19(6):467–78.

22. Sidhu SK, Sherriff M, Watson TF. *In vivo* changes in roughness of resin-modified glass ionomer materials. Dent Mater Off Publ AcadDent Mater. 1997; 13(3):208–13.

23. Goenka S, Balu R, Sampath Kumar TS. Effects of nanocrystalline calcium deficient hydroxyapatite incorporation in glass ionomer cements. J Mech Behav Biomed Mater. 2012;7:69–76.

24. Rajabzadeh G, Salehi S, Nemati A, Tavakoli R, Solati Hashjin M. Enhancing glass ionomer cement features by using the HA/YSZ nanocomposite: a feed forward neural network modelling. J Mech Behav Biomed Mater. 2014;29:317–27.

25. Randall RC, Wilson NH. Glass-ionomer restoratives: a systematic review of a secondary caries treatment effect. J Dent Res. 1999;78(2):628–37.

26. Wiegand A, Buchalla W, Attin T. Review on fluoride-releasing restorative materials–fluoride release and uptake characteristics, antibacterial activity and influence on caries formation. Dent Mater Off Publ AcadDent Mater. 2007;23(3):343–62.

27. Xie D, Brantley WA, Culbertson BM, Wang G. Mechanical properties and microstructures of glass-ionomer cements. Dent Mater Off Publ AcadDent Mater. 2000;16(2):129–38.

28. Shoji AK, H, Kentaro TGC. Fluoraluminosilicate glass powder for dental glass ionomer cement. 1988. US patent US 4775592 A. http://www.google.ca/patents/US4775592.

29. Xu HH, Sun L, Weir MD, Antonucci JM, Takagi S, Chow LC, et al. Nano DCPA-whisker composites with high strength and Ca and PO(4) release. J Dent Res. 2006;85(8):722–7.

30. Xu HH, Moreau JL, Sun L, Chow LC. Novel CaF(2) nanocomposite with high strength and fluoride ion release. J Dent Res. 2010;89(7):739–45.

31. De Maeyer EA, Verbeeck RM, Vercruysse CW. Reactivity of fluoride-containing calcium aluminosilicate glasses used in dental glass-ionomer cements. J Dent Res. 1998;77(12):2005–11.

32. Moshaverinia A, Ansari S, Moshaverinia M, Roohpour N, Darr JA, Rehman I. Effects of incorporation of hydroxyapatite and fluoroapatite nanobioceramics into conventional glass ionomer cements (GIC). Acta Biomater. 2008;4(2):432–40.

33. Moshaverinia A, Ansari S, Movasaghi Z, Billington RW, Darr JA, Rehman IU. Modification of

conventional glass-ionomer cements with N-vinylpyrrolidone containing polyacids, nano-hydroxy and fluoroapatite to improve mechanical properties. Dent Mater Off Publ AcadDent Mater. 2008;24(10):1381–90.

34. Lee JJ, Lee YK, Choi BJ, Lee JH, Choi HJ, Son HK, et al. Physical properties of resin-reinforced glass ionomer cement modified with micro and nano-hydroxyapatite. J Nanosci Nanotechnol. 2010; 10(8):5270–6.

35. Siddharthan A, Seshadri SK, Sampath Kumar TS. Microwave accelerated synthesis of nanosized calcium deficient hydroxyapatite. J Mater Sci Mater Med. 2004;15(12):1279–84.

36. Fadeel B, Garcia-Bennett AE. Better safe than sorry: Understanding the toxicological properties of inorganic nanoparticles manufactured for biomedical applications. Adv Drug Deliv Rev. 2010;62(3): 362–74.

37. El-Negoly SA, El-Fallal AA, El-Sherbiny IM. A New Modification for Improving Shear Bond Strength and Other Mechanical Properties of Conventional Glass-ionomer Restorative Materials. J Adhes Dent. 2014;16(1):41–7.

38. Elsaka SE, Hamouda IM, Swain MV. Titanium dioxide nanoparticles addition to a conventional glass-ionomer restorative: influence on physical and antibacterial properties. J Dent. 2011;39(9):589–98.

39. Tsuang YH, Sun JS, Huang YC, Lu CH, Chang WH, Wang CC. Studies of photokilling of bacteria using titanium dioxide nanoparticles. Artif Organs. 2008; 32(2):167–74.

40. Hook ER, Owen OJ, Bellis CA, Holder JA, O'Sullivan DJ, Barbour ME. Development of a novel antimicrobial-releasing glass ionomer cement functionalized with chlorhexidine hexametaphosphate nanoparticles. J Nanobiotechnol. 2014; 12(1):3.

41. 3M ESPE. Ketac nano: light curing glass ionomer restorative. St. Paul: 3M ESPE; 2007.

42. Curtis AR, Palin WM, Fleming GJ, Shortall AC, Marquis PM. The mechanical properties of nano-filled resin-based composites: the impact of dry and wet cyclic pre-loading on bi-axial flexure strength. Dent Mater Off Publ AcadDent Mater. 2009;25(2): 188–97.

43. de Paula AB, Fucio SB, Ambrosano GM, Alonso RC, Sardi JC, Puppin-Rontani RM. Biodegradation and abrasive wear of nano restorative materials. Oper Dent. 2011;36(6):670–7.

44. de Fucio SB, de Paula AB, de Carvalho FG, Feitosa VP, Ambrosano GM, Puppin-Rontani RM. Biomechanical degradation of the nano-filled resin-modified glass-ionomer surface. Am J Dent. 2012;25(6):315–20.

45. Guo X, Wang Y, Spencer P, Ye Q, Yao X. Effects of water content and initiator composition on photopolymerization of a model BisGMA/HEMA resin. Dent Mater Off Publ AcadDent Mater. 2008;24(6): 824–31.

46. Neelakantan P, John S, Anand S, Sureshbabu N, Subbarao C. Fluoride release from a new glass-ionomer cement. Oper Dent. 2011;36(1):80–5.

47. Moreau JL, Xu HH. Fluoride releasing restorative materials: effects of pH on mechanical properties and ion release. Dent Mater Off Publ AcadDent Mater. 2010;26(11):e227–35.

48. Perdigao J, Dutra-Correa M, Saraceni SH, Ciaramicoli MT, Kiyan VH. Randomized clinical trial of two resin-modified glass ionomer materials: 1-year results. Oper Dent. 2012;37(6): 591–601.

49. Coutinho E, Cardoso MV, De Munck J, Neves AA, Van Landuyt KL, Poitevin A, et al. Bonding effectiveness and interfacial characterization of a nano-filled resin-modified glass-ionomer. Dent Mater Off Publ AcadDent Mater. 2009;25(11):1347–57.

50. Friedl K, Hiller KA, Friedl KH. Clinical performance of a new glass ionomer based restoration system: a retrospective cohort study. Dent Mater Off Publ AcadDent Mater. 2011;27(10):1031–7.

51. Marquezan M, Osorio R, Ciamponi AL, Toledano M. Resistance to degradation of bonded restorations to simulated caries-affected primary dentin. Am J Dent. 2010;23(1):47–52.

52. Bonifacio CC, Werner A, Kleverlaan CJ. Coating glass-ionomer cements with a nanofilled resin. Acta Odontol Scand. 2012;70(6):471–7.

53. Bagheri R, Azar MR, Tyas MJ, Burrow MF. The effect of aging on the fracture toughness of esthetic restorative materials. Am J Dent. 2010;23(3): 142–6.

54. Lammeier C, Li Y, Lunos S, Fok A, Rudney J, Jones RS. Influence of dental resin material composition on cross-polarization-optical coherence tomography imaging. J Biomed Opt. 2012;17(10): 106002.

55. Moszner N, Klapdohr S. Nanotechnology for dental composites. Int J of Nanotechnology. 2004;1(1/2): 130–56.

56. Klapdohr S, Moszner N. New inorganic components for dental filling composites. Monatshefte fur Chemie. 2005;136:21–45.

57. Wilson KS, Zhang K, Antonucci JM. Systematic variation of interfacial phase reactivity in dental nanocomposites. Biomaterials. 2005;26(25): 5095–103.

58. Chen MH, Chen CR, Hsu SH, Sun SP, Su WF. Low shrinkage light curable nanocomposite for dental restorative material. Dent Mater Off Publ AcadDent Mater. 2006;22(2):138–45.

59. Furman B, Rawls HR, Wellinghoff S, Dixon H, Lankford J, Nicolella D. Metal-oxide nanoparticles for the reinforcement of dental restorative resins. Crit Rev Biomed Eng. 2000;28(3–4):439–43.

60. Wang Y, Lee JJ, Lloyd IK, Wilson Jr OC, Rosenblum M, Thompson V. High modulus nanopowder reinforced dimethacrylate matrix composites for dental cement applications. J Biomed Mater Res A. 2007; 82(3):651–7.

61. Tian M, Gao Y, Liu Y, Liao Y, Hedin NE, Fong H. Fabrication and evaluation of Bis-GMA/TEGDMA dental resins/composites containing nano fibrillar silicate. Dent Mater Off Publ AcadDent Mater. 2008;24(2):235–43.

62. Xia Y, Zhang F, Xie H, Gu N. Nanoparticle-reinforced resin-based dental composites. J Dent. 2008;36(6):450–5.

63. Zhang K, Melo MA, Cheng L, Weir MD, Bai Y, Xu HH. Effect of quaternary ammonium and silver nanoparticle-containing adhesives on dentin bond strength and dental plaque microcosm biofilms. Dent Mater Off Publ AcadDent Mater. 2012;28(8):842–52.

64. Zhang K, Li F, Imazato S, Cheng L, Liu H, Arola DD, et al. Dual antibacterial agents of nano-silver and 12-methacryloyloxydodecylpyridinium bromide in dental adhesive to inhibit caries. J Biomed Mater Res B Appl Biomater. 2013;101(6):929–38.

65. Sideridou ID, Karabela MM, Vouvoudi E. Physical properties of current dental nanohybrid and nanofill light-cured resin composites. Dent Mater Off Publ AcadDent Mater. 2011;27(6):598–607.

66. Endo T, Finger WJ, Kanehira M, Utterodt A, Komatsu M. Surface texture and roughness of polished nanofill and nanohybrid resin composites. Dent Mater J. 2010;29(2):213–23.

67. Taha DG, Abdel-Samad AA, Mahmoud SH. Fracture resistance of maxillary premolars with class II MOD cavities restored with Ormocer, Nanofilled, and Nanoceramic composite restorative systems. Quintessence Int. 2011;42(7):579–87.

68. Tabatabaei MH, Sadrai S, Bassir SH, Veisy N, Dehghan S. Effect of food stimulated liquids and thermocycling on the monomer elution from a nano-filled composite. Open Dent J. 2013;7:62–7.

69. Eren D, Bektas OO, Siso SH. Three different adhesive systems; three different bond strength test methods. Acta Odontol Scand. 2013;71(3–4):978–83.

70. Vanderlei A, Passos SP, Ozcan M, Bottino MA, Valandro LF. Durability of adhesion between feldspathic ceramic and resin cements: effect of adhesive resin, polymerization mode of resin cement, and aging. J Prosthodont Off J Am Coll Prosthodont. 2013;22(3):196–202.

71. Ernst CP, Brandenbusch M, Meyer G, Canbek K, Gottschalk F, Willershausen B. Two-year clinical performance of a nanofiller vs a fine-particle hybrid resin composite. Clin Oral Investig. 2006;10(2):119–25.

72. Polydorou O, Konig A, Hellwig E, Kummerer K. Long-term release of monomers from modern dental-composite materials. Eur J Oral Sci. 2009;117(1):68–75.

73. da Silva JM, da Rocha DM, Travassos AC, Fernandes Jr VV, Rodrigues JR. Effect of different finishing times on surface roughness and maintenance of polish in nanoparticle and microhybrid composite resins. Eur J Esthet Dent Off J Eur Acad Esthet Dent. 2010;5(3):288–98.

74. Feitosa VP, Fugolin AP, Correr AB, Correr-Sobrinho L, Consani S, Watson TF, et al. Effects of different photo-polymerization protocols on resin-dentine muTBS, mechanical properties and cross-link density of a nano-filled resin composite. J Dent. 2012;40(10):802–9.

75. Egilmez F, Ergun G, Cekic-Nagas I, Vallittu PK, Lassila LV. Short and long term effects of additional post curing and polishing systems on the color change of dental nano-composites. Dent Mater J. 2013;32(1):107–14.

76. Al Kheraif AA, Qasim SS, Ramakrishnaiah R, Ihtesham ur R. Effect of different beverages on the color stability and degree of conversion of nano and microhybrid composites. Dent Mater J. 2013; 32(2):326–31.

77. Takano T, Tasaka A, Yoshinari M, Sakurai K. Fatigue strength of Ce-TZP/Al2O3 nanocomposite with different surfaces. J Dent Res. 2012;91(8):800–4.

78. Fischer J, Stawarczyk B, Sailer I, Hammerle CHF. Shear bond strength between veneering ceramics and ceria-stabilized zirconia/alumina. J Prosthet Dent. 2010;103(5):267–74.

79. Xu HH, Weir MD, Sun L, Moreau JL, Takagi S, Chow LC, et al. Strong nanocomposites with Ca, PO(4), and F release for caries inhibition. J Dent Res. 2010;89(1):19–28.

80. Nakabayashi N, Kojima K, Masuhara E. The promotion of adhesion by the infiltration of monomers into tooth substrates. J Biomed Mater Res. 1982;16(3):265–73.

81. Hosoya Y, Tay FR. Hardness, elasticity, and ultrastructure of bonded sound and caries-affected primary tooth dentin. J Biomed Mater Res B Appl Biomater. 2007;81(1):135–41.

82. Nakabayashi N, Nakamura M, Yasuda N. Hybrid layer as a dentin-bonding mechanism. J Esthet Dent. 1991;3(4):133–8.

83. Poitevin A, De Munck J, Van Landuyt K, Coutinho E, Peumans M, Lambrechts P, et al. Critical analysis of the influence of different parameters on the microtensile bond strength of adhesives to dentin. J Adhes Dent. 2008;10(1):7–16.

84. Breschi L, Mazzoni A, Ruggeri A, Cadenaro M, Di Lenarda R, De Stefano Dorigo E. Dental adhesion review: aging and stability of the bonded interface. Dent Mater Off Publ AcadDent Mater. 2008;24(1):90–101.

85. Osorio R, Yamauti M, Osorio E, Roman JS, Toledano M. Zinc-doped dentin adhesive for collagen protection at the hybrid layer. Eur J Oral Sci. 2011;119(5):401–10.

86. Sano H, Takatsu T, Ciucchi B, Horner JA, Matthews WG, Pashley DH. Nanoleakage: leakage within the hybrid layer. Oper Dent. 1995;20(1):18–25.

87. Hashimoto M, Tay FR, Ohno H, Sano H, Kaga M, Yiu C, et al. SEM and TEM analysis of water degradation of human dentinal collagen. J Biomed Mater Res B Appl Biomater. 2003;66(1):287–98.

88. Hebling J, Pashley DH, Tjaderhane L, Tay FR. Chlorhexidine arrests subclinical degradation of dentin hybrid layers *in vivo*. J Dent Res. 2005;84(8):741–6.

89. Spencer P, Wang Y. Adhesive phase separation at the dentin interface under wet bonding conditions. J Biomed Mater Res. 2002;62(3):447–56.

90. Kim JS, Cho BH, Lee IB, Um CM, Lim BS, Oh MH, et al. Effect of the hydrophilic nanofiller loading on the mechanical properties and the microtensile bond strength of an ethanol-based one-bottle dentin adhesive. J Biomed Mater Res B Appl Biomater. 2005;72(2):284–91.

91. Lohbauer U, Wagner A, Belli R, Stoetzel C, Hilpert A, Kurland HD, et al. Zirconia nanoparticles prepared by laser vaporization as fillers for dental adhesives. Acta Biomater. 2010;6(12):4539–46.

92. Osorio E, Toledano M, Yamauti M, Osorio R. Differential nanofiller cluster formations in dental adhesive systems. Microsc Res Tech. 2012;75(6):749–57.

93. Denehy GE. A direct approach to restore anterior teeth. Am J Dent. 2000;13(Spec No):55D–9.

94. Hickel R, Dasch W, Janda R, Tyas M, Anusavice K. New direct restorative materials. FDI Commission Project. Int Dent J. 1998;48(1):3–16.

95. Hosseinalipour M, Javadpour J, Rezaie H, Dadras T, Hayati AN. Investigation of mechanical properties of experimental Bis-GMA/TEGDMA dental composite resins containing various mass fractions of silica nanoparticles. J Prosthodont Off J Am Coll Prosthodont. 2010;19(2):112–7.

96. Xu HH, Moreau JL, Sun L, Chow LC. Nanocomposite containing amorphous calcium phosphate nanoparticles for caries inhibition. Dent Mater Off Publ AcadDent Mater. 2011;27(8):762–9.

97. Van Landuyt KL, Snauwaert J, De Munck J, Peumans M, Yoshida Y, Poitevin A, et al. Systematic review of the chemical composition of contemporary dental adhesives. Biomaterials. 2007;28(26):3757–85.

98. Kasraei ShA M, Khamverdi Z, Khalegh Nejad S. Effect of nanofiller addition to an experimental dentin adhesive on microtensile bond strength to human dentin. J Dent (Tehran). 2009;6:91–6.

99. Osorio E, Toledano M, Aguilera FS, Tay FR, Osorio R. Ethanol wet-bonding technique sensitivity assessed by AFM. J Dent Res. 2010;89(11):1264–9.

100. Tay FR, Moulding KM, Pashley DH. Distribution of nanofillers from a simplified-step adhesive in acid-conditioned dentin. J Adhes Dent. 1999;1(2):103–17.

101. de Monredon-Senani S, Bonhomme C, Ribot F, Babonneau F. Covalent grafting of organoalkoxysilanes on silica surfaces in water-rich medium as evidenced by Si-29 NMR. J Sol-gel Sci Techn. 2009;50(2):152–7.

102. Cheng L, Weir MD, Zhang K, Xu SM, Chen Q, Zhou X, et al. Antibacterial nanocomposite with calcium phosphate and quaternary ammonium. J Dent Res. 2012;91(5):460–6.

103. Cheng L, Weir MD, Xu HH, Antonucci JM, Kraigsley AM, Lin NJ, et al. Antibacterial amorphous calcium phosphate nanocomposites with a quaternary ammonium dimethacrylate and silver nanoparticles. Dent Mater Off Publ AcadDent Mater. 2012;28(5):561–72.

104. Li F, Weir MD, Chen J, Xu HH. Comparison of quaternary ammonium-containing with nano-silver-containing adhesive in antibacterial properties and cytotoxicity. Dent Mater Off Publ AcadDent Mater. 2013;29(4):450–61.

105. Morones JR, Elechiguerra JL, Camacho A, Holt K, Kouri JB, Ramirez JT, et al. The bactericidal effect of silver nanoparticles. Nanotechnology. 2005;16(10):2346–53.

106. Zhang K, Cheng L, Imazato S, Antonucci JM, Lin NJ, Lin-Gibson S, et al. Effects of dual antibacterial agents MDPB and nano-silver in primer on microcosm biofilm, cytotoxicity and dentine bond properties. J Dent. 2013;41(5):464–74.

107. Cheng L, Zhang K, Weir MD, Liu H, Zhou X, Xu HH. Effects of antibacterial primers with quaternary ammonium and nano-silver on Streptococcus mutans impregnated in human dentin blocks. Dent Mater Off Publ AcadDent Mater. 2013;29(4):462–72.

108. Skrtic D, Antonucci JM, Eanes ED, Eichmiller FC, Schumacher GE. Physicochemical evaluation of bioactive polymeric composites based on hybrid amorphous calcium phosphates. J Biomed Mater Res. 2000;53(4):381–91.

109. Melo MA, Cheng L, Zhang K, Weir MD, Rodrigues LK, Xu HH. Novel dental adhesives containing nanoparticles of silver and amorphous calcium phosphate. Dent Mater Off Publ AcadDent Mater. 2013;29(2):199–210.

110. Melo MA, Cheng L, Weir MD, Hsia RC, Rodrigues LK, Xu HH. Novel dental adhesive containing antibacterial agents and calcium phosphate nanoparticles. J Biomed Mater Res B Appl Biomater. 2013;101(4):620–9.

111. Besinis A, van Noort R, Martin N. Infiltration of demineralized dentin with silica and hydroxyapatite nanoparticles. Dent Mater Off Publ AcadDent Mater. 2012;28(9):1012–23.

112. Cai YR, Tang RK. Calcium phosphate nanoparticles in biomineralization and biomaterials. J Mater Chem. 2008;18(32):3775–87.

113. Sadat-Shojai M, Atai M, Nodehi A, Khanlar LN. Hydroxyapatite nanorods as novel fillers for improving the properties of dental adhesives: Synthesis and application. Dent Mater Off Publ AcadDent Mater. 2010;26(5):471–82.

114. Earl JS, Wood DJ, Milne SJ. Nanoparticles for dentine tubule infiltration: an *in vitro* study. J Nanosci Nanotechnol. 2009;9(11):6668–74.

115. Kajita M, Hikosaka K, Iitsuka M, Kanayama A, Toshima N, Miyamoto Y. Platinum nanoparticle is a useful scavenger of superoxide anion and hydrogen peroxide. Free Radic Res. 2007;41(6):615–26.

116. Salamanca-Buentello F, Persad DL, Court EB, Martin DK, Daar AS, Singer PA. Nanotechnology and the developing world. PLoS Med. 2005;2(5):e97.

117. Hoshika S, Nagano F, Tanaka T, Ikeda T, Wada T, Asakura K, et al. Effect of application time of colloidal platinum nanoparticles on the microtensile bond strength to dentin. Dent Mater J. 2010; 29(6):682–9.

118. Nagano F, Selimovic D, Noda M, Ikeda T, Tanaka T, Miyamoto Y, et al. Improved bond performance of a dental adhesive system using nano-technology. Bioed Mater Eng. 2009;19(2–3):249–57.

119. Hoshika S, Nagano F, Tanaka T, Wada T, Asakura K, Koshiro K, et al. Expansion of nanotechnology for dentistry: effect of colloidal platinum nanoparticles on dentin adhesion mediated by 4-META/MMA-TBB. J Adhes Dent. 2011;13(5):411–6.

120. Toledano M, Sauro S, Cabello I, Watson T, Osorio R. A Zn-doped etch-and-rinse adhesive may improve the mechanical properties and the integrity at the bonded-dentin interface. Dent Mater Off Publ AcadDent Mater. 2013;29(8):e142–52.

121. Osorio R, Yamauti M, Osorio E, Ruiz-Requena ME, Pashley DH, Tay FR, et al. Zinc reduces collagen degradation in demineralized human dentin explants. J Dent. 2011;39(2):148–53.

122. de Souza AP, Gerlach RF, Line SR. Inhibition of human gingival gelatinases (MMP-2 and MMP-9) by metal salts. Dent Mater Off Publ AcadDent Mater. 2000;16(2):103–8.

123. Toledano M, Yamauti M, Ruiz-Requena ME, Osorio R. A ZnO-doped adhesive reduced collagen degradation favouring dentine remineralization. J Dent. 2012;40(9):756–65.

124. Sakharov DV, Lim C. Zn protein simulations including charge transfer and local polarization effects. J Am Chem Soc. 2005;127(13):4921–9.

125. Atai M, Solhi L, Nodehi A, Mirabedini SM, Kasraei S, Akbari K, et al. PMMA-grafted nanoclay as novel filler for dental adhesives. Dent Mater Off Publ AcadDent Mater. 2009;25(3):339–47.

126. Solhi L, Atai M, Nodehi A, Imani M. A novel dentin bonding system containing poly(methacrylic acid) grafted nanoclay: synthesis, characterization and properties. Dent Mater Off Publ AcadDent Mater. 2012;28(10):1041–50.

127. Solhi L, Atai M, Nodehi A, Imani M, Ghaemi A, Khosravi K. Poly(acrylic acid) grafted montmorillonite as novel fillers for dental adhesives: synthesis, characterization and properties of the adhesive. Dent Mater Off Publ AcadDent Mater. 2012;28(4): 369–77.

128. Moraes RR, Garcia JW, Barros MD, Lewis SH, Pfeifer CS, Liu J, et al. Control of polymerization shrinkage and stress in nanogel-modified monomer and composite materials. Dent Mater Off Publ AcadDent Mater. 2011;27(6):509–19.

129. Moraes RR, Garcia JW, Wilson ND, Lewis SH, Barros MD, Yang B, et al. Improved dental adhesive formulations based on reactive nanogel additives. J Dent Res. 2012;91(2):179–84.

130. Cheng YJ, Zeiger DN, Howarter JA, Zhang XR, Lin NJ, Antonucci JM, et al. *In situ* formation of silver nanoparticles in photocrosslinking polymers. J Biomed Mater Res B. 2011;97B(1):124–31.

131. Madler L, Krumeich F, Burtscher P, Moszner N. Visibly transparent & radiopaque inorganic organic composites from flame-made mixed-oxide fillers. J Nanopart Res. 2006;8:323–33.

132. Chen Q, Zhao Y, Wu W, Xu T, Fong H. Fabrication and evaluation of Bis-GMA/TEGDMA dental resins/composites containing halloysite nanotubes. Dent Mater Off Publ AcadDent Mater. 2012;28(10): 1071–9.

133. Chen L, Xu C, Wang Y, Shi J, Yu Q, Li H. BisGMA/TEGDMA dental nanocomposites containing glyoxylic acid-modified high-aspect ratio hydroxyapatite nanofibers with enhanced dispersion. Biomed Mater. 2012;7(4):045014.

134. Beyth N, Yudovin-Farber I, Bahir R, Domb AJ, Weiss EI. Antibacterial activity of dental composites containing quaternary ammonium polyethylenimine nanoparticles against Streptococcus mutans. Biomaterials. 2006;27(21):3995–4002.

135. Yudovin-Farber I, Beyth N, Nyska A, Weiss EI, Golenser J, Domb AJ. Surface characterization and biocompatibility of restorative resin containing nanoparticles. Biomacromolecules. 2008;9(11): 3044–50.

136. Beyth N, Houri-Haddad Y, Baraness-Hadar L, Yudovin-Farber I, Domb AJ, Weiss EI. Surface antimicrobial activity and biocompatibility of incorporated polyethylenimine nanoparticles. Biomaterials. 2008;29(31):4157–63.

137. Beyth N, Yudovin-Farber I, Perez-Davidi M, Domb AJ, Weiss EI. Polyethyleneimine nanoparticles incorporated into resin composite cause cell death and trigger biofilm stress *in vivo*. Proc Natl Acad Sci U S A. 2010;107(51):22038–43.

138. Langhorst SE, O'Donnell JN, Skrtic D. *In vitro* remineralization of enamel by polymeric amorphous calcium phosphate composite: quantitative microradiographic study. Dent Mater Off Publ AcadDent Mater. 2009;25(7):884–91.

139. Regnault WF, Icenogle TB, Antonucci JM, Skrtic D. Amorphous calcium phosphate/urethane methacrylate resin composites. I. Physicochemical characterization. J Mater Sci Mater Med. 2008;19(2): 507–15.

140. Moreau JL, Sun L, Chow LC, Xu HH. Mechanical and acid neutralizing properties and bacteria inhibition of amorphous calcium phosphate dental nanocomposite. J Biomed Mater Res B Appl Biomater. 2011;98(1):80–8.

141. Moreau JL, Weir MD, Giuseppetti AA, Chow LC, Antonucci JM, Xu HH. Long-term mechanical durability of dental nanocomposites containing amorphous calcium phosphate nanoparticles. J Biomed Mater Res B Appl Biomater. 2012;100(5):1264–73.

142. Cheng L, Weir MD, Xu HH, Kraigsley AM, Lin NJ, Lin-Gibson S, et al. Antibacterial and physical properties of calcium-phosphate and calcium-fluoride nanocomposites with chlorhexidine. Dent Mater Off Publ AcadDent Mater. 2012;28(5):573–83.

143. Janardhanan R, Karuppaiah M, Hebalkar N, Rao TN. Synthesis and surface chemistry of nano silver particles. Polyhedron. 2009;28(12):2522–30.

144. Kasraei S, Azarsina M. Addition of silver nanoparticles reduces the wettability of methacrylate and silorane-based composites. Braz Oral Res. 2012;26(6):505–10.

145. Pillalamarri SK, Blum FD, Tokuhiro AT, Bertino MF. One-pot synthesis of polyaniline – metal nanocomposites. Chem Mater. 2005;17(24):5941–4.

146. Fan C, Chu L, Rawls HR, Norling BK, Cardenas HL, Whang K. Development of an antimicrobial resin–a pilot study. Dent Mater Off Publ AcadDent Mater. 2011;27(4):322–8.

147. Durner J, Obermaier J, Draenert M, Ilie N. Correlation of the degree of conversion with the amount of elutable substances in nano-hybrid dental composites. Dent Mater Off Publ AcadDent Mater. 2012;28(11):1146–53.

148. Tavassoli Hojati S, Alaghemand H, Hamze F, Ahmadian Babaki F, Rajab-Nia R, Rezvani MB, et al. Antibacterial, physical and mechanical properties of flowable resin composites containing zinc oxide nanoparticles. Dent Mater Off Publ AcadDent Mater. 2013;29(5):495–505.

149. Ling L, Xu X, Choi GY, Billodeaux D, Guo G, Diwan RM. Novel F-releasing composite with improved mechanical properties. J Dent Res. 2009;88(1):83–8.

150. Xu HHK, Moreau JL, Sun LM, Chow LC. Strength and fluoride release characteristics of a calcium fluoride based dental nanocomposite. Biomaterials. 2008;29(32):4261–7.

151. Weir MD, Moreau JL, Levine ED, Strassler HE, Chow LC, Xu HHK. Nanocomposite containing CaF2 nanoparticles: thermal cycling, wear and long-term water-aging. Dent Mater. 2012;28(6):642–52.

152. Leung D, Spratt DA, Pratten J, Gulabivala K, Mordan NJ, Young AM. Chlorhexidine-releasing methacrylate dental composite materials. Biomaterials. 2005;26(34):7145–53.

153. Anusavice KJ, Zhang NZ, Shen C. Controlled release of chlorhexidine from UDMA-TEGDMA resin. J Dent Res. 2006;85(10):950–4.

154. Xu HHK, Martin TA, Antonucci JM, Eichmiller FC. Ceramic whisker reinforcement of dental resin composites. J Dent Res. 1999;78(2):706–12.

155. Xu HH, Eichmiller FC, Smith DT, Schumacher GE, Giuseppetti AA, Antonucci JM. Effect of thermal cycling on whisker-reinforced dental resin composites. J Mater Sci Mater Med. 2002;13(9):875–83.

156. Dickens SH, Flaim GM, Takagi S. Mechanical properties and biochemical activity of remineralizing resin-based Ca-PO4 cements. Dent Mater Off Publ AcadDent Mater. 2003;19(6):558–66.

157. Xu HH, Weir MD, Sun L, Takagi S, Chow LC. Effects of calcium phosphate nanoparticles on Ca-PO4 composite. J Dent Res. 2007;86(4):378–83.

158. Xu HH, Weir MD, Sun L. Nanocomposites with Ca and PO4 release: effects of reinforcement, dicalcium phosphate particle size and silanization. Dent Mater Off Publ AcadDent Mater. 2007;23(12):1482–91.

159. Hannig C, Hannig M. Natural enamel wear–a physiological source of hydroxylapatite nanoparticles for biofilm management and tooth repair? Med Hypotheses. 2010;74(4):670–2.

160. Venegas SC, Palacios JM, Apella MC, Morando PJ, Blesa MA. Calcium modulates interactions between bacteria and hydroxyapatite. J Dent Res. 2006;85(12):1124–8.

161. Lezaja M, Veljovic DN, Jokic BM, Cvijovic-Alagic I, Zrilic MM, Miletic V. Effect of hydroxyapatite spheres, whiskers, and nanoparticles on mechanical properties of a model BisGMA/TEGDMA composite initially and after storage. J Biomed Mater Res B Appl Biomater. 2013;101:1469–76.

162. Habekost LV, Camacho GB, Lima GS, Ogliari FA, Cubas GB, Moraes RR. Nanoparticle loading level and properties of experimental hybrid resin luting agents. J Prosthodont Off J Am Coll Prosthodont. 2012;21(7):540–5.

163. Schmidt H. Organically modified silicates by the sol-gel process. Mat Res Soc Symp Proc. 1984;32:327–35.

164. Schmidt H. Inorganic-organic composites by sol-gel techniques. J Sol-Gel Sci Technology. 1994;1:217–231.

165. Wolter H, Storch W, Ott H. New inorganic/organic copolymers (ormocers) for dental applications. Mater Res Soc Symp Proc. 1994;346:143–9.

166. Moszner N, Gianasmidis A, Klapdohr S, Fischer UK, Rheinberger V. Sol-gel materials 2. Light-curing dental composites based on ormocers of cross-linking alkoxysilane methacrylates and further nano-components. Dent Mater Off Publ AcadDent Mater. 2008;24(6):851–6.

167. Mahmoud SH, El-Embaby AE, AbdAllah AM, Hamama HH. Two-year clinical evaluation of ormocer, nanohybrid and nanofill composite restorative systems in posterior teeth. J Adhes Dent. 2008;10(4):315–22.

168. Efes BG, Dorter C, Gomec Y. Clinical evaluation of an ormocer, a nanofill composite and a hybrid composite at 2 years. Am J Dent. 2006;19(4):236–40.

169. Efes BG, Dorter C, Gomec Y, Koray F. Two-year clinical evaluation of ormocer and nanofill composite with and without a flowable liner. J Adhes Dent. 2006;8(2):119–26.

170. Mejare I, Lingstrom P, Petersson LG, Holm AK, Twetman S, Kallestal C, et al. Caries-preventive effect of fissure sealants: a systematic review. Acta Odontol Scand. 2003;61(6):321–30.

171. Imazato S, Kinomoto Y, Tarumi H, Torii M, Russell RRB, McCabe JF. Incorporation of antibacterial monomer MDPB into dentin primer. J Dent Res. 1997;76(3):768–72.

172. Fan Y, Townsend J, Wang Y, Lee EC, Evans K, Hender E, et al. Formulation and characterization of antibacterial fluoride-releasing sealants. Pediatr Dent. 2013;35(1):E13–8.

173. Kang ES, Jeon YC, Jeong CM, Huh JB, Yun MJ, Kwon YH. Effect of solution temperature on the mechanical properties of dual-cure resin cements. J Advan Prosthod. 2013;5(2):133–9.

174. Rekow ED, Harsono M, Janal M, Thompson VP, Zhang G. Factorial analysis of variables influencing stress in all-ceramic crowns. Dent Mater Off Publ AcadDent Mater. 2006;22(2):125–32.

175. Lu C, Wang R, Mao S, Arola D, Zhang D. Reduction of load-bearing capacity of all-ceramic crowns due to cement aging. J Mech Behav Biomed Mater. 2013;17:56–65.

176. Scherrer SS, Quinn JB, Quinn GD, Wiskott HW. Fractographic ceramic failure analysis using the replica technique. Dent Mater Off Publ AcadDent Mater. 2007;23(11):1397–404.

177. Ali SM. Prediction of the service life of dental porcelains by the measurements of post-indentation slow crack growth. Dent Mater J. 1993;12(1):45–53.

178. Cesar PF, Soki FN, Yoshimura HN, Gonzaga CC, Styopkin V. Influence of leucite content on slow crack growth of dental porcelains. Dent Mater Off Publ AcadDent Mater. 2008;24(8):1114–22.

179. Zhu SM, Fahrenholtz WG, Hilmas GE. Influence of silicon carbide particle size on the microstructure and mechanical properties of zirconium diboride-silicon carbide ceramics. J Eur Ceram Soc. 2007;27(4):2077–83.

180. Niihara K. New design concept of structural ceramic – ceramic nanocomposites. J Ceram Soc Jpn. 1991;99:974–82.

181. Zhang J, Liao Y, Li W, Zhao Y, Zhang C. Microstructure and mechanical properties of glass-infiltrated Al_2O_3/ZrO_2 nanocomposites. J Mater Sci Mater Med. 2012;23(2):239–44.

182. Allen AJ, Long GG, Kerch HM, Krueger S, Skandan G, Hahn H, et al. Sintering studies of nanophase ceramic oxides using small angle scattering. Ceram Int. 1996;22(4):275–80.

183. Chevalier J, Gremillard L, Deville S. Low-temperature degradation of Zirconia and implications for biomedical implants. Annu Rev Mater Res. 2007;37:1–32.

184. Yang SF, Yang LQ, Jin ZH, Guo TW, Wang L, Liu HC. New nano-sized Al_2O_3-BN coating 3Y-TZP ceramic composites for CAD/CAM-produced all-ceramic dental restorations. Part I. Fabrication of powders. Nanomed Nanotechnol Biol Med. 2009;5(2):232–9.

185. Garvie RC, Hannink RH, Pascoe RT. Ceramic Steel. Nature. 1975;258(5537):703–4.

186. Tsukuma K, Shimada M. Strength, fracture-toughness and vickers hardness of Ceo2-stabilized tetragonal Zro2 polycrystals (Ce-Tzp). J Mater Sci. 1985;20(4):1178–84.

187. Tsukuma K. Mechanical-properties and thermal-stability of Ceo2 containing tetragonal zirconia polycrystals. Am Ceram Soc Bull. 1986;65(10):1386–9.

188. Nawa M, Bamba N, Sekino T, Niihara K. The effect of TiO2 addition on strengthening and toughening in intragranular type of 12Ce-TZP/Al2O3 nanocomposites. J Eur Ceram Soc. 1998;18(3):209–19.

189. Nawa M, Nakamoto S, Sekino T, Niihara K. Tough and strong Ce-TZP/alumina nanocomposites doped with titania. Ceram Int. 1998;24(7):497–506.

190. Losic D, Mitchell JG, Voelcker NH. Diatomaceous lessons in nanotechnology and advanced materials. Adv Mater. 2009;21(29):2947–58.

191. Santoyo-Salazar J, Gonzalez G, Schabes-Retchkiman PS, Ascencio JA, Tartaj-Salvador J, Chavez-Carvayar JA. Synthesis and characterisation of YSZ-Al2O3 nanostructured materials. J Nanosci Nanotechnol. 2006;6(7):2103–9.

192. Chevalier J, Taddei P, Gremillard L, Deville S, Fantozzi G, Bartolome JF, et al. Reliability assessment in advanced nanocomposite materials for orthopaedic applications. J Mech Behav Biomed Mater. 2011;4(3):303–14.

193. Silva NRFA, Bonfante EA, Rafferty BT, Zavanelli RA, Martins LL, Rekow ED, et al. Conventional and modified veneered zirconia vs. metalloceramic: fatigue and finite element analysis. J Prosthodont Implant Esthet Reconstruct Dent. 2012;21(6):433–9.

194. Kelly JR. Ceramics in restorative and prosthetic dentistry. Annu Rev Mater Sci. 1997;27:443–68.

195. Theocharopoulos A, Chen X, Wilson RM, Hill R, Cattell MJ. Crystallization of high-strength nanoscale leucite glass-ceramics. Dent Mater. 2013;29(11):1149–57.

196. Chang J, Da Silva JD, Sakai M, Kristiansen J, Ishikawa-Nagai S. The optical effect of composite luting cement on all ceramic crowns. J Dent. 2009;37(12):937–43.

197. Uno M, Kurachi M, Wakamatsu N, Doi Y. Effects of adding silver nanoparticles on the toughening of dental porcelain. J Prosthet Dent. 2013;109(4):241–7.

198. Cutler RA, Mayhew RJ, Prettyman KM, Virkar AV. High-toughness Ce-Tzp/Al2o3 ceramics with improved hardness and strength. J Am Ceram Soc. 1991;74(1):179–86.

199. Fischer J, Stawarczyk B. Compatibility of machined Ce-TZP/Al2O3 nanocomposite and a veneering ceramic. Dent Mater Off Publ AcadDent Mater. 2007;23(12):1500–5.

200. Fischer J, Stawarczyk B, Trottmann A, Hammerle CHF. Impact of thermal properties of veneering ceramics on the fracture load of layered Ce-TZP/A nanocomposite frameworks. Dent Mater. 2009;25(3): 326–30.
201. Terui Y, Sato K, Goto D, Hotta Y, Tamaki Y, Miyazaki T. Compatibility of Ce-TZP/AI(2)O(3) nanocomposite frameworks and veneering porcelains. Dent Mater J. 2013;32(5):839–46.
202. Anusavice KJ, Kakar K, Ferree N. Which mechanical and physical testing methods are relevant for predicting the clinical performance of ceramic-based dental prostheses? Clin Oral Implants Res. 2007;18 Suppl 3:218–31.
203. Helkimo E, Carlsson GE, Helkimo M. Bite force and state dentition. Acta Odontol Scand. 1976;35: 297–303.
204. Belli R, Geinzer E, Muschweck A, Petzchelt A, Lohbauer U. Mechanical fatigue degradation of ceramics versus composites for dental restorations. Dent Mater Off Publ Acad Dent Mater. 2014;30: 424–32.
205. Guth JF, Zuch T, Zwinge S, Engels J, Stimmelmayr M, Edelhoff D. Optical properties of manually and CAD/CAM-fabricated polymers. Dent Mater J. 2013;32(6):865–71.
206. Koller M, Arnetzl GV, Holly L, et al. Lava ultimate resin nano ceramic for CAD/CAM: customization case study. Int J Comput Dent. 2012;15:159–64.
207. Robinson C, Connell S, Kirkham J, Shore R, Smith A. Dental enamel – a biological ceramic: regular substructures in enamel hydroxyapatite crystals revealed by atomic force microscopy. J Mater Chem. 2004;14(14):2242–8.

Remineralizing Nanomaterials for Minimally Invasive Dentistry

9

Xu Zhang, Xuliang Deng, and Yi Wu

Abstract

Modern dentistry advocates early prevention of tooth decay and minimally invasive management of dental caries (minimally invasive dentistry, MID). Remineralization is an important therapeutic method in MID. At the moment, traditional remineralizing agents and methods are not adapted to the requirements of MID. Recent studies indicate that the development of nanomaterials, especially biomimetic ones, as remineralizing agents, provides novel remineralizing strategies for MID. Here, we review the progress of the development of remineralizing nanomaterials for different applications in MID. Some nanomaterials, including calcium fluoride, hydroxyapatite, and amorphous calcium phosphate in nanoscales, are incorporated into restorative materials such as composite resins, glass ionomers, and adhesive systems. These dental materials play a remineralizing role through releasing fluoride calcium and phosphate ions. Other nanomaterials composed of stabilizers and amorphous calcium phosphates (ACP), such as nanocomplexes of casein phosphopeptides (CPP) and ACP, polyacrylic acid (PAA)-ACP, polyaspartic acid (PASP)-ACP, and phosphorylated chitosan (Pchi)-ACP, provide a biomimetic remineralizing strategy by mimicking biomineralization processes, which could *de novo* form dental hard tissues through nonclassical crystallization pathways. However, it is unpractical to restore small clinically visible cavities with

X. Zhang, PhD (✉) • Y. Wu, MSc
Department of Endodontics, School and Hospital
of Stomatology, Tianjin Medical University,
No. 12 Observatory Road, Heping District, Tianjin
300070, People's Republic of China
e-mail: zhxden@gmail.com

X. Deng, PhD
Geriatric Dentisty, Peking University School
and Hospital of Stomatology, Beijing, China

A. Kishen (ed.), *Nanotechnology in Endodontics: Current and Potential Clinical Applications*,
DOI 10.1007/978-3-319-13575-5_9, © Springer International Publishing Switzerland 2015

nanomaterials reviewed in this chapter at present. Most of the research covered in this chapter focuses primarily on laboratory tests. Future comprehensive research with respect to clinical applicability is required before employing remineralizing nanomaterials routinely in clinical practices.

Abbreviations

ACP	Amorphous calcium phosphates
CaF_2	Calcium fluoride
CMC	Carboxymethyl chitosan
CPP	Casein phosphopeptides
DMP1	Dentin matrix protein
DPP	Dentin phosphoprotein also known as DMP2 or phosphophoryn
EDX	Energy-dispersive X-ray spectroscopy
F	Fluoride
FAP	Fluorapatite $Ca_{10}(PO_4)_6F_2$
$f\beta$-TCP	Funcionalized β-TCP
GTR	Guided tissue remineralization
HAP	Hydroxyapatite
IP	Ionic activity product
K_{sp}	The solubility product
MID	Minimally invasive dentistry
NaF	Sodium fluoride
nano-CaF_2	CaF_2 nanoparticle
NCPs	Noncollagenous proteins
n-FHA	Nanofluorohydroxyapatite
n-HAP	Nano-sized HAP
OCP	Octacalcium phosphate
PAA	Polyacrylic acid
PASP	Polyaspartic acid
Pchi	Phosphorylated chitosan
PEO	Ethylene oxide
PILP	Polymer-induced liquid-precursor
PVPA	Polyvinylphosphonic acid
R	Gas constant $8.314 \, J \cdot K^{-1} \, mol^{-1}$
S	Supersaturation
SAED	Selected area electron diffraction
SEM	Scanning electron microscopy
STMP	Sodium trimetaphosphate
T	Absolute temperature
TEM	Transmission electron microscope
TPP	Sodium tripolyphosphate
β-TCP	Beta tricalcium phosphate $Ca_3(PO_4)_2$

9.1 Introduction

Dental caries remains to be the most common oral disease in many developing as well as industrialized countries. Thus, attempts to prevent caries progression in the developed world have been of prime focus in order to reduce the prevalence of caries. It is well established that dental caries is a dynamic disease process caused by the unbalance between demineralization and remineralization processes [1]. Traditionally, dentists use an "extension for prevention" surgical approach to manage dental caries, with G.V. Black cavity designs specified for each lesion type [2]. Minimally invasive dentistry (MID), a modern evidence-based approach to caries management, has evolved based on the emergence of atraumatic restorative techniques in the 1970s. It is a treatment that conserves maximum natural tooth tissue and restores the appearance, function, and aesthetics [3–5]. The conception of MID gradually replaces the status of "extension for prevention."

The main components of MID include assessment of the risk of disease, early caries diagnosis, remineralization for prevention and treatment of early caries, use of special restorations, dental materials and equipment for MID, and surgical intervention only when required or only after the disease has not been controlled [6–11]. Since remineralization is to achieve mineralization or restore mineral loss within a demineralized matrix/tooth, it is an ideal method to treat the caries that does not require surgical intervention. Additionally, remineralization contributes to prevent early caries and recurrent caries and treat dentin hypersensitivity.

Remineralization of enamel is mainly based on the growth of residual crystals. However, it is difficult to repair the enlarged defects of enamel caries by traditional remineralization agents and their deliveries, such as fluoride and fluoride-containing dentifrice. It is also demonstrated that the methods applied for enamel remineralization, such as the usage of fluoride, facilitated dentin remineralization, and the remineralization mechanism, are similar in both tissues [12, 13]. However, compared with remineralization of enamel, remineralization of dentin using fluoride is less effective. This could be ascribed to the fact that fluoride mainly remineralizes residual crystals in dental lesions, which act as seed crystals, but the residual crystals are lacking in dentin lesions where a more organic matrix exists [1]. The organic matrix in dentin is composed of Type I collagen and NCPs (noncollagenous proteins), such as DMP1 (dentin matrix protein) and DPP (dentin phosphoprotein, also known as DMP2 or phosphophoryn) with highly phosphorlyated serine and threonine residues [14]. The inorganic mineral (mainly calcium-deficient carbonate-containing hydroxyapatite (HAP)) in dentin is embedded in the organic matrix. Some studies indicated that some NCPs are inhibitors of mineralization and if removed, dentin remineralization will be enhanced [15]. Therefore, the process of dentin remineralization is more complex than that of enamel, which is more challenging in MID. Novel remineralizing agents (materials) and techniques should be developed to fulfill the remineralizing requirements of MID.

So far, fluoride treatment remains to be the best established remineralizing strategy for early enamel caries. Enamel remineralization using fluoride and its delivery methods have been studied extensively. Various forms of calcium phosphates that provide calcium and phosphate ions for remineralization are vital for remineralizing agents. These agents can be added to restorative materials or directly applied on tooth surface to perform remineralizing functions. The addition of the remineralizing agents in conventional size to restorative materials results in poor functional performance/aesthetics, unsatisfactory biocompatibility, and poor workability. On the contrary, due to the small size and high surface area of the nanoparticles, nanoscales of remineralizing materials are able to release high levels of mineral ions at a low filler level, which has no impact on the incorporation of reinforcing (but nonreleasing) fillers in the same material. In addition, since the well-sized nanomaterials are similar to the scale of the natural building blocks of dental hard tissues, they can facilitate mineralization of collagen and form the materials with enamel-like microstructure. Therefore, nanomaterials could *de novo* repair early caries lesions and thus can protect diseased hard tissue from further demineralization to form visible cavities.

Inspired by the rationale of biomineralization of dental hard tissues, it is possible to mimic this "natural" mechanism to accomplish remineralization of collagen in demineralized dentin. This methodology is termed as biomimetic mineralization, imitating the natural process of mineralization to realize remineralization [16]. The advantage of biomimetic mineralization is that it simulates the natural formation process of mineral crystals on the surface of organs without using special equipments and strict conditions [16]. Some biomimetic remineralizing agents have been developed, such as nanocomplexes of stabilizer and amorphous calcium phosphate. These nanocomplexes can remineralize enamel lesions [17]. Combined with phosphorylated collagen, these nanocomplexes also can remineralize dentin lesions through mineralization of collagen in dentin [18].

In this review, we focus on recent remineralizing nanomaterials showing ideals of MID to manage caries. These materials include calcium fluoride, amorphous calcium phosphate (ACP), beta tricalcium phosphate (β-TCP), hydroxyapatite (HAP) nanoparticles, nanocomplexes of casein phosphopeptides, and amorphous calcium phosphate (CPP-ACP). Particularly, other biomimetic nanocomplexes of stabilizer and amorphous calcium phosphate will be discussed. These nanomaterials include polyacrylic acid (PAA)-ACP, polyaspartic acid (PASP)-ACP, and phosphorylated chitosan (Pchi)-ACP that are novel remineralizing agents based on the nonclassical nucleation theory and the guided tissue remineralization (GTR) strategy. The rationale of remineralization effects of these materials on enamel and dentin will be discussed and summarized.

9.2 Structure and Compositions of Dental Hard Tissues and Caries

9.2.1 Enamel

The dental hard tissue is composed of enamel, dentin, and cementum [19]. The enamel is an outer layer of 1–3 mm thickness, which covers and protects dentin and pulp cavity [20]. In mature enamel, approximately 96 wt% of enamel consists of crystalline HAP, the rest of the composition being the organic materials (ca. 0.6 % by weight) and water (ca. 3.4 % by weight) [20]. The average size of HAP crystallites in enamel is about 30 nm thick, 60 nm wide, and several microns long [21]. Precise spatial arrangement of these fibers leads to a superstructural organization [22], which can be described as several hierarchical levels from nano- to microscales [23, 24]. Since enamel is acellular and independent of nutrient supply, once formed, mature enamel cannot self-repair. Dental cavities start appearing on a small scale, which can be made reversible without using aggressive excavation procedures, and thus at that point, they should be effectively tackled. Naturally, enamel lesions could be remineralized by the epitaxial growth of residual crystals (as nucleation sites) or restore the partially demineralized crystals with the healthy saliva containing calcium and phosphate ions, which provides a supersaturated environment with respect to HAP. If saliva is not enough for remineralization of enamel, enhancing remineralizing treatment should be used, such as providing topical sources of calcium and phosphate ions in oral environments. It is proposed that enamel tissues *de novo* form through interaction between nanospheres and nanofibers of amelogenin, the main protein in the developing enamel matrix. These nanospheres and nanofibers formed by self-assembly of amelogenin at the nanoscale regulate the nucleation, growth, and morphology of apatite crystals in enamel [25]. Therefore, the ideal remineralization strategy is mimicking the genesis of this complex tissue, thus offering restoration of teeth to their natural forms [26].

9.2.2 Dentin

The underlying dentin is a porous, calcified tissue, which forms the major bulk of the tooth structure. It is thought that the flexibility of dentin may help prevent the brittle enamel from fracturing. Compared to enamel, dentin tissue consists of more organic components (ca. 21 % by weight) [27]. The organic matrix of dentin is mainly composed of Type I collagen (ca. 92 % by weight) and noncollagenous proteins (NCPs) [27]. The inorganic components of dentin are primarily calcium-deficient carbonate-containing HAPs [27]. Dentin contains 20 % water by volume, and it is in bound or unbound states [27]. Water molecules hydrate the organic matrix and occupy the interfibrillar space. Dentin is structurally divided into peritubular dentin and intertubular dentin. Peritubular dentin is a highly mineralized tissue containing more inorganic components and less organic matrix. Intertubular dentin is mostly composed of Type I collagen fibrils ranging from 50 to 200 nm in diameter surrounded by nanocrystalline apatites [28]. Some reports showed that the apatite crystallites in dentin are plate like in shape, ca. 50–60 nm long, and up to 3.5 nm thick [29], which are smaller than those of enamel. According to the current understanding of biomineralization of dentin, the process of biomineralization involves the interaction of NCPs with collagen fibrils and minerals [30]. NCPs work as nucleator and inhibitor to accomplish mineralization of collagen. The functional groups of NCPs are crucial for heterogeneous nucleation in biomineralization.

According to the location with respect to the collagen fibrils, apatite crystallites are classified into extrafibrillar minerals and intrafibrillar minerals. Extrafibrillar mineral is located in the spaces separating the collagen fibrils [31–33], and the intrafibrillar one is largely in the gap regions of the fibrils extending between tropocollagen molecules [31–33]. It was observed that the demineralization of the extrafibrillar minerals is faster than that of the intrafibrillar minerals in vitro [34]. The collagen matrix in the intertubular dentin could inhibit demineralization as the crystals remaining in extensively demineralized

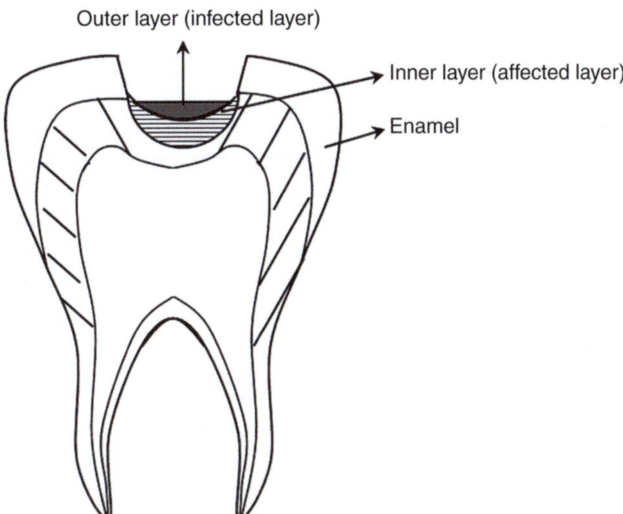

Fig. 9.1 The different zones of dentin caries

regions are mainly ones located in intimate association with the collagen fibrils. Some studies showed that collagen fibrils in carious dentin cannot remineralize unless residual mineral crystals remain in the demineralized lesion [12, 13, 35]. Thus, the effect of traditional remineralization in dentin is dependent on the amount of residual crystals. In addition, it should be noted that the mechanical property of dentin is related to intrafibrillar mineralization of collagen [36]. Therefore, in order to remineralize dentin effectively and obtain ideal mechanical properties, the collagen in dentin lesions should be completely remineralized, including both extra- and intrafibrillar mineralization. Inspired by the role of NCPs in mineralization of collagen, biomimetic remineralization methods would be developed.

9.3 Remineralization Treatment in Minimally Invasive Dentistry

Minimally invasive dentistry embodies at least five principles: remineralization of early lesions, reduction in cariogenic bacteria and elimination of the risk of further demineralization and cavitation, minimum surgical intervention of cavitated lesions, repair rather than replacement of defective restorations, and disease control [37]. A goal of MID is to manage noncavitated carious lesions noninvasively through remineralization to prevent disease progression and improve aesthetics, strength, and function of teeth. In recent decades, the therapeutic importance of remineralization has been generally accepted. In MID, remineralization or arrest of lesions should be utilized to a maximum since there is no real substitute for natural tooth structure.

Early in caries development, caries may affect only enamel. At this time, remineralization treatment and improvement of oral hygiene would stop carious development and recover carious lesions. Once the extent of decay reaches the deeper layer of dentin, dentin caries occurs. Remineralization of dentin is also significant for management of dentin caries. However, it is difficult to accomplish remineralization of deep caries in clinical practice because the caries process has passed a "point of no return" [38]. Carious dentin is usually described in terms of two altered layers in clinical practice, an outer layer (infected layer) and an inner carious layer (affected layer) (Fig. 9.1). The outer layer is contaminated with bacteria in clinical caries, and the collagen fibers are degraded, which is not considered for remineralization, while the inner layer is bacteria free with limited denaturation of the collagen, which could be remineralized [39]. In clinical practice, it is recommended that the outer (infected) layer of dentin caries should be removed during cavity preparation, and the inner layer is conserved for bonding of restorative material, though this is an invasive methodology [40].

Although roots of teeth have a very thin layer of cementum over a large layer of dentin, root caries is mainly caused by demineralization of dentin. Root caries is widespread and increasing in aged adults because pathology and treatment of periodontal diseases expose root surfaces and make them prone to caries attack. It is technically difficult to treat root caries because the lesions tend to "wrap around" the cervical margins of the teeth so that optimal restorative materials are lacking [41]. Therefore, remineralization of dentin is also significant for treatment of root caries. Different in vitro root caries models have been developed for remineralization studies [42].

The etiology of dentin hypersensitivity is ascribed to demineralization that results in exposure of dentinal tubules [15]. Thus, based on a remineralization mechanism, the formation of a mineral layer on the surface of demineralized dentin which results in mineral deposit or plugs in the dentinal tubules is an important therapeutic strategy for dentin hypersensitivity.

The clinical methodologies to deal with carious dental hard tissue can be classified into restorative and nonrestorative repairs [43, 44]. The restorative treatment needs to remove all softened and discoloring dentin to eliminate all infected tissue and create a hard tissue foundation to support a restoration. The use of hand instrumentation and a round steel bur often results in removal of healthy tissue and accidental exposure of the pulp. Moreover, the notion has generally been accepted that restorative intervention is likely the beginning of a long sequence of rerestorations often leading to crowns and implants irrespective of how well the first filling is prepared [44]. The nonrestorative repair for dental caries is in agreement with the concept of MID.

So far, remineralization of superficial enamel caries using fluoride and its derivatives has been well documented in hundreds of completed studies [44]. It has also been demonstrated that fluoride treatment can also facilitate remineralization of shallow dentin and cemetum lesions in vitro [43, 44]. In addition, it was reported that the caries lesion extending to dentin (deep lesion) could

be remineralized in vitro although this process is very slow [38]. Now, an emerging view in caries prevention and management is that restorative treatment should be delayed to provide maximum possibilities for natural lesion repair and arrest [38]. Thus, an efficient remineralization strategy for dentin caries is required to achieve a minimally invasive approach for dental caries management.

9.4 Nanomaterials for Remineralization Treatments

The essential feature in the demineralization of dental hard tissues is that a substantial number of mineral ions are removed from the apatite lattice network and its structural integrity is destroyed, leading to the permeability of the dental hard tissues. The possibility of dissolution (demineralization) and formation (remineralization) of HAP depends on the Gibbs free energy change caused by the degree of saturation with respect to HAP, as shown in Eqs. (9.1) and (9.2).

$$\text{Ca}_{10}\left(\text{PO}_4\right)_6\left(\text{OH}\right)_2 \underset{\text{Remineralization}}{\overset{\text{Demineralization}}{\rightleftarrows}} 10\text{Ca}^{2+}+6\text{PO}_4^{3-}+2\text{OH}^- \tag{9.1}$$

$$\Delta G = RT \ln\left(S\right) = RT \ln\left(\text{IP}/K_{sp}\right) \tag{9.2}$$

where S is the degree of supersaturation with respect to HAP, R the gas constant ($8.314\ \text{J}\cdot\text{K}^{-1}\ \text{mol}^{-1}$), T the absolute temperature, IP the ionic activity product, and K_{sp} the solubility product (K_{sp} for HAP: 2.34×10^{-59}) [45]. The IP value for HAP is given by $[\text{Ca}^{2+}]_{10}[(\text{PO}_4)^{3-}]_6[\text{OH}^-]_2\gamma_1^2\gamma_2^{10}\gamma_3^6$, where γ_z represents the activity coefficient of a z-valent ion. From the point of thermodynamics, if $IP<K_{sp}$ ($\Delta G<0$), the solution is unsaturated with respect to HAP, and HAP will dissolve (demineralize); if $IP>K_{sp}$ ($\Delta G>0$), the solution is supersaturated, and HAP will form again (remineralize). Since K_{sp} for FAP (fluorapatite, $\text{Ca}_{10}(\text{PO}_4)_6\text{F}_2$) is 3.16×10^{-60} [45], FAP is more resistant to acid attack than HAP. Therefore, favorable pH value, enough calcium and phosphate ions available, together with adequate fluoride ions are keys for the remineralization of dental hard tissues. Fluoride and calcium phosphate in various forms, such as calcium fluoride, tricalcium phosphate (TCP), hydroxyapatite, fluorohydroxyapatite, and amorphous calcium phosphate (ACP), as the source of fluorine, calcium, and phosphate ions for remineralization of dental hard tissues have been tested for many years. Since a small particle can release ions at higher concentrations, nanoparticles have better ion release profiles than microparticles [46]. Thus, nanoscales of calcium phosphate materials as remineralizing agents have received more attention. Since it is difficult to directly use nanomaterials to remineralize teeth in the oral environment, these materials are often added to restorative materials as inorganic fillers, such as resin composites. It should be noted that these nanomaterials are not used to reinforce the polymer but to release calcium, phosphate, and fluoride ions for remineralization of dental hard tissues.

9.4.1 Calcium Fluoride Nanoparticles

Currently, the effect of fluoride as a preventive agent on dental caries is well established. The role of fluoride in combating caries has been summarized by Wefel [47], ten Cate and Featherstone [48], and Stoodley et al. [49]. Featherstone proposed the three main mechanisms to explain the anticaries effect of fluoride. Firstly, fluoride can inhibit the metabolic and physiological pathways in the cariogenic biofilm that produces organic acids to demineralize dental hard tissues. Secondly, fluoride inhibits demineralization when it replaces ions in hydroxyapatite (HAP) of dental hard tissues to form fluorapatite (FAP) during an acid challenge; FAP is highly resistant to dissolution by acid compared with HAP. Thirdly, fluoride remineralizes dental hard tissue by attaching onto the crystal surface and then attracting calcium ions followed by phosphate ions to nucleate for new mineral formation and growth. Therefore, fluoride treatment is a very important method in minimally invasive

dentistry (MID). As remineralizing agents such as sodium fluoride (NaF) and calcium fluoride (CaF_2), fluoride has been added to many dental restorative materials (e.g., glass ionomers and resin composites) and preventive products (e.g., dentifrices and mouthwash). The deposits of calcium fluoride (CaF_2) and CaF_2-like in biofilm, as fluoride (F) reservoir, provide fluoride ions for remineralizaton in oral environment. However, the current fluoride-releasing restorative materials have poor mechanical properties or low fluoride recharge capability [50–52].

One approach to resolve this problem is to develop dental composites with fluoride nanoparticles. The CaF_2 nanoparticle (nano-CaF_2) has a 20-fold higher surface area compared with traditional CaF_2 [53, 54]. Thus, the composite with nano-CaF_2 shows high fluoride release and still maintains strength and wear resistance [54, 55]. The addition of nano-CaF_2 increases cumulative fluoride release; the composites containing 20–30 % nano-CaF_2 match the fluoride release rates of traditional glass ionomer materials [53]. Therefore, inhibiting caries by the addition of nano-CaF_2 is achieved without compromising long-term mechanical properties of composites [53, 54, 56].

9.4.2 Calcium Phosphate–Based Nanomaterials

Today, nanoparticles of hydroxyapatite (HAP), tricalcium phosphate (TCP), and amorphous calcium phosphate (ACP) have been developed as sources to release calcium/phosphate ions and increase the supersaturation of HAP in carious lesions [57–59].

Beta Tricalcium Phosphate (β-TCP, $Ca_3(PO_4)_2$)

β-TCP serves as a bioactive source of mineralizing components and has been used as a Class II device used in facilitating bone remodeling in maxillofacial procedures (FDA, 2005) and orthopedic applications (FDA, 2003). β-TCP can be functionalized with organic and/or inorganic to form funcionalized β-TCP ($f\beta$-TCP). It has been reported that the combination of fluoride and fTCP produces stronger, more acid-resistant minerals relative to fluoride, native β-TCP, or fTCP alone [60]. As a low-dose system, fTCP does not rely on high levels of calcium and phosphate to drive remineralization [60]. $f\beta$-TCP provides a barrier that prevents premature fluoride–calcium interactions and aids in mineralization when applied *via* common preparations and procedures [60]. The combination of fluoride plus fTCP has been used to remineralize enamel lesions. Karlinsey et al reported that the combination of NaF (*i.e.*, 500, 950, 1,100, or 5,000 ppm F⁻) plus fTCP in a simple aqueous solution can significantly remineralize white spot enamel lesions relative to that achievable with fluoride alone [60]. In addition, when added to commercial mouth rinse and dentifrice containing fluoride, fTCP provided significantly greater fluoride uptake and rehardening relative to a fluoride-free and controlled fluoride-only mouth rinse and dentifrice [61]. These studies demonstrate that since fTCP can enhance fluoride-based nucleation activity with subsequent remineralization driven by dietary and salivary calcium and phosphate, the combination of fluoride and fTCP appears to be a promising approach to remineralization of dental hard tissues.

Currently, nanoscale β-TCP has been used for bone tissue regeneration due to its higher compressive strength, degradation rate, osteoconductivity, and protein absorption compared to submicron β-TCP [62]. Thus, combination of fluoride and nanomaterials of β-TCP may achieve more effective remineralizing results. However, since β-TCP is often added to mouth rinse and dentifrice, the toxicity of nanoscale β-TCP should be evaluated adequately.

Hydroxyapatite (HAP) Nanoparticles

Synthetic HAP is a biocompatible material, and nano-sized HAP (n-HAP) is similar to the apatite crystal of tooth enamel in morphology and crystal structure. Therefore, it is logical to consider n-HAP as compound substitute for the natural mineral constituent of enamel, with which defects of dental enamel would be repaired.

It has been reported that n-HAP particles can remineralize initial submicrometer enamel caries [63, 64]. If the dimensions of the n-HAP particles are adapted to the scale of the submicrometer- and nano-sized defects, the reparation of the enamel surface can be greatly improved by using these n-HAP particles. It is shown that the basic building blocks of enamel are 20–40 nm HAP nanoparticles [65]. *In vitro* data indicate that n-HAP with a size of 20 nm fits well with the dimensions of the nanodefects on the enamel surface caused by acidic erosion [64]. Under *in vitro* conditions, these n-HAP particles can strongly attach to the demineralized enamel surface and inhibit further acidic attack [64]. Thus, the use of well-sized n-HAP particles similar to the scale of the natural building blocks of enamel could *de novo* repair early carious lesions and thus can protect them from further demineralization to form visible cavities. In the other study, an enamel-like nanocrystal layer with 10 µm thickness in small cavities was achieved *in vitro* by pasting fluoride-substituted HAP on the enamel within 15 min, but this process was carried out under pH 3.5 and high concentrations of hydrogen peroxide [66]. In view of the real conditions of the oral cavity and potential toxicity of n-HAP, the effect of direct use of n-HAP particles on remineralization of enamel should be further investigated and confirmed in a clinical trial.

n-HAP powder can be also added to dental restorative materials for remineralization effects and improvement of mechanical properties due to its excellent biocompatibility and bioactivity [67, 68]. For instance, compared with micro-HA added to glass ionomer cement, 10 % n-HAP particles (60–100 nm) are incorporated in resin-modified glass ionomer cement, which results in an increased resistance to demineralization and acceptable bonding strength with the only drawback of exceeding the clinically suitable maximum setting time [69–71]. Furthermore, the addition of n-HAP and nanofluorohydroxyapatite (n-FHA) to glass ionomer cements increases the compressive, diametral tensile, and biaxial flexural strength of glass ionomer cements [72, 73]. Besides, the glass ionomer cement containing n-FHA has the potential to increase the amount of fluoride release [74].

Nanoparticles of HAP have been incorporated into toothpastes or mouth-rinsing solutions to facilitate the remineralization of demineralized enamel or dentin by depositing HAP nanoparticles in the lesions. Commercially available dental prophylactic products containing biomimetic carbonate hydroxyl apatite nanoparticles have been used to fill microdefects on demineralized enamel or dentin surfaces and proved to be effective *in vitro* after a 10 min application. However, these promising effects need a clinical study to support them. In addition, the toothpastes with either spheroidal or needle-like particles of n-HAP show better remineralization effect on demineralized enamel than sodium fluoride solutions [75]. However, the *in vitro* study simulating the real conditions of oral cavity or an *in vivo* study is needed to further test to prove the remineralization effects of these toothpastes.

Recently, some studies indicated that biomimetic synthesis of hierarchically organized enamel-like structures composed of n-HAP would be an ideal approach to repair enamel microcavities. In the presence of organic additives [76–85] or by using various hydrothermal conditions, the *in vitro* formation of enamel-like microstructures can be achieved. Formation of enamel-like structures in presence of amelogenin, a major extracellular matrix protein in physiological enamel development, has been well documented. Amelogenin oligomers mediate the self-assembly of oriented parallel needle-like apatite bundles to form nano- and microstructured materials, which is compositionally and morphologically similar to natural enamel [25, 76, 78–81, 83, 84, 86–88]. Amelogenin remineralizes etched enamel surfaces by forming a mineral layer containing needle-like fluoridated HAP crystals with dimensions of 35 nm [80]. Additionally, self-assembling anionic β-sheet peptides, mainly composed of glutamic acid and glutamine, form fibrillar networks as scaffolds to be mineralized and could enhance remineralization and inhibit demineralization of the enamel [82]. Surfactants also can work as micelles or microemulsions to mimic the biomineralization process during the formation of enamel [84]. HAP nanorods modified with monolayers of

surfactants can self-assemble into a prism-like enamel structure due to specific surface characteristics [84].

Although some promising *in vitro* results were obtained, the stability and the mechanical properties of the n-HAP and the enamel-like materials are not sufficient for tooth restorations, and the long time (from several hours to days) for the formation of the mineral structures also limits their clinical application [76, 81]. Therefore, besides remineralization functions, further research should improve the properties of the materials related to clinical operations, thus providing clinically conceivable biomimetic tooth repair.

Amorphous Calcium Phosphate (ACP) Nanoparticles

Amorphous calcium phosphate (ACP) is the initial solid phase precipitating from a highly supersaturated solution with respect to calcium phosphate, which is firstly described by Aaron S. Posner in the mid-1960s [89]. The morphology of ACP particles is shown as small spheroidal particles in the nanoscale (40–100 nm). Owing to its excellent bioactivity, high cell adhesion, adjustable biodegradation rate, and good osteoconduction, ACP has been widely applied in biomedical fields, especially in orthopedic and dental fields [90–93]. Since ACP can convert readily to stable crystalline phases such as octacalcium phosphate (OCP) or HAP, it is difficult to directly use ACP to remineralize dental hard tissues unless stabilized in some way. Therefore, like the nanomateirals of CaF_2 and HAP mentioned above, ACP nanoparticles, as source of calcium and phosphate ions, have also been added to composite resins, ionomer cements, and adhesives. Taking advantage of the ability of ACP to release calcium and phosphate ions, these composites, especially in the acidic oral environment, present remineralization effects on dental hard tissues to prevent secondary caries after restorations. A study using in situ caries models of humans indicated that nanoACP-containing nanocomposites prevented demineralization at the restoration–enamel margins, producing lower enamel mineral loss compared with the control composite [94]. This result could be attributed to the oral biofilm exposed to nanoACP with higher calcium and phosphorus concentrations than that exposed to the control composite [94]. This high local concentration at the surface thus stimulates precipitation and deposition into tooth structures as apatite mineral. The remineralizing potential of ACP composites can be improved by introducing Si or Zr elements during low-temperature synthesis of the filler [95]. Si and Zr ACPs increased the duration of mineral ion release by slowing down the intracomposite ACP to HAP conversion [96].

Although ACP-containing composites show remineralization ability, these composites exhibit inferior mechanical properties, durability, and water sorption characteristics due to the addition of ACP [97]. These problems could be attributed to the uncontrolled aggregation of ACP nanoparticles along with poor interfacial interaction between them [98]. Currently, stabilizing and coupling agents are used to stabilize and disperse ACP nanoparticles in the composites. It was found that anionic surfactants can stabilize the amorphous solid phase against the conversion to apatite during the precipitation of ACP; the particle size of ACP was also moderately reduced. The hydrophilic polyethylene oxide (PEO) is water compatible due to its multiple hydrogen bonding interactions with water molecules and stabilizes ACP nanoparticles by multiple chelation. Thus, the incorporated PEO in ACP fillers can prevent ACP nanoparticles from aggregating and affect the water content of the ACP-containing composites, which eventually will impact both ion release kinetics and mechanical stability of composites [99].

It has been suggested that ACP works as a precursor to bioapatite and as a transient phase in biomineralization [100]. This process is thought to be mediated by noncollagenous proteins, such as amelogenin, dentin matrix protein (DMP1), and dentin phosphophoryn (DPP, DMP2) with highly phosphorylated serine and threonine. They are biological stabilizers by chelating calcium ions to control the transformation of ACP to HAP. Therefore, it is possible to develop a biomimetic remineralizing strategy for reparation of teeth caries by mimicking the biomineralization process. In the next section, the development of nanocomplexes of stabilizers and ACP will be reviewed.

9.4.3 Nanocomplexes of Stabilizers and Amorphous Calcium Phosphate

Some *in vitro* studies indicate that some proteins and their derivatives and analogues, such as polymers and poly(amino acid) macromolecules mimicking the functional domain of these proteins, could stabilize calcium/phosphate ions as nanocomplexes of protein/amorphous calcium phosphate (ACP) in solution [101–103]. Casein phosphopeptides (CPP) obtained from milk is such an analogue of the proteins involved in biomineralization of teeth.

Nanocomplexes of Casein Phosphopeptides (CPP) and Amorphous Calcium Phosphate (ACP)

The four sequesters of casein phosphopeptides are Bos α_{S1}-casein X-5P (f59-79), Bos β-casein X-4P (f1-25), Bos α_{S2}-casein X-4P (f46-70), and Bos α_{S2}-casein X-4P (f1-21). All the peptides contain the sequence motif -Pse-Pse-Pse-Glu-Glu-, and the major peptides of the preparation are Bos α_{S1}-casein X-5P (f59-79) (ab. α_{S1}(59–79)) and Bos β-casein X-4P (f1-25) (ab. β (1–25)). It is proposed that the CPP binds to the spontaneously forming ACP nanoclusters under alkaline conditions (e.g. pH 9.0), producing a metastable colloid of nanocomplexes of CPP-ACP. From the stoichiometric and cross-linking analyses [104], the stabilized nanocoplexes of CPP-ACP complex have unit formula of $[\alpha_{S1}(59–79)(ACP)_7]_6$ and $[\beta(1–25)(ACP)_8]_6$ [105]. A "closed complex" model of $\beta(1–25)$ complexed with alkaline amorphous calcium phosphate has recently been proposed [104, 105]. This model indicated that all the charged residues of CPP significantly interact with the alkaline calcium phosphate core particle. The hydrodynamic radii of $\beta(1–25)$-ACP complex were estimated at 1.526 ± 0.044 nm at pH 6.0 and 1.923 ± 0.082 nm at pH 9.0.

Reynolds et al. developed a routine preparation of nanocomplexes of CPP-ACP by titrating calcium ions, phosphate ions, and hydroxide ions at pH 9.0 into CPP solutions, followed by purifying with filtration and drying. The nanocomplexes of CPP-ACP have been trademarked as Recaldent™ and added to sugar-free gum and dental professional products (GC Tooth Mousse™). Currently, this product is used for the prevention and treatment of early caries and the treatment of tooth sensitivity, especially after in-office bleaching procedures, ultrasonic scaling, hand scaling, and root planing. It was demonstrated that CPP-ACP binds well to *Streptococcus mutans* in oral biofilm, thereby providing a mineral ion reservoir to release free calcium and phosphate for inhibiting demineralization and enhancing subsequent remineralization [106]. The affinity of CPP-ACP to biofilm could attributed to calcium of CPP-ACP competing the calcium binding sites of biofilm, which will decrease the amount of calcium bridge between bacteria and the acquired pellicle, and among bacteria themselves [107].

In an *in vitro* caries model, Reynolds et al. investigated the effects of CPP-ACP solutions on remineralization of artificial lesions in human third molars [108]. With 10 day remineralization treatment, 1.0 % CPP-ACP (pH 7.0) solution recovered 63.9±20.1 % of mineral loss [108]. In addition, the combination of CPP-ACP and fluoride has additive remineralizing effects on carious lesions [109]. A recent clinical trial compared the remineralizing effect of a sugar-free gum containing 18.8 mg CPP-ACP with that of a sugar-free gum not containing CPP-ACP on enamel [110]. After in situ remineralisation by the CPP-ACP–containing gum, the lesions with subsequent acid attack showed demineralization beneath the remineralized zone, indicating that the remineralized mineral by CPP-ACP was more resistant to acid challenge [17].

Nanocomplexes of CPP-ACP provide a new effective remineralization method for minimally invasive management of dental caries. Although Reynolds did not emphasize that CPP-ACP is a biomimetic product, but the mechanism of the CPP-stabilized ACP formation is similar to that of phosphorylated protein-stabilized ACP, which transforms into HAP crystal in the process of biomineralization of dental hard tissues. Therefore, inspired from CPP-ACP and understandings on the roles of acidic noncollagenous proteins, such as phosphoproteins, in biomineralization of dental hard tissues, we obtain a strategy for development of novel biomimetic-remineralizing agents for MID: finding analogues of acidic noncollagenous proteins that are capable of stabilizing ACP.

Biomimetic Nanocomplexes of Stabilizer and Amorphous Calcium Phosphate (ACP)

Up to the present, transient amorphous mineral phases have been found in biomineral systems in different phyla of the animalia kingdom [111]. For example, ACP has been reported to form as a precursor phase of carbonated hydroxyapatite in Chiton teeth [112]. Also, an ACP phase has been observed in the newly formed bony zebrafish fin rays [113]. A comprehensive analysis of the mineral phases in the early secretory enamel of the mouse's mandibular incisor indicated that the outer, younger, early secretory enamel contained a transient disordered ACP phase. The disordered ACP phase is a precursor of crystalline hydroxyapatite and transforms into the final apatitic crystalline mineral with time [111]. It was also suggested that the transient ACP phase can directly deposit inside the gap regions of collagen fibrils during bone and dentin maturation [113].

It is proposed that a variety of acidic noncollagenous proteins are involved in the biomineralization of ACP to HAP [114, 115]. Since the acidic noncollagenous proteins contain aspartic acid and glutamic acid–rich domains, they may act as nucleators or inhibitors, growth modifiers, anchoring molecules, or scaffolds for mineral deposition [116–118]. The functional domains of these acidic noncollagenous proteins can be mimicked by some polyelectrolytes and poly(amino acid) macromolecules containing phosphoryl or carboxyl groups, such as polyacrylic acid (PAA) and polyaspartic acid. These biomimetic materials can stabilize ACP and reduce these amorphous phases to nanoscale [119], known as biomimetic nanocomplexes of stabilizer and ACP. This biomimetic process, also known as "polymer-induced liquid precursor" (PILP) [120, 121], has been used to synthesize nanoscale ACP to biomimetically mineralize type I collagen [120, 121]. This biomimetic process has been reported to be independent of ion solubility products and relatively insensitive to changes in pH and osmolarity, which is difficult to be explained by classical crystallization theory. Accordingly, nonclassical crystallization theory (pathway) was proposed to describe the biomimetic mineralization process based on PILP [122]. In the nonclassical pathway, inorganic nanocrystals coated/stablized with organic molecules can form larger mesocrystals via self-assembly and crystallographic alignment. These mesocrystals work as intermediates for the formation of single macroscopic crystals.

Although remineralization of enamel lesions is conceivably achieved and evaluated, the remineralization of dentin is still debatable [15, 38, 123]. Although the deposition of interfibrillar/extrafibrillar apatite minerals can contribute to the increase of the gray intensity value of samples after remineralization, without intrafibrillar mineralization, interfibrillar/extrafibrillar mineralization of dentin alone does not result in a highly mineralized collagen matrix [36] and cannot be regarded as true remineralization of demineralized dentin. Thus, the microradiography as a convincing method for determining the degree of remineralization of enamel [15, 38] has recently been challenged [124].

More recently, based on the rules of nonclassical crystallization pathways, guided tissue remineralization (GTR) was proposed [18]. In their studies, presence of polyacrylic acid (PAA) as calcium phosphate–binding matrix protein analogues or metastable ACP nanoprecursors formed in a Portland cement/phosphate-containing fluid system [18]. Interestingly, only both in the presence of polyvinylphosphonic acid (PVPA) or PAA and sodium trimetaphosphate (STMP) or sodium tripolyphosphate (TPP) as collagen-binding matrix phosphoprotein analogues, the intrafibrillar remineralzation of collagen in dentin was achieved [101, 102]. In guided tissue remineralization, the PAA-based biomimetic analogue is employed as a sequestration agent to stabilize nanoscale ACP that can infiltrate the internal water compartments of a collagen fibril. The phosphorus-based analogue of matrix phosphoproteins binds to the collagen via electrostatic interaction or chemical phosphorylation reactions to attract these nanoprecursors of ACP to the gap zones between the collagen molecules [125]. The self-assembly of nanoprecursors of ACP and their subsequent transformation into nanocrystals of HAP would result in the hierarchical and intrafibrillar remineralization of collagen in demineralized dentin. Therefore, GTR with biomimetic remineralizing materials is potentially useful in the remineralization of partially demineralized caries-affected dentin.

The concepts of CPP-ACP and GTR provide a novel strategy for remineralization treatment in MID. In the concept of CPP-ACP, for remineralization of enamel, one analogue is required to mimic the sequestration function of matrix proteins to stabilize ACP, while in the GTR concept, two polyanionic analogues mimicking the sequestration and templating functions of matrix proteins are involved to stabilize ACP and direct ACP nanoprecursors into collagen molecules for remineralization of dentin. Thus, the analogues stabilizing ACP can aid to remineralize both enamel and dentin.

Presently, studies have proposed phosphorylated chitosan (Pchi) and carboxymethyl chitosan (CMC) as calcium phosphate–binding matrix protein analogues to stabilize ACP to form nanocomplexes of Pchi-ACP or CMC-ACP. Chitosan is a linear copolymer of glucosamine and N-acetyl glucosamine in a $\beta 1 \rightarrow 4$ linkage obtained by N-deacetylation of chitin. Chitosan and its derivatives have emerged as a new class of novel biomaterials due to their versatile biological activity, excellent biocompatibility, and complete biodegradability [126–128]. Among these derivatives of chitosan, phosphorylated chitosan (Pchi) exhibits bactericidal [129], biocompatible, bioabsorbable [130–135], and metal-chelating properties [136, 137]. It was shown that the chelating ability of phosphate groups of Pchi allowed immobilized Pchi molecules to bind with calcium ions to form nucleating sites [138], while free Pchi molecules showed an inhibitory effect on the formation of calcium phosphate deposits in solution [36]. Thus, Pchi could be used to stabilize ACP to form the nanocomplexes of Pchi-ACP.

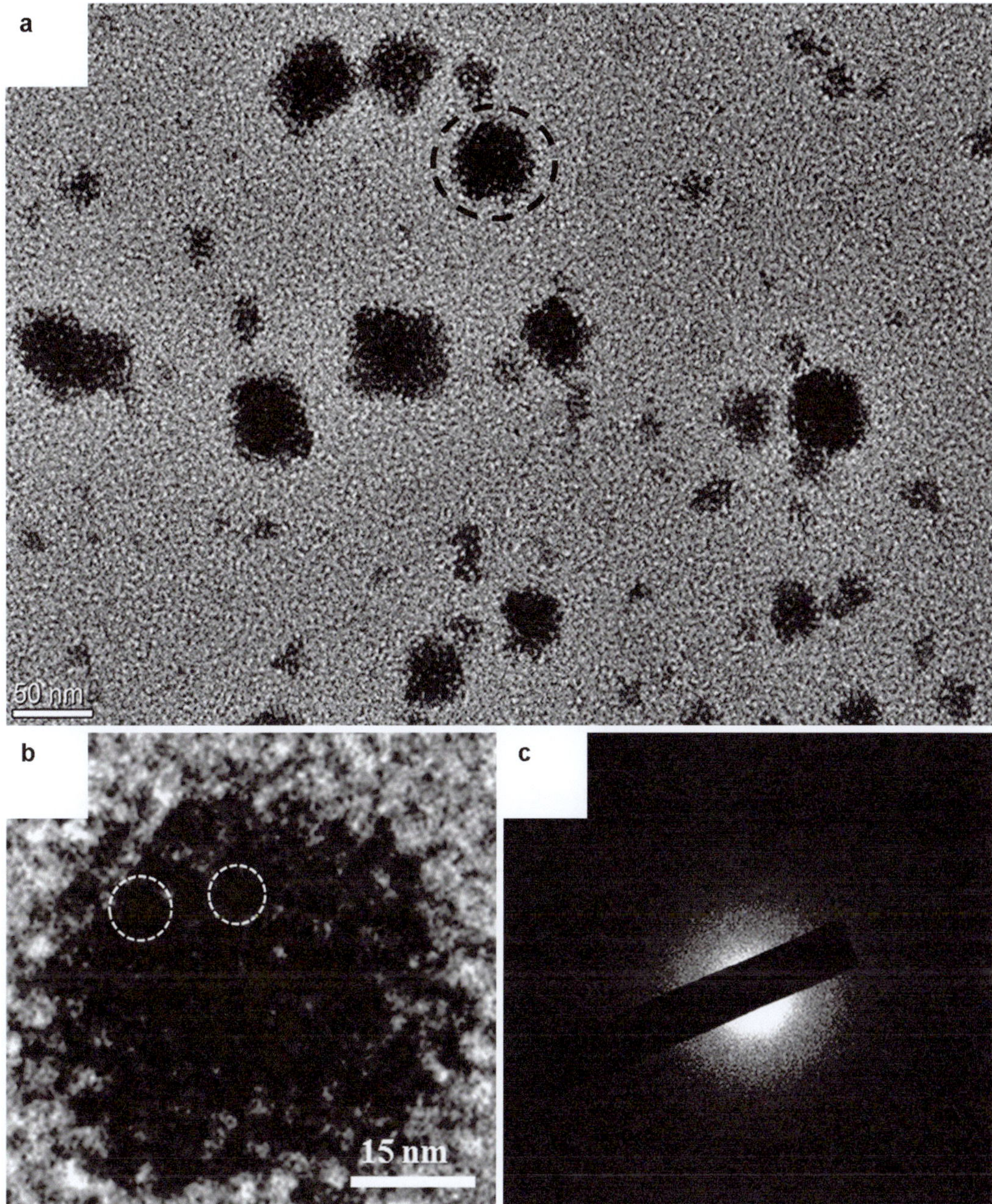

Fig. 9.2 TEM and SAED characterization of nano-complexes of Pchi-ACP. (**a**) Image of nanoparticles of ACP formed in presence of Pchi (nano-complexes of Pchi-ACP). (**b**) Higher magnification of the nanoparticle indicated by a *black dash line circle* in (a). This nanoparticle was composed of much smaller prenucleation clusters indicated by *white dash line circles*. (**c**) SAED of nanoparticles of Pchi-ACP did not show obvious dot or ring pattern characteristic of crystal structure, which indicates that its main composition is amorphous phase

The size of Pchi-ACP nanocomplex particles was determined to be less than 50 nm, and SAED results confirmed their amorphous phases (Fig. 9.2). We remineralized enamel samples with fluoride and Pchi-ACP nanocomplexes respectively. The changes in mineral profiles of

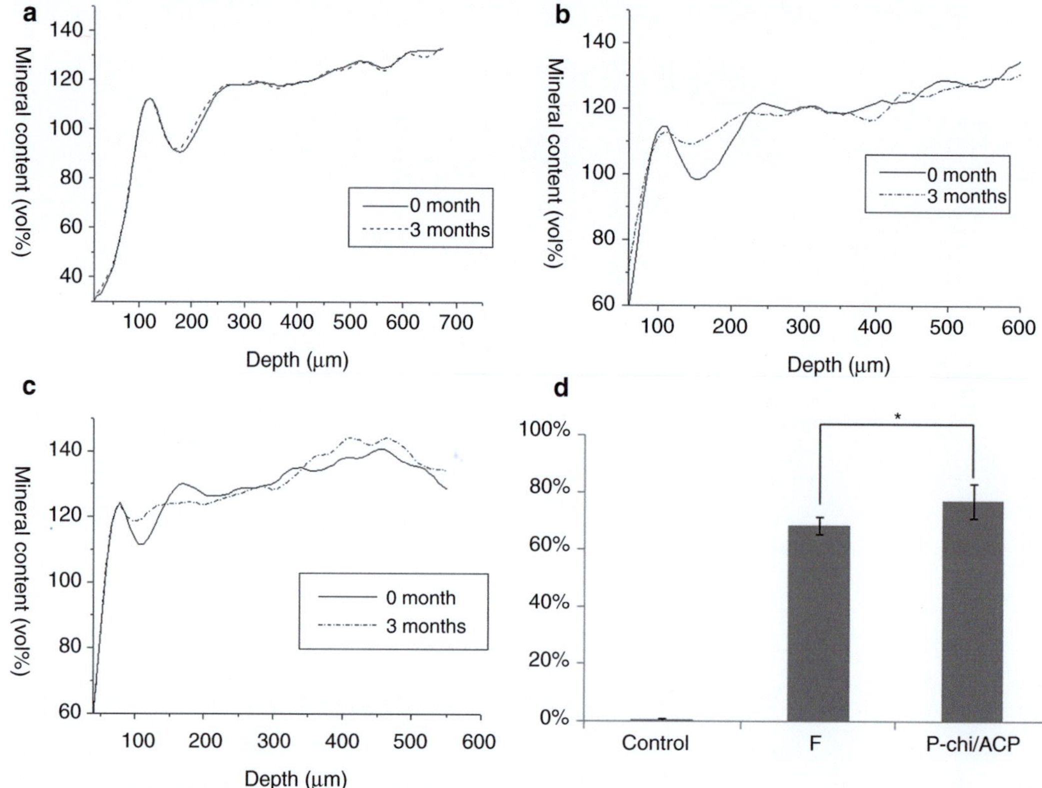

Fig. 9.3 Representative mineral profiles based Micro-CT results. (**a**) Control group; (**b**) Fluoride treatment group. (**c**) Pchi-ACP treatment group. (**d**) The remineralizing degrees of the samples in different groups. * Significantly different from F-treatment group ($p<0.05$, $n=4$, mean ± S.E.M)

representative samples showed that the control group samples did not show remineralization (Fig. 9.3a), while in the experimental groups, both fluoride and Pchi-ACP treatments led to evident changes in the mineral profiles over 3-month remineralization (Fig. 9.3b, c). The rate of remineralization of Pchi-ACP treatment was significantly higher than that of fluoride treatment ($p<0.05$) (Fig. 9.3d). In the presence of Pchi-ACP nanocomplexes, the enamel surface-adsorbed nanocomplexes of Pchi-ACP released calcium and phosphate ions, which entered the defective lattice of HAP crystals or was adsorbed onto the residual crystals to nucleate (i.e., heterogeneous nucleation) and grew into new mineral crystals. This observation was consistent with the classical nucleation theory.

An ACP phase in the incorporated nanocomplexes of Pchi-ACP may in situ transform into HAP crystals within the lesions to remineralize enamel. This assumption could be supported by the nonclassical nucleation theory [17]. Based on this theory, the nanocoplexes of Pchi-ACP could be regarded as the aggregates of prenucleation clusters (Fig. 9.2b). These aggregates (nanocomplexes of Pchi-ACP) as a transition phase gradually transformed into the final crystalline phase and thereby remineralized the lesions in demineralized enamel (Fig. 9.4). This remineralization process is similar to the natural enamel biomineralization, where coassembly of amelogenin and calcium phosphate clusters are formed initially and then transformed into a crystalline phase [133].

Pchi-ACP together with TPP was also employed to remineralize completely demineralized dentin. The SEM and EDX results showed significant degrees of remineralization (Fig. 9.5). This sandwich-like pattern indicated that the two

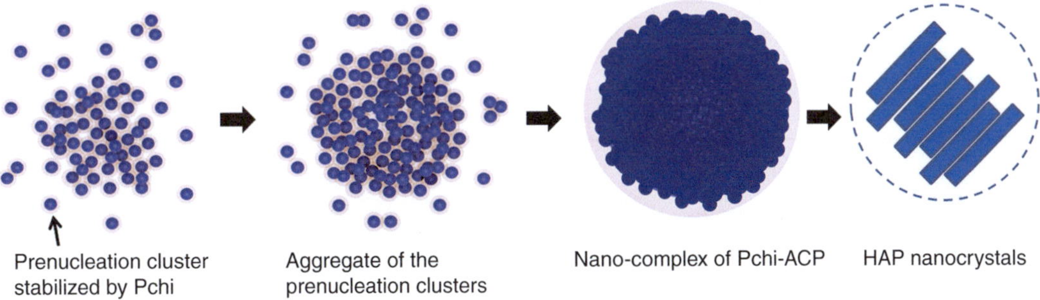

Fig. 9.4 The process of self-assembly of prenucleation clusters of ACP stabilized by Pchi into nano-complex of Pchi-ACP, and finally the nano-complexes transforming into HAP nanocrystals

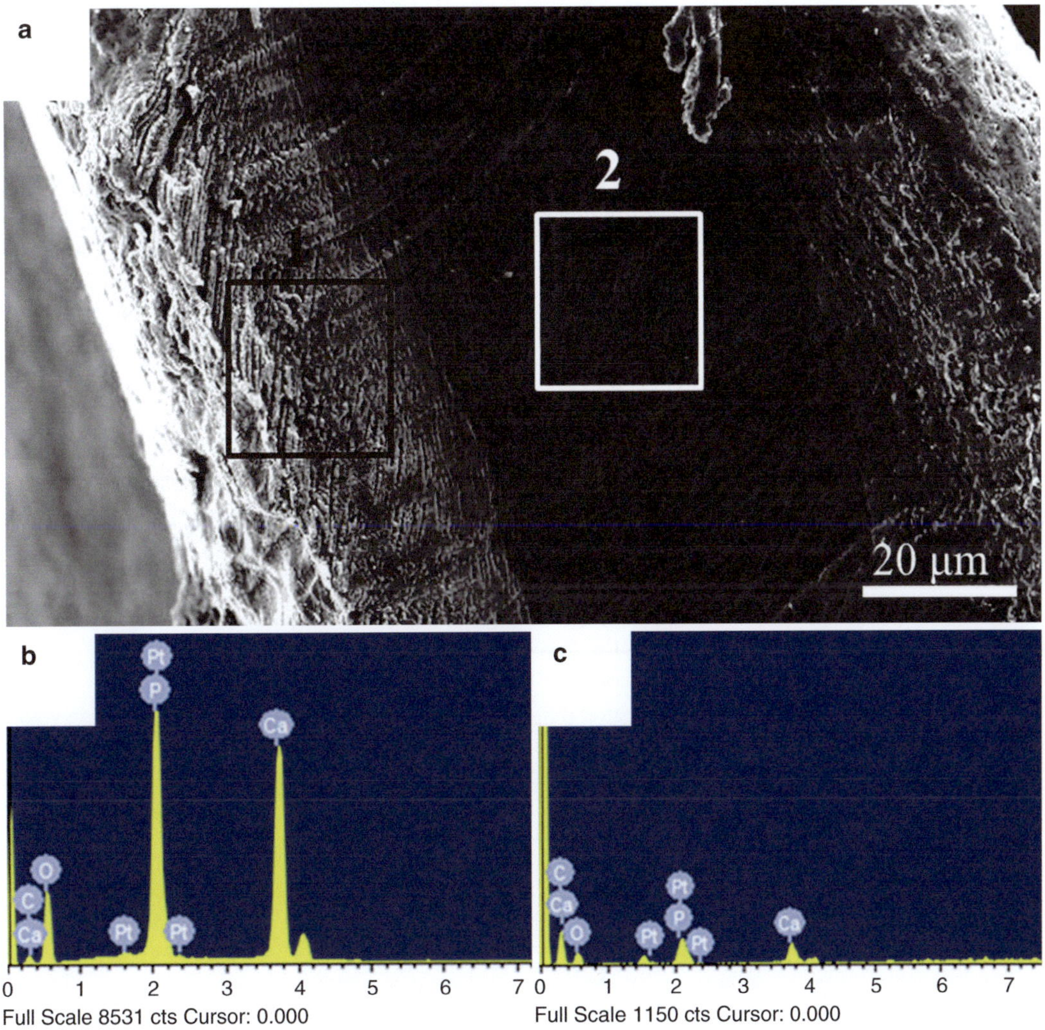

Fig. 9.5 SEM and EDX results of remineralization of completely demineralized dentin (mainly typeIcollagen). (**a**) SEM image of cross-section of remineralized dentin collagen in the presence of nano-complexes of Pchi-ACP and TPP. (**b**, **c**) EDX results corresponding to the square areas in (a) labeled with '1' and '2' indicating different contents of calcium and phosphorus in the outer layer and the middle layer, respectively

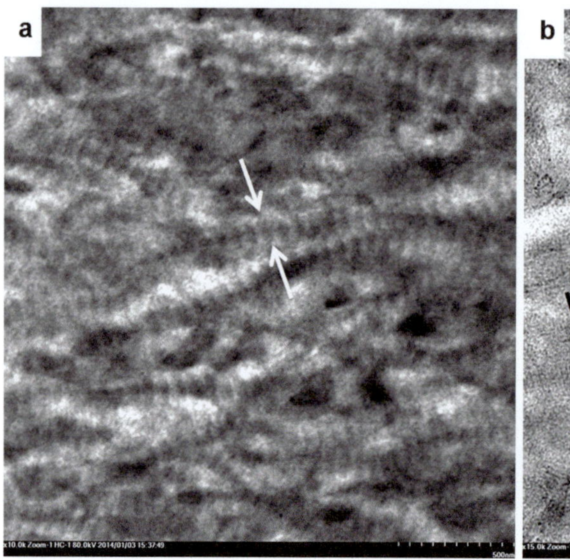
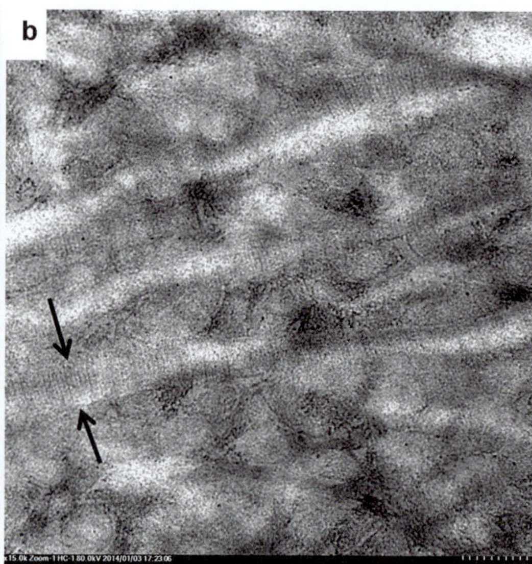

Fig. 9.6 TEM of unstained outer and middle layers of the dentin section in Fig. 9.5 (**a**) The outer layer showed characteristic type-I collagen D-bands (between *arrows*), indicating intrafibrillar minerals within collagen fibrils after remineralization. (**b**) The middle layer exhibited only faint banding characteristics (between *arrows*) due to lack of intrafibrillar minerals

outer layers were well remineralized, while the middle layer was less remineralized. This observation was further confirmed by the typical EDX result shown in Fig. 9.5b, c. The Ca/P molar ratios of the mineral crystal in the outer layers were 1.52 ± 0.03 ($n = 3$), indicating that these mineral crystals were calcium-deficient HAP, while in the middle layers, calcium and phosphorus elements were trace ingredients. This remineralizing pattern indicates that the prior remineralization of outer layers of dentin may hinder the movement of nanocomplexes of Pchi-ACP into deep dentin. TEM results indicated that in the absence of Pchi-ACP and TPP as biomimetic analogues, the fibrils in the significantly remineralized outer layers showed characteristic type-I collagen D-bands that are suggestive of intrafibrillar remineralization (Fig. 9.6a). By contrast, the fibrils in the less remineralized middle layer exhibited only faint banding characteristics due to lack of intrafibrillar minerals (Fig. 9.6b).

A current study indicates that nanocomplexes of stabilizer and ACP itself can remineralize enamel lesions; combined with phosphorylated collagen, it can remineralze dentin lesions. Most importantly, the remineralizing rationale of biomimetic nanomaterials is based on a particle-mediated, bottom-up mineralization strategy, which mimics the process of biomineralization. The biomimetic method may provide a new strategy for remineralization of enamel and dentin and could break the traditional notion that collagen matrix in dentin caries cannot be remineralized.

9.5 Concluding Remarks

Nanotechnology has been applied to the development of dental materials with better properties and anticaries or remineralizing potential. Most remineralizing nanomaterials with remineralizing function, such as CaF_2, β-TCP, HAP, and ACP, cannot be directly applied in oral environments. They are usually incorporated into restorative materials such as composite resins, glass ionomer, and adhesive systems play a remineralizing role through releasing calcium and phosphate ions from restorative materials. Other biomimetic remineralizing nanomaterials, such as nanocomplexes of CPP-ACP, PAA-ACP, and Pchi-ACP, also show remineralizing effects on carious lesions of enamel and dentin. Furthermore,

CPP-ACP is commercially available now and applied in clinics. The remineralizing mechanism of the biomimetic nanomaterials is based on a particle-mediated, bottom-up mineralization strategy, which is different from conventional remineralization techniques. These biomimetic nanomaterials could *de novo* form dental hard tissues through nonclassical crystallization pathways (i.e., kinetically driven proteins or their analogue-modulated pathways for lowering the activation energy barrier for crystal nucleation via sequential steps of phase transformations as depicted in Fig. 9.1) rather than depending on the epitaxial growth of enough residual seed crystallites (the classical crystallization pathway). Therefore, remineralizing nanomaterials, especially biomimetic nanomaterials and their remineralizing rationale, such as GTR conception, would enrich remineralizing methods in MID and thus broaden its dimensions.

With the development of remineralizing nanomateirals for MID, the line between preventive and restorative treatments is promising to be erased. However, it is unpractical to restore small clinically visible cavities with nanomaterials reviewed in this chapter at present. Most of the research covered in this chapter focused primarily on laboratory tests. Extensive future research with respect to the clinical applicability of remineralizing nanomaterials is required for their successful application in clinical practices.

References

1. Featherstone JDB. The science and practice of caries prevention. J Am Dent Assoc. 2000;131:887–99.
2. Osborne JW, Summitt JB. Extension for prevention: is it relevant today? Am J Dent. 1998;11(4):189–96.
3. Christensen GJ. The advantages of minimally invasive dentistry. J Am Dent Assoc. 2005;136(11):1563–5.
4. White JM, Eakle WS. Rationale and treatment approach in minimally invasive dentistry. J Am Dent Assoc. 2000;131(9):1250–2.
5. Rainey JT. Understanding the applications of micro-dentistry. Compend Contin Educ Dent. 2001;2(11A):1018–25.
6. Ettinger RL. Restoring the ageing dentition: repair or replacement? Int Dent J. 1990;40(5):275–82.
7. Hewlett ER, Mount GJ. Glass ionomers in contemporary restorative dentistry – a clinical update. J Calif Dent Assoc. 2003;31(6):483–92.
8. Mount GJ, Ngo H. Minimal intervention: a new concept for operative dentistry. Quintessence Int. 2000;31(8):527–33.
9. Mount GJ, Hume WR. A new cavity classification. Aust Dent J. 1998;43(3):153–9.
10. Pitts NB. Are we ready to move from operative to non-operative/preventive treatment of dental caries in clinical practice? Caries Res. 2004;38(3):294–304.
11. Tyas MJ, Anusavice KJ, Frencken JE, Mount GJ. Minimal intervention dentistry – a review. FDI Commission Project 1–97. Int Dent J. 2000;50(1):1–12.
12. Levine RS, Rowles SL. Further studies on the remineralization of human carious in vitro. Arch Oral Biol. 1973;18:1351–6.
13. Daculsi G, Kerebel B, Le Cabellec MT, Kerebel LM. Qualitative and quantitative data on arrested caries in dentine. Caries Res. 1979;13:190–202.
14. Anne G, Arthur V. Phosphorylated proteins and control over apatite nucleation, crystal growth and inhibition. Chem Rev. 2008;108(11):4670–93.
15. Kawasaki K, Ruben J, Stokroos I, Takagi O, Arends J. The remineralization of EDTA-treated human dentine. Caries Res. 1999;33:275–80.
16. Zhang HL, Liu JS, Yao ZW, Yang J, Pan LZ, Chen ZQ. Biomimetic mineralization of electrospun poly(lactic-co-glycolic acid)/multi-walled carbon nanotubes composite scaffolds in vitro. Mater Lett. 2009;63:2313–6.
17. Cross KJ, Huq NL, Reynolds EC. Casein phosphopeptides in oral health-chemistry and clinical applications. Curr pharm des. 2007;13(8):793–800.
18. Tay FR, Pashley DH. Guided tissue remineralisation of partially demineralised human dentine. Biomaterials. 2008;29(8):1127–37.
19. Jones FH. Teeth and bones: applications of surface science to dental materials and related biomaterials. Surf Sci Rep. 2001;42:75–205.
20. Hermann E, Petros GK, Konstantinos DD, Oleg SP. Principles of demineralization: Modern strategies for the isolation of organic Frameworks Part II. Decalcification. Micron. 2009;40:169–93.
21. Simmelink JW. Histology of enamel. In: Avery JK, editor. Oral development and histology. New York: Thieme Medical Publishers Inc.; 1994.
22. Habelitz S, Marshall SJ, Marshall GW, Balooch M. Mechanical properties of human dental enamel on the nanometer scale. Arch Oral Biol. 2001;46:173–83.
23. Hansma P, Turner P, Drake B, Yurtsev E, Proctor A, Mathews P, Lulejian J, Randall C, Adams J, Jungmann R, Garza-de-Leon F, Fantner G, Mkrtchyan H, Pontin M, Weaver A, Brown MB, Sahar N, Rossello R, Kohn D. The bone diagnostic instrument II: indentation distance increase. Rev Sci Instrum. 2008;79:064303.
24. Weiner S, Traub W. Bone structure: from angstroms to microns. FASEB J. 1992;6:879–85.
25. Moradian-Oldak J. Amelogenins: assembly, processing and control of crystal morphology. Matrix Biol. 2001;20(5):293–305.

26. Uskoković V, Bertassoni LE. Nanotechnology in dental sciences: moving towards a finer way of doing dentistry. Materials. 2010;3(3):1674–91.

27. Piesco NP. Histology of dentine. In: Avery JK, editor. Oral development and histology. New York: Thieme Medical Publishers Inc.; 1994.

28. Steve W, Arthur V, Elia B, Talmon A, Jerry WD, Boris S, Farida S. Peritubular dentin formation: crystal organization and the macromolecular constitutes in human teeth. J Struct Biol. 1999;126:27–41.

29. Johanssen E. Microstructure of enamel and dentine. J Dent Res. 1964;43:1007–9.

30. Veis A. A window on biomineralization. Science. 2005;37:1419–20.

31. Arsenault AL. Crystal-collagen relationships in calcified turkey leg tendons visualized by selected-area dark field electron microscopy. Calcif Tissue Int. 1988;43:202–12.

32. Traub W, Arad T, Weiner S. Threedimensional ordered distribution of crystals in turkey tendon collagen fibers. Proc Natl Acad Sci U S A. 1989;86:9822–6.

33. Landis WJ, Hodgens KJ, Arena J, Song MJ, McEwen BF. Structural relations between collagen and mineral in bone as determined by high voltage electron microscopic tomography. Microsc Res Tech. 1996;33:192–202.

34. Selvig KA. Ultrastructural changes in human dentine exposed to a weak acid. Archiv Oral Biol. 1968;13:719–34.

35. Tveit AB, Selvig KA. In vitro recalcification of dentine demineralized by citric acid. Scand J Dent Res. 1981;89:38–42.

36. Jäger I, Fratzl P. Mineralized collagen fibrils: a mechanical model with a staggered arrangement of mineral particles. Biophys J. 2000;79:1737–46.

37. Murdoch-Kinch CA, McLean ME. Minimally invasive dentistry. J Am Dent Assoc. 2003;134(1):87–95.

38. ten Cate JM. Remineralization of caries lesions extend into dentine. J Dent Res. 2001;80:1407–11.

39. Sakoolnamarka R, Burrow MF, Kubo S, Tyas MJ. Morphological study of demineralized dentine after caries removal using two different methods. Aust Dent J. 2002;47(2):116–22.

40. Arnold WH, Konopka S, Kriwalsky MS, Gaengler P. Morphological analysis and chemical content of natural dentin carious lesion zones. Ann Anat. 2003;185:419–24.

41. Clarkson BH, Feagin FF, McCurdy SP, Sheetz JH, Speirs R. Effects of phosphoprotein moieties on the remineralization of human root caries. Caries Res. 1991;25:166–73.

42. McIntyre JM, Featherstone JD, Fu J. Studies of dental root surface caries. 1: comparison of natural and artificial root caries lesions. Aust Dent J. 2000;45:24–30.

43. Nyvad B, Rejerskov O. Active root surface caries converted into inactive caries as a response to oral hygiene. Scand J Dent Res. 1986;94:281–4.

44. ten Cate JM. Remineralization of deep enamel dentine caries lesions. Aust Dent J. 2008;53:281–5.

45. Alauddin SS, Greenspan D, Anusavice KJ, Mecholsky J. In vitro human enamel remineralization using bioactive glass containing dentifrice. J Dent Res. 2005;84:2546.

46. Cheng L, Weir MD, Xu HHK, Kraigsley AM, Lin NJ, Lin-Gibson S, Zhou X. Antibacterial and physical properties of calcium–phosphate and calcium–fluoride nanocomposites with chlorhexidine. Dent Mater. 2012;28(5):573–83.

47. Wefel JS. Effects of fluoride on caries development and progression using intra-oral models. J Dent Res. 1990;69:626.

48. Ten Cate JM, Featherstone JDB. Mechanistic aspects of the interactions between fluoride and dental enamel. Crit Rev Oral Biol Med. 1991;2(3):283–96.

49. Stoodley P, Wefel J, Gieseke A, von Ohle C. Biofilm plaque and hydrodynamic effects on mass transfer, fluoride delivery and caries. J Am Dent Assoc. 2008;139(9):1182–90.

50. Kirsten GA, Takahashi MK, Rached RN, Giannini M, Souza EM. Microhardness of dentin underneath fluoride-releasing adhesive systems subjected to cariogenic challenge and fluoride therapy. J Dent. 2010;38(6):460–8.

51. Cenci MS, Tenuta LMA, Pereira-Cenci T, Del Bel Cury AA, Ten Cate JM, Cury JA. Effect of microleakage and fluoride on enamel-dentine demineralization around restorations. Caries Res. 2008;42(5):369–79.

52. Mousavinasab SM, Meyers I. Fluoride release and uptake by glass ionomer cements, compomers and giomers. Res J Biol Sci. 2009;4(5):609–16.

53. Xu HHK, Moreau JL, Sun L, Chow LC. Strength and fluoride release characteristics of a calcium fluoride based dental nanocomposite. Biomaterials. 2008;29(32):4261–7.

54. Xu HHK, Moreau JL, Sun L, Chow LC. Novel CaF$_2$ nanocomposite with high strength and fluoride ion release. J Dent Res. 2010;89(7):739–45.

55. Sun L, Chow LC. Preparation and properties of nano-sized calcium fluoride for dental applications. Dent Mater. 2008;24(1):111–6.

56. Weir MD, Moreau JL, Levine ED, Strassler HE, Chow LC, Xu HH. Nanocomposite containing CaF$_2$ nanoparticles: thermal cycling, wear and long-term water-aging. Dent Mater. 2012;28(6):642–52.

57. Zhou H, Bhaduri S. Novel microwave synthesis of amorphous calcium phosphate nanospheres. J Biomed Mater Res B Appl Biomater. 2012;100(4):1142–50.

58. Sun L, Chow LC, Frukhtbeyn SA, Bonevich JE. Preparation and properties of nanoparticles of calcium phosphates with various Ca/P ratios. J Res Natl Inst Stan. 2010;115(4):243.

59. Xu HHK, Moreau JL, Sun L, Chow LC. Nanocomposite containing amorphous calcium phosphate nanoparticles for caries inhibition. Dent Mater. 2011;27(8):762–9.

60. Karlinsey RL, Pfarrer AM. Fluoride plus functionalized β-TCP a promising combination for robust remineralization. Advan Dent Res. 2012;24(2):48–52.

61. Karlinsey RL, Mackey AC. Solid-state preparation and dental application of an organically modified calcium phosphate. J Mater Sci. 2009;44(1):346–9.

62. Ramay HRR, Zhang M. Biphasic calcium phosphate nanocomposite porous scaffolds for load-bearing bone tissue engineering. Biomaterials. 2004;25(21):5171–80.

63. Roveri N, Rimondini L, Palazzo B, Iafisco M, Battistella E, Foltran I, Lelli M. Synthetic biomimetic carbonate-hydroxyapatite nanocrystals for enamel remineralization. Advan Mater Res. 2008;47:821–4.

64. Li L, Pan H, Tao J, Xu X, Mao C, Gu X, Tang R. Repair of enamel by using hydroxyapatite nanoparticles as the building blocks. J Mater Chem. 2008;18(34):4079–84.

65. Cai Y, Liu Y, Yan W, Hu Q, Tao J, Zhang M, Tang R. Role of hydroxyapatite nanoparticle size in bone cell proliferation. J Mater Chem. 2007;17(36):3780–7.

66. Yamagishi K, Onuma K, Suzuki T, Okada F, Tagami J, Otsuki M, Senawangse P. Materials chemistry: a synthetic enamel for rapid tooth repair. Nature. 2005;433(7028):819.

67. Moshaverinia A, Roohpour N, Chee WWL, Schricker SR. A review of powder modifications in conventional glass-ionomer dental cements. J Mater Chem. 2011;21(5):1319–28.

68. Zhang H, Darvell BW. Mechanical properties of hydroxyapatite whisker-reinforced bis-GMA-based resin composites. Dent Mater. 2012;28(8):824–30.

69. Goenka S, Balu R, Sampath Kumar TS. Effects of nanocrystalline calcium deficient hydroxyapatite incorporation in glass ionomer cements. J Mech Behav Biomed Mater. 2012;7:69–76.

70. Wang QS, Wang Y, Li R, Zhao MM, Sun JJ, Gao Y. Effects of light-initiation agent on mechanical properties of light-cured nano-hydroxyapatite composite for dental restoration. Appl Mech Mater. 2012;138:1012–6.

71. Lee JJ, Lee YK, Choi BJ, Lee JH, Choi HJ, Son HK, Kim SO. Physical properties of resin-reinforced glass ionomer cement modified with micro and nano-hydroxyapatite. J Nanosci Nanotechnol. 2010;10(8):5270–6.

72. Moshaverinia A, Ansari S, Movasaghi Z, Billington RW, Darr JA, Rehman IU. Modification of conventional glass-ionomer cements with N vinylpyrrolidone containing polyacids, nano-hydroxy and fluoroapatite to improve mechanical properties. Dent Mater. 2008;24(10):1381–90.

73. Moshaverinia A, Ansari S, Moshaverinia M, Roohpour N, Darr JA, Rehman I. Effects of incorporation of hydroxyapatite and fluoroapatite nanobioceramics into conventional glass ionomer cements (GIC). Acta Biomater. 2008;4(2):432–40.

74. Lin J, Zhu J, Gu X, Wen W, Li Q, Fischer-Brandies H, Mehl C. Effects of incorporation of nano-fluorapatite or nano-fluorohydroxyapatite on a resin-modified glass ionomer cement. Acta Biomater. 2011;7(3):1346–53.

75. Zhang JX, Meng XC, Li XY, Lv KL. Remineralization effect of the nano-HA toothpaste on artificial caries. Key Eng Mater. 2007;330:267–70.

76. Wang L, Guan X, Yin H, Moradian-Oldak J, Nancollas GH. Mimicking the self-organized microstructure of tooth enamel. J Phys Chem C. 2008;112(15):5892–9.

77. Roveri N, Palazzo B, Iafisco M. The role of biomimetism in developing nanostructured inorganic matrices for drug delivery. Expert Opin Drug Deliv. 2008;5(8):861–77.

78. Wang L, Guan X, Du C, Moradian-Oldak J, Nancollas GH. Amelogenin promotes the formation of elongated apatite microstructures in a controlled crystallization system. J Phys Chem C. 2007;111(17):6398–404.

79. Fan Y, Sun Z, Wang R, Abbott C, Moradian-Oldak J. Enamel inspired nanocomposite fabrication through amelogenin supramolecular assembly. Biomaterials. 2007;28(19):3034–42.

80. Fan Y, Sun Z, Moradian-Oldak J. Controlled remineralization of enamel in the presence of amelogenin and fluoride. Biomaterials. 2009;30(4):478–83.

81. Tao J, Pan H, Zeng Y, Xu X, Tang R. Roles of amorphous calcium phosphate and biological additives in the assembly of hydroxyapatite nanoparticles. J Phys Chem B. 2007;111(47):13410–8.

82. Kirkham J, Firth A, Vernals D, Boden N, Robinson C, Shore RC, Aggeli A. Self-assembling peptide scaffolds promote enamel remineralization. J Dent Res. 2007;86(5):426–30.

83. Fowler CE, Li M, Mann S, Margolis HC. Influence of surfactant assembly on the formation of calcium phosphate materials—a model for dental enamel formation. J Mater Chem. 2005;15(32):3317–25.

84. Chen H, Clarkson BH, Sun K, Mansfield JF. Self-assembly of synthetic hydroxyapatite nanorods into an enamel prism-like structure. J Coll Interface Sci. 2005;288(1):97–103.

85. Palazzo B, Walsh D, Iafisco M, Foresti E, Bertinetti L, Martra G, Roveri N. Amino acid synergetic effect on structure, morphology and surface properties of biomimetic apatite nanocrystals. Acta Biomater. 2009;5(4):1241–52.

86. Iijima M, Moradian-Oldak J. Control of apatite crystal growth in a fluoride containing amelogenin-rich matrix. Biomaterials. 2005;26(13):1595–603.

87. He G, Dahl T, Veis A, George A. Dentin matrix protein 1 initiates hydroxyapatite formation in vitro. Connect Tissue Res. 2003;44(1):240–5.

88. Veis A. A window on biomineralization. Science. 2005;307(5714):1419–20.

89. Boskey AL. Amorphous calcium phosphate: the contention of bone. J Dent Res. 1997;76(8):1433–6.

90. Yan-Bao L, Dong-Xu L, Wen-Jian W. Amorphous calcium phosphates and its biomedical application. J Inorg Mater. 2007;22(5):775–82.

91. Sun W, Zhang F, Guo J, Wu J, Wu W. Effects of amorphous calcium phosphate on periodontal ligament cell adhesion and proliferation in vitro. J Med Biol Eng. 2008;28(2):107–10.

92. Li Y, Kong F, Weng W. Preparation and characterization of novel biphasic calcium phosphate powders (α-TCP/HA) derived from carbonated amorphous calcium phosphates. J Biomed Mater Res B Appl Biomater. 2009;89(2):508–17.
93. Dorozhkin SV. Calcium orthophosphates as bioceramics: state of the art. J Funct Biomater. 2010;1(1):22–107.
94. Melo MAS, Weir MD, Rodrigues LKA, Xu HH. Novel calcium phosphate nanocomposite with caries-inhibition in a human in situ model. Dent Mater. 2013;29:231–40.
95. Jie Z, Yu L, Wei-bin S, Hai Z. Amorphous calcium phosphate and its application in dentistry. Chem Cent J. 2011;5:40.
96. Skrtic D, Antonucci JM, Eanes ED. Effect of the monomer and filler systems on the remineralizing potential of bioactive dental composites based on amorphous calcium phosphate. Polym Advan Technol. 2001;12(6):369–79.
97. O'donnell JNR, Skrtic D, Antonucci JM. Amorphous calcium phosphate composites with improved mechanical properties1. J Bioact Compat Pol. 2006;21(3):169–84.
98. Schumacher GE, Antonucci JM, O'Donnell JNR, Skrtic D. The use of amorphous calcium phosphate composites as bioactive basing materials: their effect on the strength of the composite/adhesive/dentin bond. J Am Dent Assoc. 2007;138(11):1476.
99. Antonucci JM, Liu DW, Skrtic D. Amorphous calcium phosphate based composites: effect of surfactants and poly (ethylene oxide) on filler and composite properties. J Disper Sci Technol. 2007;28(5):819–24.
100. Eanes ED. Amorphous calcium phosphate: thermodynamic and kinetic considerations. In: Amjad Z editor. Calcium Phosphates in Biological and Industrial Systems. Springer International Publishing AG, Part of Springer Science+Business Media. 1998:21–39.
101. Liu Y, Kim YK, Dai L, Li N, Khan SO, Pashley DH, Tay FR. Hierarchical and non-hierarchical mineralisation of collagen. Biomaterials. 2011;32(5):1291–300.
102. Liu Y, Li N, Qi Y, Niu LN, Elshafiy S, Mao J, Tay FR. The use of sodium trimetaphosphate as a biomimetic analog of matrix phosphoproteins for remineralization of artificial caries-like dentin. Dent Mater. 2011;27(5):465–77.
103. Burwell AK, Thula-Mata T, Gower LB, Habeliz S, Kurylo M, Ho SP, Marshall GW. Functional remineralization of dentin lesions using polymer-induced liquid-precursor process. PLoS One. 2012;7(6):e38852.
104. Cross KJ, Huq NL, Reynolds EC. Anticariogenic peptides. In: Mine Y, Shahidi F, editors. Nutraceutical proteins and peptides in health and disease. Hoboken: CRC Press; 2005. p. 335–51.
105. Cross KJ, Huq NL, Palamara J, Perich J, Reynolds EC. Physicochemical characterization of casein phosphopeptide-amorphous calcium phosphate nanocomplexes. J Biol Chem. 2005;280:15362–9.
106. Rose RK. Binding characteristics of Streptococcus mutans for calcium and casein phosphopeptide. Caries Res. 2000;34:427–31.
107. Rose RK. Effects of an anticariogenic casein phosphopeptide on calcium diffusion in streptococcal model dental plaques. Arch Oral Biol. 2000;45:569–75.
108. Reynolds EC. Remineralization of enamel subsurface lesions by casein phosphopeptide-stabilized calcium phosphate solutions. J Dent Res. 1997;76:1587–95.
109. Reynolds EC. Calcium phosphate-based remineralization systems: scientific evidence? Aus Dent J. 2008;53:268–73.
110. Iijima Y, Cai F, Shen P, Walker G, Reynolds C, Reynolds EC. Acid resistance of enamel subsurface lesions remineralized by a sugar-free chewing gum containing casein phosphopeptideamorphous calcium phosphate. Caries Res. 2004;38:551–6.
111. Beniash E, Metzler RA, Lam RS, Gilbert PU. Transient amorphous calcium phosphate in forming enamel. J Struct Biol. 2009;166:133–43.
112. Lowenstam HA, Weiner S. Transformation of amorphous calcium phosphate to crystalline dahllite in the radular teeth of chitons. Science. 1985;227:51–3.
113. Mahamid J, Sharir A, Addadi L, Weiner S. Amorphous calcium phosphate is a major component of the forming fin bones of zebrafish: Indications for an amorphous precursor phase. Proc Natl Acad Sci. 2008;105:12748–53.
114. Combes C, Rey C. Amorphous calcium phosphates: synthesis, properties and uses in biomaterials. Acta Biomater. 2010;6:3362–78.
115. Pan HH, Liu XY, Tang RK, Xu HY. Mystery of the transformation from amorphous calcium phosphate to hydroxyapatite. Chem Comm. 2010;46:7415–20.
116. George A, Sabsay B, Simonian PA, Veis A. Characterization of a novel dentin matrix acidic phosphoprotein. Implications for induction of biomineralization. J Biol Chem. 1993;268:12624–30.
117. Hunter GK, Hauschka PV, Poole AR, Robsenberg LC, Goldberg HA. Nucleation and inhibition of hydroxyapatite formation by mineralized tissue proteins. Biochem J. 1996;317:59–64.
118. He G, Ramachandran A, Dahl T, George S, Schultz D, Cookson D, George A. Phosphorylation of phosphophoryn is crucial for its function as a mediator of biomineralization. J Biol Chem. 2005;280(39):33109–14.
119. Liou SC, Chen SY, Liu DM. Manipulation of nanoneedle and nanosphere apatite/poly(acrylic acid) nanocomposites. J Biomed Mater Res B Appl Biomater. 2005;73:117–22.

120. Olszta MJ, Odom DJ, Douglas EP, Gower LB. A new paradigm for biomineral formation: mineralization via an amorphous liquid-phase precursor. Connect Tissue Res. 2003;44 Suppl 1:326–34.

121. Gower LB, Olszta MJ, Douglas EP, Munisamy S, Wheeler DL. Biomimetic organic/inorganic composites, processes for their production, and methods of use. US patent application 20060204581, 2006.

122. Niederberger M, Cölfen H. Oriented attachment and mesocrystals: non-classical crystallization mechanisms based on nanoparticle assembly. Phys Chem Chem Phys. 2006;8:3271–87.

123. Saito T, Yamauchi M, Crenshaw MA. Apatite induction by insoluble dentin collagen. J Bone Miner Res. 1998;13:265–70.

124. Kinney JH, Habelitz S, Marshall SJ, Marshall GW. The importance of intrafibrillar mineralization of collagen on the mechanical properties of dentin. J Dent Res. 2003;82:957–61.

125. Dai L, Liu Y, Salameh Z, Khan S, Mao J, Pashley DH, Tay FR. Can caries-affected dentin be completely remineralized by guided tissue remineralization? Dent Hypotheses. 2011;2(2):74.

126. Rinaudo M. Chitin and chitosan: properties and applications. Prog Polym Sci. 2006;31:603–32.

127. Malette W, Quigley H. Adickes E. In: Muzzarelli RAA, Jeuniaux C, Gooday GW, editors. Chitin in nature and technology. New York: Plenum Press; 1986. p. 435–42.

128. Huang M, Fang Y. Preparation, characterization, and properties of chitosan-g-poly(vinyl alcohol) copolymer. Biopolymers. 2006;81:160–6.

129. Sakairi N, Shirai A, Miyazaki S, Tashiro H, Tsuji Y, Kawahara H, Yoshida T, Tokura S. Synthesis and properties of chitin phosphate. Kobunshi Ronbunshu. 1998;55:212–6.

130. Jayakumara R, Rajkumarb M, Freitas H, Selvamuruganc N, Nair SV, Furuikec T, Tamuraa H. Preparation, characterization, bioactive and metal uptake studies of alginate/phosphorylated chitin blend films. Int J Biol Macromol. 2009;44:107–11.

131. Wang XH, Zhu Y, Feng QL, Cui FZ, Ma JB. Responses of osteo- and fibroblast cells to phosphorylated chitin. Bioact Compat Polym. 2003;18:135–46.

132. Wang X, Ma J, Wang Y, He B. Bone repair in radii and tibias of rabbits with phosphorylated chitosan reinforced calcium phosphate cements. Biomaterials. 2002;23:4167–76.

133. Wang X, Ma J, Feng QL, Cui FZ. Skeletal repair in rabbits with calcium phosphate cements incorporated phosphorylated chitin. Biomaterials. 2002;23:459–4600.

134. Abarrategi A, Moreno-Vicente C, Ramos V, Aranaz I, Sanz Casado JV, López-Lacomba JL. Improvement of porous beta-TCP scaffolds with rhBMP-2 chitosan carrier film for bone tissue application. Tissue Eng Part A. 2008;14(8):1305–19.

135. Weir MD, Xu HH. Osteoblastic induction on calcium phosphate cement-chitosan constructs for bone tissue engineering. J Biomed Mater Res A. 2010;94(1):223–33.

136. Jayakumar R, Nwe N, Tokura S, Tamura H. Sulfated chitin and chitosan as novel biomaterials. Int J Biol Macromol. 2007;40:175–81.

137. Zhang X, Yang P, Yang WT, Chen JC. The bio-inspired approach to controllable biomimetic synthesis of silver nanoparticles in organic matrix of chitosan and silver-binding peptide (NPSSLFRYLPSD). Mater Sci Eng C Biomim Mater Sens Syst. 2008;28:237–42.

138. Xu Z, Neoh KG, Lin CC, Kishen A. Remineralization of partially demineralized dentine substrate using phosphorylated chitosan. J Biomed Mater Res B Appl Mater. 2011;98B(1):150–9.

Index

A. Kishen (ed.), *Nanotechnology in Endodontics: Current and Potential Clinical Applications,*
DOI 10.1007/978-3-319-13575-5, © Springer International Publishing Switzerland 2015